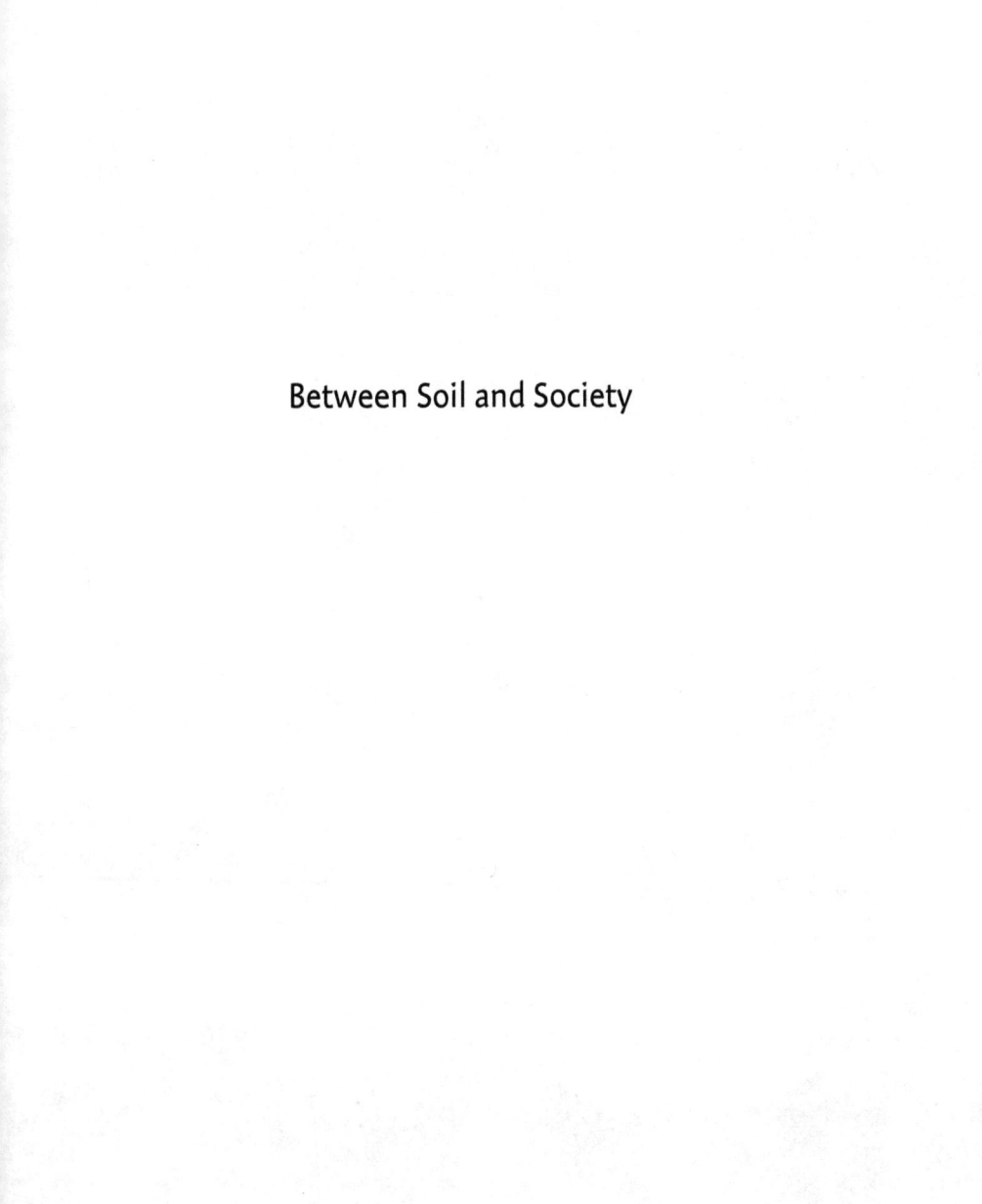

Between Soil and Society

BETWEEN SOIL and SOCIETY

Legislative History and
Political Development of
Farm Bill Conservation Policy

JONATHAN COPPESS

University of Nebraska Press   Lincoln

© 2024 by Jonathan Coppess

All rights reserved

The University of Nebraska Press is part of a land-grant institution with campuses and programs on the past, present, and future homelands of the Pawnee, Ponca, Otoe-Missouria, Omaha, Dakota, Lakota, Kaw, Cheyenne, and Arapaho Peoples, as well as those of the relocated Ho-Chunk, Sac and Fox, and Iowa Peoples.

Publication of this volume was assisted by University of Illinois, College of ACES, Hatch Act project number ILLU-470-379-7003440.

Library of Congress Cataloging-in-Publication Data
Names: Coppess, Jonathan, author.
Title: Between soil and society: legislative history and political development of farm bill conservation policy / Jonathan Coppess.
Description: Lincoln: University of Nebraska Press, 2024. | Includes bibliographical references and index.
Identifiers: LCCN 2023029213
ISBN 9781496225146 (hardback)
ISBN 9781496238580 (epub)
ISBN 9781496238597 (pdf)
Subjects: LCSH: Agricultural laws and legislation—United States. | Agriculture and politics—United States. | Agriculture and state—United States. | Agricultural conservation—Law and legislation—United States. | Soil conservation—Law and legislation—United States. | Farms—Government policy—United States. | BISAC: NATURE / Environmental Conservation & Protection | HISTORY / United States / 20th Century
Classification: LCC KF1681 .C67 2024 |
DDC 343.7307/6—dc23/eng/20231214
LC record available at https://lccn.loc.gov/2023029213

Designed and set in Minion Pro by L. Welch.

To my parents,
both retired now.
Mom was an innovative and dedicated
high school librarian;
Dad was an environmentally minded farmer
who took conservation seriously.

CONTENTS

Acknowledgments ix

PART 1. INTRODUCTION

1. Of Farming, Conservation, and the Farm Bill: An Introduction   3

2. Of Congress and the Conservation Question: A Preliminary Discussion   16

PART 2. DUST BOWL ORIGINS

3. Out of Dust, Sharecropping, and the Supreme Court: An Origin's Backstory   37

4. Reform amid Ruin: Lessons from the 74th Congress   56

PART 3. THE SOIL BANK SAGA

5. Out of Surplus and Policy Failure: The Rise of the Soil Bank   85

6. Southern Sabotage: The Swift Demise of the Soil Bank   105

PART 4. THE FOOD SECURITY ACT OF 1985

7. From Dust to Dust: The Seventies Interlude and Backstory   137

8. A New Foundation for Conservation Policy: The Food Security Act of 1985   153

PART 5. MODERN CONSERVATION POLICY DEVELOPMENTS

9. Modern Developments: Farm Bill Conservation Policy after 1985   175

10. Of Congress and the Conservation Question: A Working Theory and Closing Argument   195

    Notes   221
    Bibliography   275
    Index   327

ACKNOWLEDGMENTS

While I have dedicated this book to my parents, I want to begin the acknowledgments by thanking my wife, Susan, and our incredible children, Abigail and Warren. I am certain there are people who write books with some version of ease and grace; I am not one of those people, and the burdens of any struggles in writing can be born by those closest in proximity. So, I want to thank Susan, Abigail, and Warren for their strong support, their love, and their patience and understanding—especially in those moments when I was stuck or struggled. What you have brought to my life cannot be measured, and my appreciation has no bounds. I love you and will work to the end of my days to try and leave the world a better place for you.

I also want to thank my colleagues and collaborators in ACE, ACES, and across the campus and system of the University of Illinois. As I completed this book, I reflected on an amazing nine years at this outstanding institution. I count myself as fortunate to work amid the brilliance and dedication, to learn from so many experts, and to have the opportunity to explore intellectual curiosities with so many talented people. I've often likened myself to a kid in a candy store. I appreciate the many opportunities so far and look forward to those to come.

I also want to acknowledge and thank the Washington DC crowd, of whom there are too many to name individually. While some of us have left, the connections remain—and remain important and greatly appreciated. If you have ever bounced around bad ideas or talked policy and politics with me over a glass of whiskey or a pint of beer, thank you. I have learned much from so many talented and dedicated people in Washington and in the farm policy family. I hope that this work contributes in some meaningful way.

A huge thanks goes to the multiple reviewers of the different versions of the manuscript for this book. I struggled to pull it together. The very thorough and thoughtful comments, suggestions, and critical evaluations of the material were extraordinarily helpful and much appreciated. To that end, I would also like to extend my appreciation to Adam Sheingate at Johns Hopkins University for some incredibly productive, thought-provoking discussions and suggestions that helped immensely as I worked through aspects of this book.

Thank you to Bridget Barry, Sara Springsteen, and the team at University of Nebraska Press for their great work in making this book a reality. I also want to thank Kerin Tate for superb editing of the manuscript and for making that difficult task less so. Finally, thank you to Susan Certo for putting together the index. I really appreciate all the incredible work and professionalism that goes into completing a project like this.

Finally, I want to thank Mom and Dad for all that they have provided me over the years. From my love of reading to my interests in the environment and agricultural conservation, I dedicated this book to you as a small token of my appreciation. You developed in me the drive to achieve, a strong work ethic, and the grit to push through when things get tough—all of which proved necessary in completing this project.

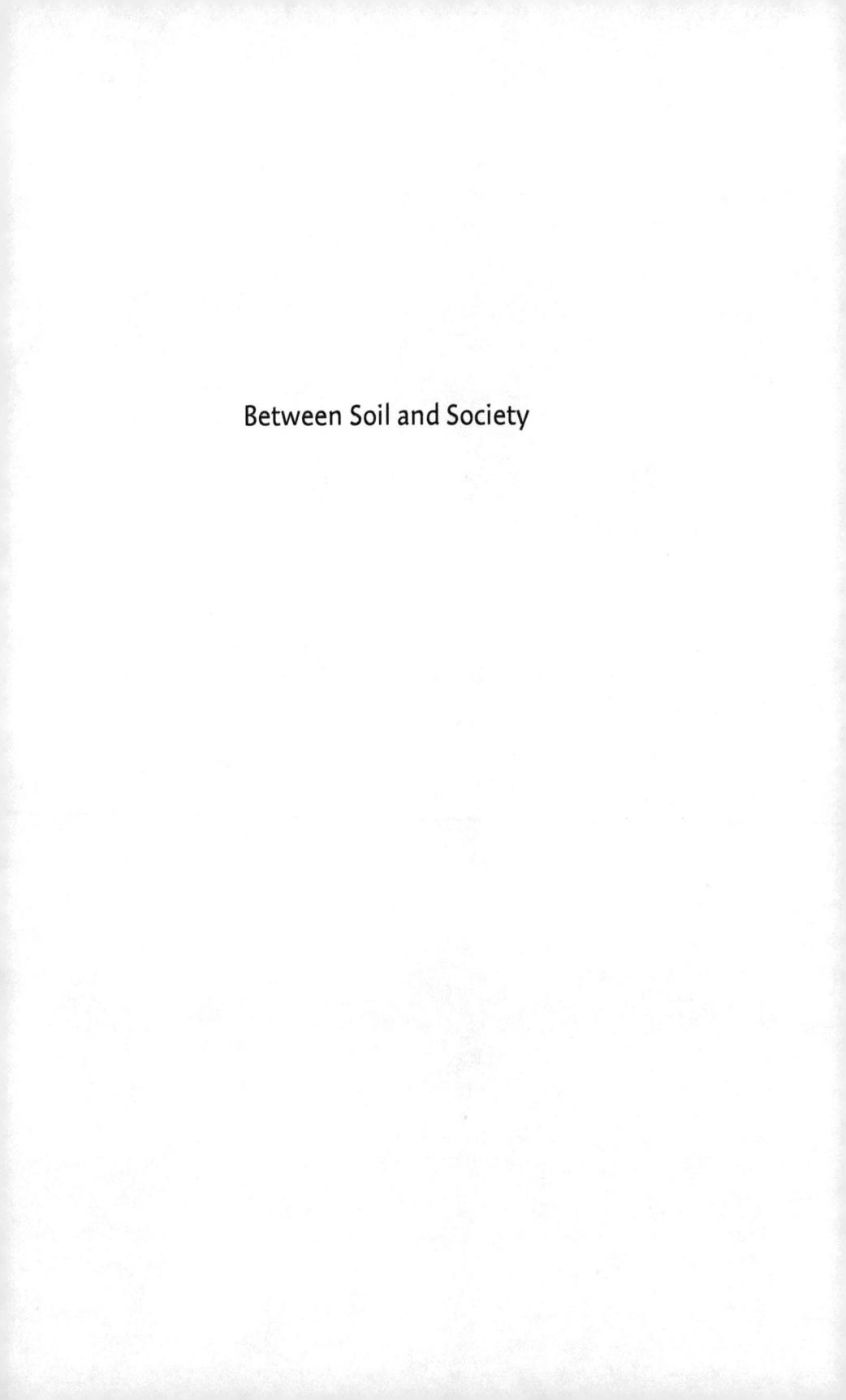

Between Soil and Society

PART 1   Introduction

1

Of Farming, Conservation, and the Farm Bill

AN INTRODUCTION

Between soil and society there exists more than ten thousand years of history, time during which people have built civilizations and torn them down. In many ways, agriculture makes bread from stones, and by doing so, agriculture provides the origin story for human society and forms its foundation.[1] To scientists, soil is the "human-nature hybrid" in which a living, functioning ecosystem is applied to food production. Society's food system requires, among many things, the production of staple commodities—wheat for cereals and breads, for example, and corn or other crops for animal feed—which is accomplished by farming. The connections between soil and society are deep and profound. Soil has underwritten human developments spanning culture, government, economies, markets, and technology. Today, farming's footprint stretches across roughly one-third of the earth's terra firma, but it all depends on a relatively thin layer of soil, an inch of which may take a thousand years to replace. Buried in the soil's memory are also portents and warnings, stories of mismanagement by societies willing to degrade the soil and its myriad valuable complexities in service of short-term economic gains. The consequences can be vast and drastic. The demise and collapse of a civilization can be calculated as the amount of time it takes people to waste the soil resource.[2]

Soils are necessary to sustain food production; therefore, it would stand to reason that society would prioritize maintaining healthy, productive soils. A society that governs itself would want to make doing so among the utmost priorities. Logically correct; however, reality is not so simple. There are vast distances between soil and society. Soil is a means to an end for society. Soil and society operate on fundamentally different time horizons. It can take a long time for degradation to truly threaten society, and repairing damage takes even longer, but society's needs are immediate and persistent. If efforts to avoid degradation are perceived as having too high a cost or as presenting unacceptable risk in the short term, then those efforts are problematic for society. While societies should arguably and logically prioritize the preservation and conservation of soil to ensure continued food production in the future, a society is unlikely to do so if it means sacrificing the present. Even the perception of a present sacrifice for an unknown future is likely beyond the political pale.

Natural resource challenges, such as those involving soil, confound humankind in the most visceral, persistent ways. No matter how it is managed or conducted—no matter how conscientious the farmer or the landowner—the production of agricultural commodities has consequences for natural resources and the environment. To grow crops, farmers need land. The history of American agriculture has witnessed the clearing of vast forests and the breaking of millions of acres of native sod; wetlands have been drained and rivers rerouted, aquifers and other reservoirs of water consumed. Each year's efforts at producing a crop can (but do not have to) involve intense tillage practices that turn the soil, leaving this extraordinary resource vulnerable to the elements. So much of the midwestern winter landscape, for example, is a barren brown of cultivated topsoil left vulnerable to the ravages of winter winds, followed by the flush of melting snow, and then rain each spring. These are self-inflicted problems in farm fields, but annual production cycles, farmer practices, increasing population demands, and unpredictable weather magnify the challenges. Soil erosion and nutrient loss provide arguably the clearest examples.

Soil erosion is the most prominent consequence historically. It is often the most visible and tangible, as soils are washed or blown from farm fields and deposited elsewhere, sometimes at great distance. Erosion is typically due to soils blown into the air by wind or carried away by water (rain or snowmelt) and is defined as soil loss that exceeds the rate at which natural processes can replace what was lost. The U.S. Department of Agriculture's Natural Resources Conservation Service (NRCS) uses a soil loss tolerance rate (T) that it defines as the "maximum rate of annual soil loss that will permit crop productivity to be sustained economically and indefinitely on a given soil."[3] It can take three hundred to one thousand years to replace an inch of eroded soil, and USDA recently estimated 4.63 tons per acre of soil erosion from croplands, or about 1.7 billion tons of lost soil per year; an acre of soil one inch deep will weigh about 167 tons. Soil erosion does occur naturally, but it is mostly the product of human activity. Soil erosion is the result of intensive agricultural production and farming practices such as tillage (plowing), which turns soil over and leaves it open, vulnerable to the weather. Erosion also depends on natural circumstances such as soil characteristics, the energy or power of the water and wind, the characteristics of the land, and whether there is plant cover and how much. In America, progress has been made, but we struggle to do better.[4]

Another critical consequence due to modern production agricultural practices is water degradation resulting from nutrient or fertilizer losses. Fertilizers feed growing plants, but the nutrients lost from farm fields degrade water quality. In much of the midwestern Corn Belt, a complex system of subsurface tiles and ditches drain former swampland so that the fields can be farmed. This massive and effective drainage system, coupled with intensive synthetic fertilizer application, pollutes local drinking water and other water resources; it is generally not regulated. One example of the unintended consequences is the persistent development of a dead zone in the Gulf of Mexico. Erosion makes the situation worse because fertilizer applications likely increase to offset yield losses due to erosion. Adding problems, much of our commercial and synthetic fertilizer production today involves fossil fuels and intense mining. Therefore, as we erode soil, we need more fossil fuels to produce fertilizers to compensate. Increased fertilizer application results in still larger losses, when excess nutrients are exported from farm fields by drainage systems and water resources. The impacts of climate change are likely to make these problems even worse, and we are, in reality, only at the beginning of our efforts to address these problems.[5]

Natural resource problems are more likely to build slowly and over longer time horizons. They are also more likely to be externalized, at least until the point of collapse is reached. Soil erosion and nutrient loss again provide primary and clear examples because some portion of the cost of individual decisions is often paid by other farmers and the general public. Lost topsoil impacts neighboring fields, is deposited in public waterways, and eventually requires a public cost to clean up; collectively, enough farmers wasting their topsoil can put at risk the nation's productive capabilities. The farmer's decision to overapply nitrogen may improve that farmer's yields in the short-term, but it also results in excess nitrate that is exported to public waters, where it can require costly equipment to clean the water for a community's drinking needs, or it produces a dead zone that impacts those who make a living by the waters. They provide powerful examples because of their scale and scope. Addressing soil erosion and nutrient losses presents complexities and difficulties *every* crop year. They are also interconnected through many inherent human factors.[6]

The consequences of farming encompass soil erosion, nutrient losses, water quality degradation, water shortages from irrigation, and habitat loss or destruction. Billions of humans, however, rely on agriculture for food. This reality risks making the environmental consequences of producing it a secondary concern at best. Humans cannot unreasonably curtail farming due to the environmental consequences farm practices can produce. Instead, we search for solutions that can be implemented within the farming and food system to decrease or avoid the consequences without losing our capacity to produce food. This is conservation. Aldo Leopold was a leading voice on conservation, providing some of the more profound and well-known statements on the matter. "A land ethic," Leopold wrote in 1949, "reflects the existence of an ecological conscience, and this in turn reflects a conviction of individual responsibility for the health of the land." Furthermore, he defined health as "the capacity of the land for self-renewal," and "conservation is our effort to understand and preserve this capacity."[7] But this, too, is complex and burdened by many layers of complications.

Conservation is rooted in an understanding that natural resources are finite and can be lost, exhausted, or destroyed, but that nature also possesses an ability to renew and rebuild or restore itself. It requires effort to reduce, prevent, and address the conflicts between food production and environmental consequences. Conservation also presents costs, both financial and managerial, and it can both increase and decrease risks. The farmer or landowner actively undertakes measures to minimize deterioration of the resource. Conservation is also neither purely public nor completely private. Natural resource degradation will result in decreased production, which, in turn, is an economic loss to the producer. Large-scale and widespread degradation will likely lead to significant losses for society. Conservation can also increase productivity and the capacity of the natural resource, making improvements through investments beyond simply preventing or decreasing deterioration. It is often connected to economic matters that emphasize maintaining or maximizing levels of production on the land. Such perspectives often tolerate loss or degradation only insofar as it does not reduce or limit productive capability. There are potential risks and controversies in the balancing act between the economics for the farmer and the inherent value of natural resources for their own sake.[8]

Society struggles to overcome the temporal, as well as the other challenges for prioritizing the preservation of soil and other natural resources. Therefore, conservation and natural resource problems are matters that necessitate public policy, collective decisions by government for the general welfare and in the public's interests.[9] The path explored in this book is a history at the intersection of farming, the environment, and public policymaking. It reviews the political development of agricultural conservation policy when Congress has enacted it in farm bills.[10]

The Farm Bill: A Brief Background

"Farm bill" is the common term for omnibus federal legislation that authorizes a variety of programs in the food and agricultural sector. Modern iterations of a farm bill combine support for farmers (payments, crop insurance, trade development, farm loans, and specialty crop assistance) with assistance to steward public resources (conservation and forestry), food assistance to low-income individuals and families both domestic and abroad (the Supplemental Nutrition Assistance Program [SNAP] and international food donations or assistance), rural economic development, and research and education. The current farm bill is the Agricultural Improvement Act of 2018 (2018 Farm Bill).[11] It has been standard since the 1970s that farm bills are written as five-year authorizations, and the 2018 Farm Bill is scheduled to expire with the 2023 calendar, crop, and fiscal years.

Farm program payments (Title I), conservation programs (Title II), and crop insurance constitute the vast majority of federal funds typically expended on farmers, although they do not constitute the entirety of federal farm policy. These policies are also not the only forms of direct assistance to producers, but they are the largest and most prominent. Two of these three primary policy categories provide economic assistance to farmers (subsidy payments and subsidized insurance), but only one assists directly with conservation. As measured by federal spending, generally between $5 billion and $7 billion is spent each year for subsidy payments in Title I. In addition, between $8 billion and $10 billion each year is expended for crop insurance, most of which subsidizes the premium cost for the farmer. By comparison, conservation programs are the smallest of the three, spending between $4 billion and $6 billion each year.[12]

Crop insurance is by far the largest federal program for farmers in terms of participants, acres impacted, and federal outlays. Through this federally subsidized insurance program, the farmer purchases the crop insurance policy and receives an insurance indemnity if there is a loss on the crop insured. The federal subsidy reduces the cost of the policy premium to the farmer; generally, the subsidy consumes more than 80 percent of the total program's costs. USDA reports approximately 330 million acres of total cropland used to produce crops in the United States each year. As many as 400 million to nearly 445 million acres are insured in any given crop year, and the total insurance liability for the program has exceeded $130 billion in recent years. Nearly 1.2 million policies are sold and earning premium each year. Finally, crop insurance is directly connected to farm risks such as yield losses or price declines during the crop year.[13]

The farm programs in Title I of the Farm Bill provide direct cash payments to a subset of farmers, mostly those with a history of planting the basic bulk commodities (e.g., corn, soybeans, wheat, rice, cotton, and peanuts) and dairy producers. Payments are predominantly triggered by low prices for that select group of crops, but farmers do not have to plant the crop for which they receive the payment. Farmers also do not need to have suffered any actual loss to receive a payment. These subsidy programs can be understood to operate as approximations of loss using national averages and historical price or revenue experiences. Any connection to farm realities or risks (including losses) is tenuous at best; in operation it is often minimal or nonexistent.

The combined $13 billion to $17 billion each year for crop insurance and farm subsidy payments goes to the same subset of bulk commodity farmers because a farmer receiving a subsidy payment is also likely to have purchased subsidized crop insurance, receiving an indemnity if there were actual losses. It is not uncommon to receive both subsidy payments and crop insurance indemnities, and both programs are subject to criticism and opposition. Arguably the critical challenge for these farm support policies is explaining and justifying the return to the taxpayer for these massive investments. It has grown increasingly difficult for farm subsidy programs, especially in recent decades, considering that even a family farmer is typically wealthier than the average American family. Today's farms are far larger and more sophisticated than traditional or historic farms; the farm in reality departs drastically from the more mythologized concepts of it.[14]

Agricultural conservation policy and programs are mostly contained in Title II of a farm bill. Of historical note, all are amendments to the base statutory text enacted by Congress in the Food Security Act of 1985. The $4 billion to $6 billion spent annually on today's conservation programs is roughly one-third of that spent on farm support policies. That total is also somewhat misleading because conservation spending is allocated across multiple programs with different priorities or policy goals. In general, all farm bill conservation programs can be placed in one of two categories: assistance for working lands conservation, or payments to retire land from production.

Working lands conservation programs pay farmers directly to adopt, maintain, improve, or advance conservation on their farms and within their production practices. The Environmental Quality Incentives Program (EQIP) provides cost-share payments for specific conservation practices adopted by the farmer. The Conservation Stewardship Program (CSP) provides annual payments in return for the farmer successfully increasing or expanding the conservation efforts on the farm. The Regional Conservation Partnership Program (RCPP) blends various authorities for supporting conservation practices, with public funding matched by private partnership contributions. It is the newest conservation program.

By comparison, retirement or reserve programs pay to remove farmland from agricultural production. The Conservation Reserve Program (CRP) operates through a long-term rental agreement with the landowner and provides an annual rental payment. The programs emphasize enrollment of environmentally sensitive land (whole or partial fields), seeking to keep it out of intensive annual production for at least a decade or more. The Agricultural Conservation Easement Program (ACEP) operates through the purchase of a long-term or permanent easement on farmland, such as to restore or protect wetlands.[15]

History and Development of Farm Bill Conservation Policy: Overview as Introduction

In *The Fault Lines of Farm Policy*, I traced the history and legislative development of direct assistance to American farmers, particularly the farm subsidy programs.[16] In some ways, this book is a companion to *Fault Lines*, and my goal is to incorporate that history and development without repeating it. This book builds upon the earlier discussion and provides a more complete history of the development of federal legislation for American farmers. But that is only a partial explanation.

One of the central arguments underpinning *Fault Lines* is that the history of farm bills provides useful case studies of Congress and the legislative process. Congress has debated roughly two dozen farm bills since 1933, offering at least as many lessons in politics and American political development—not least of which are those about coalition-building and maintenance—in the post–New Deal era.[17] Even the most cursory comparison between the histories of farm programs and farm conservation programs exposes substantial differences that raise inescapable questions. The two categories of farm policy and assistance cover the same years, and both pay roughly the same constituencies or interests, but they experienced different developmental histories. Given the central importance of soil to society, and the value of conserving this resource, a reasonable conclusion would be that the United States should prioritize conservation, but the disparities explored in this history contradict that conclusion and beg questions that form the core of this book.

Congress has legislated on matters critical to the agricultural sector since the founding, but the most direct involvement began in 1933. That year, Congress enacted legislation to provide direct benefits to some American farmers and to attempt to better manage the production of basic agricultural commodities in this country, effectively enacting the first farm bill. The first federal law to substantively address soil erosion from farming was enacted in 1935 and was followed by the Soil Conservation and Domestic Allotment Act of 1936. Both were enacted as emergency responses to catastrophe and crisis: the massive dust storms that came to be known as the Dust Bowl during the economic crisis that was the Great Depression. The work by Congress and key officials at USDA during this era forms the traditional origin story for farm bill conservation policy. Chapter 3 will provide a backstory connecting the Dust Bowl and the Great Depression to matters that help better explain the political development. The backstory is particularly important to the congressional effort for the 1936 legislation, which is covered in chapter 4. Combined, the two chapters offer a more complete perspective on the origin of agricultural conservation policy.

From 1936, twenty years passed before Congress returned to conservation policy in a meaningful way in a farm bill. Chapter 5 provides the backstory and legislative history of the Agricultural Act of 1956, which contained two programs that were versions of conservation policy and known as the Soil Bank. Chapter 6 details the relatively rapid demise of the Soil Bank and how it came to be that President Eisenhower, who had proposed the Soil Bank to Congress, requested that Congress terminate the program after barely three years in operation. Combined, these two chapters provide a more complete perspective on this second attempt at agricultural conservation policy and the purposeful strategy to destroy the Soil Bank.

After the demise of the Soil Bank, a quarter of a century passed before enactment of the landmark Food Security Act of 1985, which provided the foundation for modern conservation policy in farm bills. Chapter 7 explores the backstory for the 1985 Act. It delves into the events and developments between the Soil Bank's demise and the 1985 Act's enactment. It also reviews the rise of the modern environmental movement in the 1970s because it coincided with drastic changes in American agriculture. Farming in the 1970s led to a recycling of the twin crises experienced in the 1930s: the combination of environmental and economic collapse in the farm sector that consumed much of the 1980s. Chapter 8 details the legislative developments contained in the 1985 Act.

Finally, chapter 9 traces the legislative development of the modern suite of conservation programs in the farm bill, from the 1985 Act to the Agricultural Improvement Act of 2018 and the Inflation Reduction Act of 2022. It traces the increased spending on conservation as well as the increase in the number of programs. This discussion involves the development of working lands conservation programs to supplement or complement the more traditional land retirement programs. It also contains a specific episode when a new, innovative conservation program was drastically curtailed by partisans; innovation harmed and limited, albeit not to the extent of the Soil Bank. Chapter 10 concludes the book with an effort to derive lessons from the history and apply them to matters of Congress and American political development.

Throughout most of the history of direct federal assistance to farmers, Congress has prioritized subsidies and other payments over assistance for conserving and preserving the soil resource. In some eras the disparity is vast. It is also understated because most conservation policy has been designed more as a complement or supplement to farm support policy rather than for natural resources or environmental outcomes. Even today, taxpayers invest less than one-third of the funding for farmers in conserving natural resources.

Conservation policy has also been purposefully limited by Congress. The modern CRP has been subject to an acreage cap that Congress has reduced multiple times to achieve savings or appease opponents of the program. The main working lands program, EQIP, has operated under a fixed amount of funding that has steadily increased since 1996, but it remains limited by Congress, with consequences. In fiscal year 2020, USDA was able to fund only 27 percent of eligible program applications that it received. This has been a pattern in which tens of thousands of farmers seeking assistance with conservation are turned away from the programs.[18] Most instructive, innovations in conservation policy such as the Soil Bank in 1956 and the Conservation Security Program (CSP) in 2002 were purposefully sabotaged or restricted and damaged by Congress. These constraints or attacks on conservation policy have not had their equal in the treatment of farm program payments.

If soil is so fundamental and essential to society, and if a self-government system should best reflect the priorities of the governed—policies for the public interest and that promote the general welfare—why has that not been the experience in American farm policy? If, in fact, conservation policy delivers the greater public good, why has it occupied the lesser position, a relatively minor and secondary status? Why did it take more than fifty years longer to establish conservation programs in the farm bill status quo? Why has Congress consistently prioritized spending on farmers for policies and programs other than those that benefit the soil? More pointedly, why did conservation policy, with a focus on reducing the consequences of farming, fail to break through during the Dust Bowl? Why was it resurrected to help a faltering farm support policy system in the 1950s only to be sabotaged and terminated? What changed in the 1970s and 1980s to bring about a more fundamental change, albeit belatedly? And why after nearly one hundred years of experience, including two major crisis eras in which farm economic and soil erosion problems were intimately connected, has our federal farm policy apparatus failed to do better?

As noted above, these questions emanate from the history and development of farm bill conservation policy explored in this book; they become ineluctable when that history is compared to the history of farm assistance policies. Considering all that we now know and have learned, as well as our scientific and technological advancements in the last nine decades, it seems only reasonable to ask why we are not doing better. America is certainly not immune to the difficult consequences and externalities of farming that have been viscerally demonstrated throughout human history. Why haven't we, the world's oldest functioning democracy, learned these lessons or better applied them? These questions, and others along the same lines, were among the primary driving forces for me in researching and writing this book. I collectively refer to such questions as the conservation question and intend the conservation question to provide a tool for better understanding and more thoughtfully critiquing Congress, legislators, the legislative process, and the political development of policies. The core argument is that paying farmers for conservation provides more for the public good, and certainly more of a return for the taxpayers funding the payments, than payments through farm programs or subsidies.

In addition, the nature of farm bill conservation programs and the relatively straightforward relationship with farm subsidy policies serves at least two further ambitions. One is to develop perspective on both conservation policy and American government.[19] By asking questions about the development of conservation policy, the effort here also seeks to contribute something of value to the knowledge and understanding of the legislative process, of Congress, and of the political challenges in our national experiment in self-government. It is both a case study in Congress and an exploration of this one component of federal policy that seeks to develop an analytical companion to lived experience.

The other looks beyond the degradation and destruction of natural resources in farming. While those matters are critically important in their own right, they take on an even greater significance as we seek to understand and attempt to address the almost incomprehensible risks and potential cataclysms posed by climate change.[20] Climate change and its challenges to humankind are on the periphery of this work but never out of mind. From the Dust Bowl to modern dead zones and poisonous algal blooms in our public water bodies, these self-inflicted problems appear as analogues or even metaphors for climate change. If we learn any lessons from our efforts to implement conservation and address the consequences from farming, it stands to reason that they would be at least relevant to climate change discussions. This is a point even more appropriate and applicable in the realm of policy.

Between soil and society, only one is considered essential to the existence and survival of the other; soil certainly does not need human society to exist. That which a people preserve, conserve, and protect, compared with that which they waste, destroy, and neglect, defines society and the people within it. Where further advancements in conservation policy with the most potential for far-reaching natural resource and public interest outcomes are effectively restricted in Congress, the conservation question confronts us as a society. Any answers that can be formulated out of the history and development of farm bill conservation policy may provide guidance and counsel well beyond the narrow confines of farm bills. The search for answers and understanding is the driving force for this book. Encompassing political power to property law, the exploration will cut a path through legislative history and congressional deliberation across more than nine decades.[21]

In 1863 America's first secretary of agriculture used the Roman Empire to teach of the perils that humans have long known. He wrote a long discussion in his first report to Congress in which he laid much of the blame for Rome's decline and fall on the disastrous combination of an increasing monopoly over land, vast degradation of soils, and a decreasing return to labor, including the dignity afforded work. Rome's rise to greatness, he concluded, was due to the honorable treatment given to the cultivation of soil and to labor. He wrote, "Changing this splendid basis of prosperity, permanency, and power, whereby, resting in the soil, Rome pierced the heavens by force of thought, she grew proud and oppressive; the reins of power slipped from the hands of the middle classes; labor became disreputable; the soil a monopoly, and the masses of the people reckless, unpatriotic, and degraded." These developments, he continued, were "destructive to intelligent, interested, and really productive agriculture" and brought ruin to the Roman Empire.[22] Meant as a cautionary tale during the Civil War, his discussion remains striking and profound to this day. It serves to launch the exploration of conservation policy history that unfolds in the pages that follow. That exploration begins in chapter 2 with a discussion of Congress.

2

Of Congress and the Conservation Question

A PRELIMINARY DISCUSSION

The United States spends approximately $20 billion each year for farmers in subsidy payments, crop insurance, and conservation; most of the funding is spent on the same subset of farmers. All three are products of the 1930s and the New Deal, but only farm subsidy programs have occupied center stage in politics and development. Fifty-two years separated the first farm bill in 1933 and the farm bill that established conservation as part of the status quo in 1985. In terms of politics, all three are provided to the same interest group or political faction, but federal payments to conserve natural resources such as soil, water, and habitat arguably provide at least the expectation of a return to the general public and taxpayer. The extent to which this is achieved or is effective and meaningful, as well as any value to the taxpayer, is debatable; the basic point, however, stands.[1] The American taxpayer spends the most on policies that provide the least return and the least on the policies that provide the most return, especially over the long run.

    The mechanism by which society makes decisions is the system of government through which the words of law should reflect the priorities of the citizenry. At the forefront is the U.S. Congress, the first branch of the government created by the Constitution. Congress is the branch of government vested with the legislative power, which can be understood as the power to initiate the making of law. It is the branch for which the electoral connection with the people subject to its laws is the most direct and strongest. Enacted laws by the U.S. Congress are presumed to be statements on priorities and values of those who elected the representatives and senators. Lived experience teaches that this is more ideal than practice and may exist more in the realm of theory than reality. It does not, however, answer the myriad questions that form the conservation question posed in chapter 1.

Any answer begins with recognizing that the system for Congress and the legislative power were designed to protect the status quo and complicate enacting change. The system defaults to stalemate or no outcome in most circumstances.[2] Critically, these protections are embedded in the legislative process and the congressional design. One concern is that they are likely to create important asymmetries. The design risks working in the negative and empowering factional abilities to defeat the progress of legislation and policy. While this might prevent bad policy or legislative outcomes, it could also be too much of a barrier to any outcomes, even those in the public interest.[3] If so, the design would also limit the evolution of policies and allow problems to build and worsen over time. Such concerns are at the core of what the conservation question asks.

The history of farm bills presents a series of case studies in coalition-building and maintenance in the post–New Deal political environment. From farm assistance to natural resource efforts and direct assistance to low-income Americans to help them put food on the table, farm bills can deliver for a large cross section of the American voting public. Notably, however, a farm bill does not directly provide benefits to a majority of that voting public, but rather it provides varying degrees of benefits to an amalgamation of various interests represented by the farm bill coalition. Coalition-building among interests or factions is a critical feature of the legislative process in Congress and is among the primary lessons derived from farm bills. It also raises further questions in the vein of the conservation question.

Considering Congress: The Legislative Process in Theory and Practice

Article I, Section 1 of the U.S. Constitution provides that "all legislative Powers herein granted shall be vested in a Congress of the United States, which shall consist of a Senate and a House of Representatives."[4] The Constitution is a type of legislation, albeit superior in kind or supreme in form.[5] Arguably, the best sources for understanding the intent of the framers are the essays in *The Federalist Papers*. The essays were primarily authored by James Madison and Alexander Hamilton, each of whom was a participant in the convention that drafted the Constitution. Madison, among other things, presented a detailed plan for a new government to the convention and kept a journal or record of the debates. He is often considered the "father" of the Constitution and would go on to serve as Speaker of the House, secretary of state, and the fourth president. Hamilton was a delegate to the convention from New York and would go on to serve as the treasury secretary under President George Washington.[6]

One way to understand *The Federalist Papers* is as a committee report. In general terms, committees (including conference committees) write detailed reports that explain the legislation, justify or defend its provisions and the decisions made, and advocate for passage of the legislation by the Congress. Similarly, *The Federalist Papers* were political documents, written by those with intimate knowledge of the text of the Constitution, that explained the provisions and decisions of the convention, defended them, and advocated for the Constitution's ratification. As political documents—and similar to committee reports—they are not necessarily objective or unbiased, but they are explanatory, informative, authoritative, and instructive.[7] The advocacy essays penned by Madison and Hamilton remain guiding lights on political power, Congress, and the legislative process. While the discussion that follows offers a critique of the framers' design based on the theories contained in these essays, there is no intention to ignore the important reality that the framers of the U.S. Constitution designed a system of self-government that was unique in its time. They were conflicted individuals who were certainly fallible, but there was little precedent and few models to guide their work. The enormity of the accomplishment stands in spite of its faults.

From their essays, chief among the concerns for Madison and Hamilton in the design of a national system of self-government were organized interests known as factions.[8] Madison defined faction as "a number of citizens, whether amounting to a majority or a minority of the whole, who are united and actuated by some common impulse of passion, or of interest, adversed to the rights of other citizens, or to the permanent and aggregate interests of the community." He added a clear-eyed realpolitik: the "latent causes" of faction are "sown in the nature of man" and thus unavoidable. Importantly, "removing the causes of faction" was either unrealistic (a nation where all citizens had the same opinions, passions, and interests) or unacceptable ("destroying the liberty which is essential to its existence").[9]

Hamilton focused on refuting the arguments put forward by opponents of the Constitution. The "science of politics," he wrote, "like most other sciences, has received great improvement" from which "these are wholly new discoveries, or have made their principal progress towards perfection in modern times." He considered the system they designed in the Constitution to be a "barrier against domestic faction and insurrection" that would avoid the "disorders that disfigure[d] the annals" of earlier republics. Factions were "the tempestuous waves of sedition and party rage" that by distraction or agitation were the cause of "the rapid succession of revolutions by which [previous states] were kept in a state of perpetual vibration between the extremes of tyranny and anarchy." He attacked arguments from opponents of the Constitution for "splitting ourselves into an infinity of little, jealous, clashing, tumultuous commonwealths, the wretched nurseries of unceasing discord, and the miserable objects of universal pity or contempt." The new system of government, therefore, was intended to "repress domestic faction and insurrection" and avoid the troubled fates of earlier societies.[10]

According to Madison, the key to faction was motivation (or passion) rather than size or whether it was a majority or a minority of society. A faction is motivated to achieve narrow interests or self-interests rather than the public interest.[11] A faction, as explained by Victoria Nourse, is "not only a question of legal authority, but also a matter of personal connection and incentive" with "acts of will" through the "conduct of political relationships."[12] Thus, the concern with faction was based on an understanding of the political power they would possess, from which they would be able to work their will and act on their passions. The challenge, in Madison's words, was to "make it less probable that a majority of the whole will have a common motive to invade the rights of other citizens; or if such a common motive exists, it will be more difficult for all who feel it to discover their own strength, and to act in unison with each other."[13] Unable to remove what causes faction, a goal of the system's design was to provide the means to control the effects of faction. It would require managing the political power of factions without sacrificing basic rights or liberty.

The framers understood that they were concentrating immense political power in the federal government not long after having led a rebellion against immense power concentrated in a monarchical system. They had also experienced failure in the first system of government they had designed under the Articles of Confederation. Both Hamilton and Madison emphasized the importance of dividing the powers of federal government into three branches, representing the three basic powers of government: the power to make law (legislative), the power to enforce the law (executive), and the power to render judgment under the law (judiciary).[14] They also sought to strike a balance between the federal government and the subsidiary sovereign state governments. These are the relatively familiar concepts of the separation of powers, the system of checks and balances, and elected representation. More relevant to this effort and to the conservation question were their deep concerns with the legislative power, which they divided between two chambers with different forms of representation.

The legislative power was understood as the foremost power. Those elected to the legislature are the most directly connected to the voting public. The legislative power transforms policies, ideas, and demands into the laws of the self-governed. It is the power to create the laws of the nation or to change them. Comparatively, the president executes the laws Congress has created or changed; the Supreme Court interprets and applies the laws Congress has created or changed. Only Congress has the power *to initiate* the making or creating of law, but to exercise it, the Founders designed a complicated and difficult process.[15] According to the Supreme Court, "the legislative power of the Federal Government [must] be exercised in accord with a single, finely wrought and exhaustively considered, procedure."[16] Formally, that procedure is that "every Bill which shall have passed the House of Representatives and the Senate, shall, before it become a Law, be presented to the President of the United States."[17] In practice this means that the exact same legislative text must have been passed by a vote of both the House and Senate (bicameral approval) and be presented to the president. Upon presentment, the president may approve the legislation by signing it into law, or disapprove of the legislation by vetoing it; any veto goes through a subsequent step by which a two-thirds supermajority of each House of Congress must successfully vote to override or reject the veto.

Underwriting this design was a theory about the control or management of political power by competition. Humans may well be political animals, but they play politics in factions and groups. It would be better to let the groups compete to control each other than to attempt to implement controls on them. Or, in Madison's more eloquent explanation, "Ambition must be made to counteract ambition" to implement a "policy of supplying, by opposite and rival interests, the defect of better motives."[18] Congress was the primary arena for putting these theories about power and faction into practice. Joseph Bessette has observed, "No institution, it is argued, better recreates the conflict of interests present in American society or is more suited for the peaceful resolution of this conflict through bargaining than the U.S. Congress."[19] This competition is channeled through a process understood as an iterative type of game. Success in the process, as argued by Jeremy Waldron, is an achievement in the "circumstances of politics, including the circumstances of disagreement" and, ideally, the product of public deliberation in the public interest.[20] The most obvious concern with designing a system that is based on a theory of competition among factions is whether the competition is fair and a contest of relative equals.[21]

A closer look at the design reveals many aspects that are contradictory and seemingly paradoxical. Designed to protect against faction, it also provides excessive protection for the rights of minority interests from being overpowered by majorities. Madison was concerned that large groups or more democratic systems were overly susceptible to being captured by smaller interests. He wrote that "power would be transferred to the minority" within the majority and that the "countenance of the government may become more democratic, but the soil that animates it will be more oligarchic."[22] Madison's goals included preventing the "existence of the same passion or interest in a majority at the same time," and if a majority does form, it "must be rendered . . . unable to concert and carry into effect schemes of oppression."[23] His essays read as overly concerned with controlling majorities and the risks of bad legislation, but his theories lack sufficient and realistic concern about the need for legislation in the public interest. The result was a system that provides extraordinary protections for the status quo and an extremely difficult path to enact changes to it.

Bicameralism was a critical design feature. It established a double majority requirement for legislative enactment. This requirement means that "no law or resolution can now be passed without the concurrence, first, of a majority of the people, and then, of a majority of the States."[24] Not only must legislation achieve a majority of the people and a majority of the states, it must be approved by the president, representing national consent.

Passage of legislation in the House can be understood as representing consent of a majority of the people. Members of the House are popularly elected in districts of equal size, making them the most closely connected with the will of the voters. The House was expected to be the source of power for the larger states, with greater membership connected directly to the voters. As such, Madison expected that the House would possess significant political advantages.[25] *Passage of legislation by the Senate can be understood as achieving consent of a majority of the States.*[26] Senators are elected statewide, and each state has two senators providing each state with equal representation in that chamber. This design protected the interests of the smaller or less populated states. The Senate provided an "additional impediment" to strengthen against "improper acts of legislation."[27]

Bicameralism and presentment, however, are not the whole of the legislative game. Congress has added further complexities through its rules of operation and its organization around a system of standing committees along specific jurisdictional subject matter lines.[28] All bills are referred to, or initiated by, one or more standing committees. The chair and ranking minority member of the committee manage the legislative process, beginning with debating and amending bills in a committee meeting commonly referred to as a markup.[29] Thus, before a bill can clear Madison's double majority it must first achieve a majority of the policy specialists on the committee of jurisdiction, which sets and controls the agenda for the legislation.[30]

The standing committees play an outsized role in the legislative process, and within the committee, long-standing practices and norms regarding seniority help concentrate power further. As would be expected, the first requirement for attaining seniority is reelection; continuous service in Congress and on a committee has generally been necessary to build seniority. This custom is part of, and a strong example of, the unwritten rules and customs in Congress that have impact on the legislative process, as well as important implications for policy development.[31] It allows members who are consistently reelected to amass substantial power, at least over the matters within the jurisdiction of the committee(s) upon which they sit. A key example of this will be seen in the House Committee on Agriculture over the course of much of the history explored in this book, as southern representatives in safe districts were able to hold the chairmanship of the committee and wield great power over policy for decades.

In addition, rules for consideration and debate of legislation on the House and Senate floor impact the ability to succeed against Madison's double majority and, thus, are critical to policy development. All bills in the House must be approved by the Rules Committee before being considered on the House floor. The Rules Committee establishes the rules for debate and amendment, such as setting time limits on debate and restricting the number or subject of amendments. For example, the Rules Committee can protect a bill from amendments by setting a closed rule that precludes any amendments, or it can establish an open rule that does not place any limit on amendments. As should be clear, these different outcomes can drastically shape the agenda and the bill—limiting amendments protects the committee's work and decisions, while unlimited amendments permit members outside of the committee, or those who lost at the committee level, to seek to work their will on the bill. In many ways, the rule will have a substantial impact on a bill's chances for success on the floor.[32]

For the Senate, the most prominent matters for floor consideration are the Senate Rules for unanimous consent, cloture, and filibuster.[33] In short, the Senate needs either unanimous consent to proceed on a matter, or sixty votes to invoke cloture and cut off debate. If nothing else, the Senate's rules alter the theoretical construct of Madison's double majority and create an effective supermajority of the states. Today, this supermajority hurdle creates an even bigger issue when it provides outsized power to the smallest, least populated states. Overall, these rules and the customs or practices that have built up around them have an outsized impact on a bill's chances, which, in turn, determine much about policy development.

Finally, if the House and Senate pass different versions of the legislation, the regular process for reconciling those differences is the formation of a temporary conference committee. Members of this committee are appointed by the respective party leadership in each chamber, and conferees are typically from the committee of jurisdiction. Returning the legislation to the initial drafters and bill managers further strengthens the power of those committees.[34] The rules and design of conference committees bear on their importance to the final exercise of the legislative power. For example, the conference chair and vice chair are the chairs of the House and Senate committees of jurisdiction, and voting is unique. Conference does not decide by simple majority vote because it would be too easy for one chamber or the other to stack conference; agreement in conference requires both a majority of the House conferees and a majority of the Senate conferees.[35]

Conference committee represents the decision-making delegation within Congress at its most bicameral. Those tasked by Congress with managing the legislative power have the last opportunity to draft its provisions before the bill is put to a final decision by the entire Congress.[36] In total, then, to succeed in the legislative game, a bill and its supporters must (1) achieve a majority of the specialists on the committees of jurisdiction, (2) achieve a majority of the people in the House and likely a supermajority in the Senate, (3) achieve a majority of each chamber's specialist representatives on a conference committee, (4) pass both chambers a second time, and (5) achieve the imprimatur of the nation in the form of the president's signature.

It is in the legislative process that the theory about managing political power is put to the test. In practice, the design of the process creates multiple points to stop progress of a bill. These are generally referred to as vetogates.[37] The best known is, of course, the prerogative of the president to veto legislation that has made it through Congress. The myriad vetogates in this complex system should be obvious even if actual operations seem confusing or worse. Defenders of the status quo or opponents of legislation have multiple opportunities available to them to kill or stop progress on a bill, and it is far easier to accomplish. Supporters of a bill and its authors or managers must defeat these vetogates at each level and successively.

In the simplest of terms, the legislative process operates through seven vetogates (House Committee, Senate Committee, House floor, Senate floor, Conference Committee, House floor, Senate floor) before being presented to the president for signature or veto. If the president vetoes a bill, the process to override it adds two more but with supermajority requirements. To achieve presentment to the president, legislation must win at each of the seven, but opponents need win only once at any stage. This creates powerful asymmetries. Consider that if legislation were subject to the flip of a coin with heads for success and tails for loss, it would need to land on heads seven consecutive times to get to the president; landing on tails in any flip of the coin would kill the bill.[38] While this simple analogy provides some clarifying perspectives on the process, it is also riddled with shortcomings. For one, vote counting is not a random flip of a coin; for another, the Senate today requires a supermajority.

What matters is the ways in which the system's design and the rules, customs, and practices within the process favor inaction and minority interests. If the system was designed based on a theory of competition, that design is problematic. By tilting the competition in favor of inaction, the design favors minority interests. A simple majority has the power to defeat a minority interest's demands or policies. By comparison, the system provides factions a form of negative power—the ability to stop something from being accomplished—rather than enhancing a positive power of achieving enactment. Negative power is arguably too easy and certainly requires a minimal amount of effort, skill, or strategy. It offers little in terms of actual self-government because negative power merely favors getting nothing done. It can thwart majority interests or policies with popular support, further aggrandizing power to the minority by frustrating majority demands and leaving impressions of failure, incompetence, or inability in the minds of voters. In total, this greatly benefits the smaller, better organized (and financed) factional interests. By providing minority factions with multiple opportunities to block progress, the system enhances the power of factions rather than reduces or limits them.

In reality, there is little justification and less logic in a system of self-government that cannot accomplish anything. Accordingly, the most important strategy for the legislative game and for achieving success in the circumstances of politics is the building and maintaining of political or legislative coalitions. The key to unlocking the legislative process and clearing Madison's double majority is coalition-building.[39] Coalition-building can help clear the double majority hurdle, but it doesn't resolve the larger problems in the system's design. Upon Madison's paradox we stumble again. Coalition-building could also be a majority faction both discovering its strength and acting upon it.[40] Madison's theories did not distinguish between a proper majority acting in the public's interest and an improper majority. His were concerns that a majority might oppress minority interests or compromise a minority's rights, especially over property. To be fair to Madison, it would be difficult in a self-governing system to define an improper majority or delineate how it might be acting in ways contrary to the public interest.

There can be important differences between a popular majority and a coalition that can achieve majority votes in Congress. It was Madison's view that "society itself will be broken into so many parts, interests, and classes of citizens, that the rights of individuals, or of the minority, will be in little danger from interested combinations of the majority."[41] In this view there is little concern for a popular majority. Thus, coalition-building for votes in Congress is an amalgamation of smaller or lesser factions and interests. Even here, however, the system was designed, as explained by legal scholars, such that "large groups representing diffuse interests typically will not form" and the competition is stacked in favor of negative power.[42] It is theoretically possible and even practically likely that a successful congressional coalition will not represent a majority of the population or align with majority views on policy. A farm bill coalition, for example, represents a significant minority of American citizens or voters even with more than forty million receiving benefits from SNAP. It does bring together a large cross section of Americans, as well as interests in most of the states, to achieve success against Madison's double majority hurdle.

Countermajoritarian features permeate the constitutional design, including Congress.[43] The congressional design risks disfavoring majority achievement in the public interest and empowering smaller, factional interests. If smaller groups have the advantage, then narrow factional interests can control or manipulate outcomes when the public interest demands achievements. The most effective and frequently used method for this is a form of legislative hostage-taking. A minority faction can hold hostage policies or bills demanded by, or popular with, majority interests and even the public interest, using the vetogates to the faction's benefit. More importantly, the mechanism of coalition-building requires agreement among coalitional interests and factions to achieve the necessary votes in each chamber of Congress. More often than not, if a coalition wants to achieve anything it has to acquiesce to the demands of one or more minority interests. This provides asymmetrical negotiating leverage or bargaining power that a faction can use to extract outsized benefits for itself, or to protect its narrowest interests against the demands of others in the coalition or a majority of the governed public. History provides many concerning examples.[44]

A Conservation Question Example: The 2001 Reform Amendment in the House

A direct example of the conservation question occurred on the floor of the House of Representatives in October 2001. Representative George Miller (D-CA) argued that "something has gone wrong" with farm policy because in "farm bill after farm bill after farm bill" Congress has "just shoveled the subsidies to the largest of the farmers." Then, he asked, "what is the benefit for the other half of America?"[45] At the time, the House was debating what would become the 2002 Farm Bill, and Representative Miller was among a group of members who were fighting for an amendment to modestly reform farm policy by shifting some expected spending from farm subsidy payments to conservation. The reform amendment was narrowly defeated. The amendment's defeat demonstrated that the legislative process is not kind to reforms. It also demonstrated key pieces for how and why that is the case. The 2001 debate is included with the discussion about the theory and design of Congress as an initial example of the conservation question.

An important piece of the overall background to the 2001 debate and the 2002 Farm Bill was a budget anomaly at the turn of the millennium. During a brief period of projected federal budgetary surpluses, the House and Senate Ag Committees cut a deal for an *increase* in total spending on the new farm bill by more than $73 billion over ten years. Led by Chairman Larry Combest (R-TX) and ranking Democrat Charlie Stenholm (D-TX), the House Agriculture Committee bill spent nearly 70 percent of the *increased* funds on farm program payments.[46] Accordingly, the committee's farm bill added to a status quo in which farm program spending was already more than three and a half times that of conservation spending.[47]

Representatives Sherwood Boehlert (R-NY) and Ron Kind (D-WI) led on the reform amendment to shift some of the additional funding from farm payment programs to conservation programs. It was a debate over how best to divide the estimated *increases* in spending between them. As Representative Kind explained, the amendment would "fundamentally reform agricultural policy" by taking "a little bit of the increase in subsidy payments that will go to the largest commodity producers" and investing the funds "into voluntary incentive-based land and water conservation programs," leveling the field between the two policies and benefiting more farmers in more regions of the country.[48] Representative Boehlert was more blunt. He compared the committee's work to that of a banana republic that divided up the spoils. He claimed that the committee bill was "an exercise in raw power" backed by threats from the chairman against other programs and priorities.[49]

A key argument was the fact that conservation programs were underfunded, burdened by backlogs—farmers applied and were approved but could not receive funding.[50] Representative Kind reported that 70 percent "of farmers and ranchers seeking Federal funds to improve water quality are annually rejected due to the inadequacy of funding."[51] The existing backlog for conservation programs was a measure of political support to which representatives should arguably have given great weight—farmers were effectively voting with their feet, going to USDA for assistance but being turned away because Congress underfunded the programs. Because the committee included some increase in conservation funding, however, the argument was easily countered.[52] The additional funding also offered the opportunity to make threats, such as that by ranking member Charlie Stenholm (D-TX), who warned, "Be careful what you vote for lest you might get it."[53]

Reformers put forward strong arguments and almost won.[54] The state of the farm economy in 2001 made the push for reform more challenging, but that is not to say that the reformers were ineffective or that their arguments were the wrong ones. The 2001 debate demonstrated the difficulties in pursuing reforms of the status quo against the preferences of entrenched interests. It exposed much about factional power plays in the legislative process and raised difficult questions that challenged Madison's theoretical concepts. It demonstrated the problems with the competition among factional interests and the asymmetrical power among factions within coalitions. If paying for conservation practices was also in the public's interest because it provided secondary benefits to the taxpayers that funded the payments, that version of the public interest could not prevail against the narrowest private interest in the competition for public benefits. Why couldn't the reform amendment garner enough support?

Reform was viewed as taking from traditional farm interests with powerful allies in the chair and ranking of the committee, both representing Texas cotton interests. By any honest accounting, the committee bill clearly favored not just farm program payments but specifically payments to southern cotton farmers.[55] Chairman Combest attacked the reform amendment as an attempt to "rob" the farm programs and "create divisions among regions of the country."[56] His accusations helped marshal allies in defense, all of whom framed the amendment as an actual loss of funds for their districts and farmers.[57] It did not enter their arguments that conservation funds also went to their farmers and that it was not so much a loss as a different avenue for payments.

The core of the argument put forward by those defending the Agriculture Committee's bill was that subsidizing a cheap, abundant food supply was the whole point of the farm bill.[58] Representative Bob Goodlatte (R-VA) argued, for example, that the farm bill was "dedicated to the proposition that America is a land that has been noted throughout its history for producing the greatest, most abundant, safest and most affordable food supply anywhere in the world" and that the reform amendment would "do great harm to the American workers . . . but also to the American consumer" by shifting funding to conservation.[59] Economic distress could cause farmers to go out of business, and the production of food they represent could be lost. With enough of this outcome, the food supply to the nation could be at risk. Speculative and tenuous, and often exaggerated, it has proven an effective argument with politicians uncomfortable with the risk of hungry constituents, food shortages, or even food cost spikes. In this view, subsidizing cheap food was to be in the public interest and a priority.

Conservation assistance is not connected to production, crop yields, or crop prices. The farmer adopting the practices does not get additional assistance when prices are low. At the same time, the costs of the practices are magnified in the farm's bottom line, the primary concern of the producer. Conservation cost-share policies may lack support because the payments only cover part of the total costs to the farmer for adopting conservation.[60] This is more problematic in times of farm financial distress—which was generally the case for farmers in 2001. It can also be argued that the ability to conserve soils and other natural resources depended upon sufficient income support. For example, Representative Jerry Moran (R-KS) claimed that farmers could not be "stewards of the land absent an income in which to continue farming."[61]

Conservation policy is burdened by a history in which it was generally limited to paying farmers to take land out of production. This reserve policy is most prominent in the Conservation Reserve Program (CRP), and the longer land is locked away from production, the more pronounced the issue. More funds for conservation, therefore, "would take millions of acres of land out of production" that would otherwise be producing food.[62] Given the vagaries of weather and its impact on crop yields, there is a reasonable concern that taking too much land out of production could risk supplies in a bad year such as a severe drought. But this is also easily exaggerated. It would take rather extreme combinations of weather problems coupled with an unusually large amount of acreage under conservation contract.

Another challenging argument for conservation that pays to remove acres from production is that it would harm "small towns and merchants that depend on [farmers'] business," especially during "troubled economic times," when such efforts "could precipitate a rural farm crisis."[63] To others, adding to conservation amounted to taking money from the deserving farmers and handing it over to those who did not deserve it, such as landlords.[64] These arguments have important history behind them. They also serve reminders that conservation reserve policies involve the federal government taking an interest in private lands, which presents significant political challenges. Representative Ray LaHood (R-IL), for example, argued that the "best name for this amendment is the 'land grab amendment' because this affects the idea that we can take a big chunk out of a farm bill that was delicately put together and turn it into something that can be called conservation or preserving the land."[65]

Conservation payments also come with many conditions and requirements. The payments also implicate the American system of private property rights. Conservation payments involve—albeit on a voluntary basis as a condition for receiving the payment—changes in the use of the individual's land in some form or fashion. It is here that conservation policy approaches private property rights. Representative Marion Berry (D-AR) argued that farmers did not need "one more way where the Federal Government can come and tell them what they have to do with their land" and did not need "a social engineering program."[66] Representative Mike Pence (R-IN) attacked the amendment because, he claimed, it "has been written or encouraged by the environmental lobby, rather than by actual farmers."[67] Conservation is voluntary—options for the private property owner, not deprivations of basic rights—and the voluntary adoption of conservation is in return for direct federal assistance. The proximity to land use, however, has allowed opponents to raise the specter of regulation, governmental control, or interference with private property, and this relatively light brush with private property rights adds significant political challenges for conservation policy.

Finally, the hypersensitivity around private property rights exposes potentially more troubling issues. Representative Frank Lucas (R-OK) placed as fine a point on the matter as can be expected in a public, on-the-record debate. Representative Lucas was chairman of the conservation subcommittee of the House Agriculture Committee. He warned that the amendment was "biased toward certain producers" because it "ensures that small and socially-disadvantaged farmers are awarded a contract" for conservation.[68] The particular point Representative Lucas was making likely resonated far beyond its meager space in the Congressional Record. The term "socially disadvantaged farmers" was coined by Congress to address historic discrimination in federal farm lending programs.[69]

The reform amendment did not just take funding out of farm program payments but shifted funding away from the largest, most sophisticated farmers. It treaded upon matters of equity and the glaring, persistent inequities in how federal funds are distributed by farm policies. This argument may have struck a particular political nerve to the extent that it can be interpreted to mean that the amendment posed a specific threat because it was likely to divert some of the funding to farmers who were not white. Substantial, troubling historical baggage burdened that comment, some of which will be uncovered in the historical explorations that follow. If Professor Key's observation needed further proof, or needs evidence of its relevance in modern times, Representative Lucas's statement provided it.[70]

Farm programs may well be vulnerable to charges that they are ineffective and often counterproductive, but that was not enough to reform them. Income support remains a high priority of the farm factions that receive it, especially in low price situations. It remains the priority even though it conflicts with a history of arguments that conservation investments are needed to protect the nation's productive capabilities.[71] As an example for the conservation question, the 2001 debate demonstrated the difficulties in achieving reforms but also exposed some important substantive challenges for conservation policy. To the extent it represents more of the public's interest, conservation must contend with the system's asymmetries while also addressing the concerns about land-use regulation and private property, the impacts on farm incomes and rural economies, and the potential (but exaggerated) risks to the public's interest in its food supplies.

Concluding Thoughts on Congress and the Conservation Question

There is no shortage of criticism of Congress, but much of it can lack a full understanding of the challenges designed in the legislative process. The conservation question is intended to provide a tool for better understanding and more thoughtful critique. The smallest interests exert outsized influence and control over the process. Farm bills provide prime examples in which farm factions capture disproportionate federal benefits and are effective in defeating changes or reforms. Critical to these outcomes are the protections the legislative process and congressional design provide to the status quo and the difficulties in achieving a legislative outcome. While this might prevent bad policy or legislative outcomes, it reflects a troubling pessimism about an "excess of law-making."[72] It can block policies in the public interest but also limit the evolution of policies in response to incremental challenges or problems that build and worsen over time. It limits the ability of the public interest to succeed in Congress, and this can cause a failure to adapt statutes to changing circumstances or needs. The failure to adapt allows problems to build into dangerous situations that are more difficult to address or require more drastic responses.

The legislative history and political development of conservation policies in the farm bill provide important examples that appear to contradict the impressive arguments put forward by key authors of the Constitution in *The Federalist Papers*, but not entirely. Agricultural conservation policy represents an intersection of the public interest with private property rights and narrow factional interests seeking to maximize benefits. In *The Federalist*, no. 10, James Madison warned that the "most common and durable source of factions has been the various and unequal distribution of property."[73] He knew well of what he wrote; it may be one of his most durable observations.

In addition, the development of a system of policy or law is very dependent upon its origins and prior experiences. Known as path dependency, each step is determined in some part by the steps that preceded it. In law, the concept of precedent is similar. Precedent is a previous case or decision that can serve as an example or authority for a later, similar case, with decisions aligned on the principles previously established. Both path dependence and precedent can be defined more generally as a "course of conduct once followed which may serve as a guide for future conduct" or actions.[74] Add to this understanding the clear warning from Alexander Hamilton in the introductory essay to *The Federalist Papers*. He wrote that "among the most formidable of the obstacles which the new Constitution will have to encounter" is the "obvious interest of a certain class of men in every State to resist all changes which may hazard a diminution of the power, emolument, and consequence of the offices they hold under the State establishments."[75] Path dependency is not necessarily automatic but is a result of factional interests benefiting from the design's protections of the status quo.

Accordingly, a robust understanding of factions and their power in the political processes is crucial. Such an understanding may begin with Mancur Olson's conclusion from the 1960s. He wrote that "since relatively small groups will frequently be able voluntarily to organize and act in support of their common interests, and since large groups normally will not be able to do so, the outcome will not be symmetrical" and that the smaller groups will "have disproportionate power."[76] More recently, law professor Herbert Hovenkamp added a profoundly important piece to this understanding. He wrote that the protections against faction in the Constitution are all structural and that "if the Constitution's intent was to eliminate capture with this set of structural devices, it failed." He argued that this was because it is "much easier for a focused interest group" or faction "to prevent a bill from being passed." As such, the "Constitution's constraints on bill passage are successful in limiting the power of factions if doing nothing is the baseline, and the faction wants socially harmful legislation," but these very same structural constraints "are counterproductive, however, if the public interest requires legislation but a faction opposes it."[77]

The history and development of conservation policy also demonstrate that change is possible and that the status quo is not impregnable. Dr. Christopher Bosso provided an important clue based on an observation about political power. When a faction is "forced to deploy all available weapons" in the defense of policy positions, such efforts can be costly in terms of political power. He explained that political power, "like power in any form, is most effective in its threat, or reputation, than in its actual use."[78] Throughout the history, this observation will be valuable in understanding how change succeeds and the ways in which the status quo can be altered.

Pick a metaphor; these four concepts and five quotes can serve as guideposts on the historical journey. The points contained within them can serve as handholds and footholds for the climb. When needed, any one or all of the matters can provide food for thought and nourishment along the way. Combined, they provide critical ingredients for building the foundation of a working theory on Congress, which, in turn, can answer the conservation question. For now, they conclude this chapter and launch the in-depth exploration of the legislative history and political development of farm bill conservation policies. The larger implications and issues for Congress will be revisited, reevaluated, and reassessed in chapter 10.

PART 2 Dust Bowl Origins

3

Out of Dust, Sharecropping, and the Supreme Court

AN ORIGIN'S BACKSTORY

"Last year, on the 11th of March, for the first time since white man came to America, the eastern seaboard of the United States was visited by a great dust storm which originated west of the Mississippi in the Great Plains States." This was part of the response by Dr. Hugh Hammond Bennett to a question about the dust storms that had impacted large swaths of the country during testimony to Congress on March 21, 1935. He added, "Material carried by that storm went out into the Atlantic Ocean for a distance of 300 miles or more, falling on the decks of ships." It was not an isolated event. Earlier in March, he added, a dust storm had turned the rain in Washington DC yellow "for the first time in the history of the United States."[1] On March 20, 1935, he pointed out that "it will be impossible to maintain permanent prosperity over large areas of the United States if the present rapid destruction and impoverishment of our most valuable agricultural lands by accelerated erosion are permitted to continue."[2] In his testimonies to Congress, Bennett reported that the storms had blown over 300 million tons of topsoil, "equivalent to 150,000 acres of land, 12 inches deep." Worse still, he estimated more than 50 million acres of farmland had been "essentially destroyed by wind or water erosion" and another 105 million acres had lost "practically all of the topsoil."[3] He was quite clearly sounding alarms. Considered the Father of Soil Conservation, Dr. Bennett was the leading voice on soil erosion and conservation in the Roosevelt administration and a crusader for soils over a career that spanned nearly fifty years.[4] Just over a month after his testimony, on April 27, 1935, President Franklin Delano Roosevelt signed into law an act to provide for the "protection of land resources against soil erosion," and federal policy for agricultural conservation was born.[5]

If there is an official version of the origin story for agricultural conservation policy, it tends to be built around Bennett's dramatic and timely testimony to Congress. The Dust Bowl compounded the catastrophe of the Great Depression, spurring congressional action in Roosevelt's momentous first term.[6] Roosevelt and Democrats in Congress ushered in the New Deal, a mammoth political and policy undertaking to respond to vast emergencies in a perilous era. It permanently changed the American system of self-government, altering the relation of American citizens to their government and their government to them. Dr. Ira Katznelson has written of the New Deal as a "most profound and enduring response to the challenge of navigating emergencies and managing the conflicts that are inherent in efforts to guard liberal democracy, while safeguarding its own institutions and advancing its values."[7] And while the New Deal recedes further into history, increasingly lost to the ever-shortening political memory of the nation, it remains incredibly relevant. The New Deal offers opportunities to "reexamine . . . pressing theoretical as well as political issues" with "much to teach us still," especially (and arguably) in our modern moment of economic, political, social, and environmental turmoil.[8]

From this telling, conservation policy became one part of the vast emergency response effort that was the New Deal. The heroic efforts of Dr. Bennett and his allies in their crusade to combat soil erosion, reflected in Bennett's testimony to Congress in the spring of 1935 as the soils of the Great Plains darkened the skies over Washington DC, provided the critical catalyst.[9] That there was much more to the origin is not surprising; origin stories are tricky things, and policy is no exception. The Dust Bowl narrative oversimplifies the history. It is convenient but incomplete. The 1935 Act was not the first effort for federal agricultural conservation policy, and it was certainly not the first in federal natural resource conservation policy. Earlier developments in conservation policy generally trace to the early 1900s, the Theodore Roosevelt administration, and the work of Gifford Pinchot.[10]

The 1935 Act provides a clear demarcation because it marks the starting point for the agricultural conservation policy as it now exists and is authorized by farm bills. The Dust Bowl was a key driver, but it was not the only one. A more complete origin story, therefore, includes the larger political and economic setting. It provides a backstory that brings important context and perspective to help better understand the initial developments in this direction and its limits.

Situating the 1935 Act in the Longer History

It was not as if USDA or Congress first discovered the issue of soil erosion as Oklahoma topsoil fell upon the Capitol in 1934 and 1935. There is a much longer, deeper history built under the episode that was the Dust Bowl. Soil erosion has been a known consequence of agricultural production since the beginning of human efforts to produce food; as noted in chapter 1, soil erosion from farming practices has been a significant factor in the decline and collapse of human societies and civilizations throughout recorded human history.[11] In fact, drought and dust storms were known to be a part of life on the Great Plains long before the Dust Bowl.[12] What the Dust Bowl accomplished was to drive home to Congress and USDA the consequences of policy decisions and past inactions.

Intensive farming in the Great Plains, mostly to produce wheat, was by design. It began when Congress enacted the Homestead Act of 1862, which provided 160 acres or less for free to U.S. citizens who agreed to settle and cultivate the land for five years. It was an early attempt at social reform by the agrarian utopians of the pre–Civil War era known as the "free land movement." The policy was an attempt to counter the large landholding of monopolies, corporations, speculators, and others by providing poor people with some wealth out of the public domain in the form of farmland and home. Simplistic and overloaded with reformist idealism, the homesteading policy was problematic on many levels. For one, the land considered public domain was the home of indigenous peoples, many of whom had been relocated previously. Reality also challenged the ideal, as railroads, large property owners, and speculators also benefited disproportionately. More relevant to this discussion, as settlement moved further west into the more arid regions of the Great Plains, 160 acres was simply too small to sustain a farm operation. There is little to excuse the failure to account for the well-understood challenges facing the newcomers and the basic realities of how poorly suited these taken lands were for intensive row crop farming.[13] The Dust Bowl was the intersection of various policy problems and human folly, compounded by hubris and greed.

An exceptional amount of acreage had been put under intense cultivation. And yet, breaking the native sod in the southern plains was also only part of the problem. Sod-breaking was not a singular event or a one-off environmental harm. Each year, the method of farming and the manner crops were raised potentially compounded the problems. The cultivation methods used by these farmers on these new fields drastically magnified the impact of breaking the native sod of the plains. The transplanted farmers had come from the U.S. Midwest or East, or from Europe, regions where rains were plentiful. Many were unlikely to have had any experience with the climate and soils of the region. They were required to break and farm the land and advised to strictly adhere to an intensive cultivation method that pulverized the topsoil, leaving it exposed and extremely vulnerable to the high winds and drought conditions in the region. The most notorious example was the slogan that "rain follows the plow" on the plains. Whether or not any farmers succumbed to the false allure that plowing could help produce rain, their inexperience would have made them vulnerable to mistakes or worse. Pseudoscientific theories and propaganda pushed by the railroads and other landholding interests sought to promote settlement in the region for the sake of short-term profit.[14]

Even as it was helping drive expansion deeper into the more arid regions of the Great Plains, Congress was also taking very tentative initial steps with respect to erosion. For example, in the Hatch Act of 1887, Congress established experiment stations at the land-grant universities for purposes that included the need to "preserve the fertility of our soils."[15] USDA was also moving slowly and methodically on developing an understanding of the problem. In 1894 the secretary of agriculture established the Division of Agricultural Soils to investigate soils, soil conditions, and the production of crops. It was a new line of work for USDA that by the turn of the century was resulting in annual field operations reports to Congress consisting largely of soil surveys across the United States to classify and map the types of soils.[16] Plenty of warnings were provided in these reports, and Congress had received substantial information about the state of American soils and conditions across much of the country. Expansion deeper into the Great Plains continued. USDA's soil survey work was necessary to provide scientific research into farm practices best suited to the region. Without more action by Congress, however, USDA was slow to extend what limited agricultural research and knowledge existed. Efforts to educate and assist the transplanted farmers would not have matched the federal government's efforts to give away the land to settlers rapidly plowing under the plains, nor would it be expected to keep pace with—let alone get ahead of or counter— the self-serving propaganda and marketing by railroads and speculators.

Spring 1935 was not Bennett's first warning to Congress. In 1928 he reported to the Senate Agriculture Committee that soil erosion was one of the most important problems facing American agriculture. Bennett's testimony was comprehensive. The picture he painted of the problem of soil erosion was disturbing and extensive. "In this country, about 10,000,000 acres of formerly cultivated upland already have been destroyed by gullying and about 3,000,000 acres of once rich alluvium have been ruined or seriously damaged."[17] USDA estimated an annual loss of more than $200 million from soil erosion. The damage to fields harmed farm yields and required that farmers add more and more fertilizer, but it also impacted rivers and other waterways. He detailed the losses in every part of the country and even tied it to the record flooding of the Mississippi River in 1927. He was reporting all of this in support of a bill that Senator Morris Sheppard (D-TX) had introduced for the conservation of rainfall and soils. Farmers could adopt various practices that would help, such as terracing, but USDA needed both additional authority and additional funding to address soil erosion.

The Senate Ag Committee reported the Sheppard bill favorably in May 1928.[18] The bill passed the Senate on May 23, 1928, without amendment or debate but never made it out of the House Ag Committee.[19] Senator Sheppard reintroduced the bill in 1929, and the Senate Ag Committee again reported it favorably. The Senate again passed it without amendment or debate in February 1931. The bill died in the House Ag Committee once again.[20] The information and experiences compiled by USDA's soils experts certainly helped elevate the issue, but they were unable to move Congress. The Sheppard bill twice died in the hands of the House Ag Committee for reasons never publicly explained. After all of the surveys, investigations, and stockpiling of information and experience, USDA "had no program of direct technical assistance to farmers for conserving soil and water" as late as 1933.[21] It would take catastrophes that could not be ignored to provide the catalyst for change.

The Great Depression was the first catastrophe to drive change. In October 1929 the American stock market crashed, taking down the world economy and ushering in the Great Depression. For American farmers, however, the Depression had begun nearly a decade earlier, when crop prices crashed in the wake of World War I. Depressed prices depressed farm incomes and land prices, causing many farmers to lose their farms. Multiple attempts to enact federal policy to help farmers were defeated throughout the Roaring Twenties by conservative Republican Congresses and presidencies adhering to a form of laissez-faire governing philosophy, pursuant to which markets and businesses were to be free of governmental interference (except the protective tariffs for industry). Attempts to at least begin addressing soil erosion were also victims of this governing philosophy.

Catastrophes and changes were seemingly concentrated in a relatively short span of time but had been building for decades. In hindsight the results appear foreseeable. Transplanted farmers pulverized the soils of a windswept, semiarid region, moving deeper into the more problematic southern Great Plains around the turn of the twentieth century. The initial economic and patriotic boost of World War I (1914–18) drove more transplanted farmers to follow suit, ignoring whatever good advice was available. The situation likely became a sort of feeding frenzy or snowball effect; more and more wanted in on the action as temporary exuberance fueled excess and recklessness. When the war ended, the boom times came to an abrupt stop, and the farm economy collapsed into depression. The farm depression only made matters worse. It exacerbated the bad behaviors and recklessness because financially stressed farmers worked the land harder (and expanded acreage where possible) in a desperate attempt to offset low prices and collapsed incomes. If the soil were pulverized to capture a share of the boom, it had to be so again to cover the losses in the bust. With the onset of the Great Depression in 1929 the disaster was completed. Many farmers failed or walked away from the broken land, leaving behind barren soils vulnerable to the high winds. All that was left to complete a collapse into catastrophe was a severe drought.

Beginning in 1931 a severe drought hit the Great Plains. When drought arrived, the Dust Bowl intruded upon the Depression. It lasted the better part of a decade and is still considered the "drought of record" for the United States. Strong winds whipped up the pulverized topsoil of the drought-stricken High Plains, causing dust storms from roughly 1932 into 1940. The most severe wind erosion occurred during the years 1935 to 1938, causing massive dust storms that not only ravaged the southern Great Plains but blew the soils east to be imprinted on the American psyche and memory; images of the dust storms remain indelible. To this day, the Dust Bowl is considered one of the worst environmental disasters in the nation's history and a unique political event, although dust storms in that region were not all that unique.[22]

While drought and dust storms were long part of the reality of the Great Plains, the 1930s drought had been preceded by a relatively long period of above-normal precipitation and favorable weather. It was almost a trick of nature. A stretch of good weather coincided with the tail end of homesteading, wartime demand, and a significant expansion of intensive wheat production. The dust storms that followed likewise coincided with the economic and social devastation of the Great Depression. Great Plains soils in the skies of the Eastern Seaboard of the United States provided visceral evidence of society's myriad, compounding troubles; the economic crisis combined with that of testing the outer bounds of nature's tolerance.[23]

The Political Power of Southern Democrats and a Troubled Start for Farm Policy

Congress provided one measure of the profound changes wrought by the events of this era. Republicans held large majorities in the 70th (1927–29) and 71st (1929–31) Congresses. These were the Congresses in which the Sheppard bill was buried in the House Agriculture Committee. The 1930 midterms delivered significant losses for President Hoover and his party. Control of the 72nd Congress (1931–33) was effectively split but narrowly in favor of the Democrats. Roosevelt's election in 1932 overwhelmingly defeated President Hoover and swept Democrats into power in the 73rd Congress (1933–35) with a two-hundred-seat majority. In the wake of the crash in 1929, Hoover's relatively easy 1928 election was followed by his party losing over 150 seats in the House and twenty senators.[24] In just two election cycles, the elected branches of government had gone from strong Republican control to an overwhelming Democratic juggernaut; Roosevelt and the 73rd Congress initiated the New Deal. Among the many changes that resulted, Congress began reversing the homesteading policy in 1934 after it had resulted in the breaking of as much as one hundred million acres of native vegetation. Congress also authorized the federal purchase of up to seventy-five million acres to be removed from cultivation. The Roosevelt administration also sought to use the Civilian Conservation Corps to plant millions of trees as a windbreak on the eastern edge of the Great Plains.[25]

In Congress, Roosevelt's defeat of Hoover placed the political branches of government in control of the Democratic Party. The change in partisan control, however, masked a much more important change in political power. Congressional control by Democrats brought to power southern representatives and senators, including the gavels in key committees and control of the leadership posts in both chambers.[26] It was a nearly impenetrable redoubt of power, combined as it was with one-party (white supremacist) control in the region. The South controlled Congress and held the ability to operationalize strategic vetoes over any matter of social reform that might interfere with, or could potentially threaten, the Jim Crow system. Southern political power, moreover, was built on the planter class, for whom cotton was king of an antiquated plantation system.[27]

The myriad emergency response efforts of the New Deal included the Agricultural Adjustment Act of 1933, enacted in early May. It delegated broad authority to the secretary of agriculture, Henry A. Wallace, to address the Depression in the agricultural sector. The adjustment policy was based on an economic diagnosis that the staple commodities, corn, cotton, and wheat, were oversupplied and that the surplus crops were harming prices. From this, economists and other experts designed a policy intended to better manage the national agricultural plant and align supplies with demand. They argued that this method would improve crop prices and farm incomes. The method for this was built upon acreage-based production controls, a concept designed on what some historians have called an industrial model for agriculture. In effect, the acreage reductions intended to control supplies were the agricultural equivalent to laying off factory workers.[28]

The policy's most notorious example was the USDA decision to pay cotton planters to plow under a quarter of the growing cotton crop in 1933. The starting point for farm policy was integral to the origin of conservation policy, as will be discussed further, but a critical distinction is important to understand. Acres reduced in the production controls were not necessarily acres taken out of production; farmers could switch crops, shifting acres out of the reduced crop (e.g., cotton) and plant them to another crop (e.g., feed grains such as sorghum, oats, or rye). For cotton, switching crops was part of the point. Cotton interests quickly concluded that the 1933 Act was not sufficient or was not working, and they instituted more drastic control measures in the 1934 Bankhead Cotton Control Act. The policy was a mechanism for underwriting the movement toward a more diversified southern farm sector but with substantial consequences.

On the other side of the political equation, the new administration was home to many progressive, reform-minded New Dealers to whom the Depression and the Dust Bowl provided unique opportunities for far-reaching reforms; overlapping catastrophes were arguably a mandate for change. From Depression to dust, the New Deal reform effort was a mammoth political undertaking, and the agricultural sector was not an exception. In a real sense, the New Deal was gaining power with experience, adapting to each new problem and response. Policies, programs, and bureaucracies were being put together in rapid time, seeking to address deeply entrenched, long-running, and systemic challenges—some of which exposed further problems or made them worse.

As farm policy moved from adjustment in 1933 to cotton control in 1934, some New Dealers were developing their views around environmental degradation and the need for conservation to help revitalize the rural and national economies. In 1934 Hugh Hammond Bennett sent recommendations to Roosevelt's National Planning Board that included the need for governmental acquisition of "badly eroded lands unsuited to continuing safe agricultural use and convert them into forest or grazing preserves."[29] The board reported in December 1934 on the need for national policies to better align land ownership and use with the "interest of the general public welfare, as contrasted with purely individual or group interests."[30] Not surprisingly, among the reforms was an idealistic push for "a new land policy for the nation, one designed around the idea of land-use planning" and one that acknowledged that "the so-called farm problem [w]as an outgrowth of land misuse." Reforms implicating land use or private property protections collided directly with the troubled southern power structure in Congress. As one historian of the era noted, reformers from the Midwest and East Coast "were strangely ill-prepared for tough, plantation politics" and when they were confronted by the reaction of entrenched political power, they "caved to dominant interests."[31]

The New Deal reformers stumbled into a brutal reality in the South, beginning with the acreage-based production control policies of 1933 and 1934. When USDA paid cotton planters to plow under a quarter of the growing cotton crop, it was the labor of those who had planted the crop that completed the task of plowing it under, but the need for labor decreased in proportion to the acres reduced.[32] It may have originated with economists, but acreage reduction policy was written by southerners in Congress and implemented by southern cotton interests at USDA. The contracts were between USDA and the planters. The cotton contracts were the handiwork of Oscar Johnston of Mississippi, who managed one of the largest cotton plantations in America and hundreds of Black sharecroppers. A prominent member of the planter class, he was a high-level political appointee at USDA with critical roles for the Agricultural Adjustment Administration, cotton support programs, and the newly created Commodity Credit Corporation.[33] Johnston not only drafted the critical provisions of the cotton contract to favor the planters; he also educated planters on how to avoid sharing federal benefits with tenants or sharecroppers.[34]

Here a darker side to the backstory unfolds, one that intersected with efforts to develop conservation policy. While the public record in Congress on it is sparse, the cotton sharecropper saga undoubtedly complicated policy development. It would almost certainly have limited the goals and aspirations of reformers. In Congress, southerners controlled the legislative process. They consistently bent policy to meet their demands and limited reforms deemed unacceptable or threatening.

More than almost any other major commodity, cotton farming in the New Deal era was reliant on intense, extensive hand labor throughout the season.[35] For poor Black farmers, moreover, cotton production involved a toxic debt peonage system, another form of labor exploitation in the "Shadow of Slavery."[36] Sharecropping was a system engineered by the planter class that has been described as a "peculiar form of wage payment" because it differed significantly from the typical or traditional tenancy system.[37] For example, a cash tenant pays an annual rent to farm the acres owned by the landlord, covers all costs of farming, and receives all revenue from the crop. A share-rent tenant and landlord split the costs of producing the crop, along with the proceeds from the harvested crop and the risks inherent in the crop. These systems were not cotton sharecropping.[38]

The sharecropper was not a tenant. In order to produce a cotton crop, the sharecropper borrowed from the planter for the costs (likely inflated) of all needed inputs such as seed, implements, equipment, and fertilizer. In addition to lending for crop production, the planter also provided housing, food, clothing, and similar items on credit. The sharecropper provided the labor for that crop throughout the season. Once the sharecropper harvested the crop, he delivered it to the landlord. It was the landlord, however, who sold the crop and calculated the proceeds attributed to the sharecropper. Any proceeds were first used for settling the cropper's debts, also calculated by the landlord. A sharecropper for whom the proceeds did not cover the debt remained in debt, which only increased the following crop year. If the planter decided that the crop was worth more than was owed, the remaining funds from the sale went to the sharecropper and his family. Sharecropping also, however, lacked any of the most basic features—not to mention any legal protections or rights—of a standard employer-employee relationship.

Debt was critical to every facet of the operation of the cotton sharecropper system. Contracts were a critical tool between the (more sophisticated) planter and the (often disadvantaged) sharecropper. Known as "furnish," the system reduced the sharecropper to dependency. The sharecropper possessed no legal rights in the crops grown on furnish, because those crops were produced under debt. Worse still, the sharecropper's debt could legally tie him and his family to the planter. Some southern states went so far as to make it a crime for anyone to fail to fulfill the contract. Some state laws made it all too easy to convict the debtor, and once convicted, the debtor was further degraded to convict labor; slavery in all but the semantics.[39] The planter had near-complete control and unchecked power, while at the same time he was also dependent upon the sharecropper to complete the work of producing a crop; it was the sharecropper, in fact, who knew how to produce a crop of cotton.[40] Awful and inhumane, the sharecropping system was not only brutal and exploitive but also riddled with hypocrisy and fueled by insecurities. These challenges for the planters would be magnified by federal assistance if coupled with interference from reformers.

Designed as emergency assistance for struggling farmers, the acreage-based production controls in the South provided a method by which planters could manage a self-serving exit from the exploitive sharecropping system they created. As acres were reduced—rented to the federal government in return for a payment that was not shared with them—the sharecroppers and tenants could be downgraded to create a pool of cheap labor that could be exploited without the risks inherent in furnish. Sharecroppers did the work; the planters captured the federal benefits. These federal benefits, in turn, helped the planters finance mechanization and modernization, as well as land consolidation, displacing more farm labor. Thousands of destitute sharecroppers were driven off the cotton plantations or out of cotton farming—many out of the South altogether.[41] At risk of treating summarily the depravity of this situation, those who were the worst off and possessed the least power were those driven into further poverty in the middle of the Great Depression—worse for those who were Black, given the brutal realities in the land of Jim Crow.[42]

Such an undertaking could not proceed easily, even in the South, and created new dilemmas for the planters. Foremost was the large, resident labor force—many of whom were Black and destitute, their rights curtailed by violent, racist oppression—at a time of serious unrest and depression. Any federal aid delivered directly to sharecroppers could very well have undermined the planters' control.[43] The formation of the Southern Tenant Farmers Union (STFU) was a critical example. Formed in Arkansas by both white and Black sharecroppers in July 1934, the STFU focused on the USDA cotton contracts and mistreatment at the hands of the planters.[44] By the winter of 1934–35, the issue was quickly overtaking the ability of USDA's cotton allies to deny a problem existed or cover it up. When liberal New Deal lawyers also opened a fight over the cotton contract, the matter required a swift response. On February 5, 1935, Secretary Wallace caved to the pressures being exerted by cotton interests, including those in Congress. The secretary went along with a purge of the New Deal liberal lawyers who had fought to protect sharecroppers in the cotton contracts. They were, quite simply, sacrificed to the power of the planters and southerners in Congress.[45] Power exercised at USDA and in state courts was backed by real violence on the ground and the constant threats of more.[46]

The dust storms and Bennett grabbed congressional attention after the purge over the cotton contracts at USDA. The timing also coincided with an increasing emphasis on the resettlement of poor farm families trying to scratch out a living on poor lands as part of the newly created Resettlement Administration.[47] Farm tenancy connected societal ills with some of farming's most pressing natural resource consequences, and the two overlapped in more than just the Congressional Record. By 1937 the Roosevelt administration reported to Congress that "erosion of our soil has its counterpart in erosion of our society."[48] The tough realities in that conclusion also risked conveniently blaming the tenant.

On February 11, 1935, Senator John H. Bankhead (D-AL) introduced legislation to create the Farm Tenant Homes Corporation as a mechanism to help promote farm ownership by tenants.[49] In his testimony to the Senate Agriculture Committee about the bill, Secretary of Agriculture Henry Wallace connected tenancy, sharecropping, and soil erosion. The "disintegration of the farm system in the South, particularly the plantation phase, has become progressively more rapid since the World War," explained Secretary Wallace. He went on to drastically understate the situation, adding that "operation of the cotton programs" had "injected additional complications into the usual tenant and farm-labor relationships."[50] L. C. Gray, chief of the Land Policy Section in the Agricultural Adjustment Administration, added that "tenancy is associated with serious soil erosion, exhaustion of soil fertility, the decay of buildings and other improvements" in many areas of the country.[51]

The basic reasoning was that poor tenants with little security in the tenancy had little reason and less ability to care for the land, the soil, or the property. Too focused on the short-term matters of survival and subsistence, poor tenant farmers and sharecroppers would not invest in the longer-term efforts to conserve the soil or other natural resources. This reasoning overlooked, however, the fact that the poor tenant's insecure situation was by design. Those farmers operated in a system that was unfair and loaded with disparities that were often overly beneficial to the landlords. To the extent that they were the victims of an unfair arrangement, the tenants should not have received the blame.

At the core, soil erosion and erosion of the national social contract were more than coincidence; degradation of the soil, the people, and the nation go hand in hand. This was not adequately reflected in the policies working through Congress and USDA, however. Federal policy to help tenants raised concerns about the amount of control given to the federal government over land use and farming. Agricultural conservation policy would implicate these same concerns. Southern sharecropping seemed intent on proving the first secretary of agriculture's conclusions about the fall of ancient Rome, punctuated by Great Plains dust deposited on the Roman statute adorning the Capitol dome.[52]

A First Step for Conservation:
The Soil Erosion Act of 1935

In multiple ways, 1935 was a critical juncture. The purge at USDA and initial steps on farm tenancy occurred in February. These were followed by the worst of the dust storms and Dr. Bennett's dramatic testimony to Congress in late March. Enactment of the Soil Erosion Act of 1935 was achieved in April.

The 1935 Act was enacted by acclamation. The House passed it with little debate and no recorded vote on April 1, 1935. The Senate required little debate and produced minor or technical amendments; senators agreed to the bill by unanimous consent on April 15, 1935. Congress quickly agreed on a final bill and sent it to President Roosevelt. The president signed it into law on April 27, 1935. The law focused on research and studying the problems of soil erosion, with efforts to educate farmers or landowners on how to control it. It was essentially what Congress had been unable to achieve in 1928 and 1931. Relatively narrow and focused on bureaucratic matters, the bill did not provide new federal funding.[53]

Congress recognized that "the wastage of soil and moisture resources on farm, grazing, and forest lands of the Nation, resulting from soil erosion, is a menace to the national welfare," adding "that it is hereby declared to be the policy of Congress to provide permanently for the control and prevention of soil erosion."[54] Authorship of the 1935 Act has been credited to members from New Mexico "familiar with the ravages of wind erosion, the shortage of water, the need to improve grazing conditions on badly eroded public lands, and the flooding and siltation of irrigated farmlands."[55] Much of the swift response to the dust storms can be credited to the handiwork of reformers in Roosevelt's first administration. Within the 1935 Act's development, there were hints of the push for major reforms within the crusade to save America's soils from the ill-considered actions of farmers and landowners.[56] The committee reports acknowledged that the "interest of the Nation in controlling erosion far exceeds that of the private landowner" because "land destroyed is land gone forever." The committees also acknowledged that "erosion directly threatens vast Federal investments in dams and channels and annually requires the expenditure of large sums for dredging operations."[57]

If the dust-laden winds seeded such reforms, there is little evidence in the legislative text. Congress did not communicate an intent to enact major reforms in land use or alter private property rights on behalf of the general public's welfare. Congress authorized the secretary to "acquire lands, or rights or interests therein, by purchase, condemnation, or otherwise, whenever necessary for the purposes" of the Act, but any powers over private lands were restricted to research, demonstration, and education, and only as "necessary . . . for experimental purposes, field headquarters, and engineering structures such as dams."[58] The committees also absolved the farmers and landowners, adding that they "have not been responsible for the erosion which has occurred" because in the "disposal of the public domain, settlers were encouraged to acquire the public lands and cultivate them" but without "restrictions, instructions, or advice as to methods under which the land should be used in order to protect it from erosion."[59] Rather than any of the more substantive reform goals from New Dealers, the bill was a federal hand guiding farmers and landowners toward voluntarily learning about and possibly adopting better farm practices on their rapidly eroding lands.

Arguably the most consequential provision might seem its most banal and bureaucratic. Congress created the Soil Conservation Service (SCS) and placed it under the helm of the U.S. Department of Agriculture, consolidating all federal efforts for controlling soil erosion there, including existing New Deal authorizations and funds to combat erosion and future soil erosion efforts.[60] The move marked an important victory for USDA in a bureaucratic battle.[61] Moving federal soil erosion work from the Department of the Interior to USDA was logical given that working farms and Depression-damaged farmers had contributed so substantially to the dust storms.

The 1935 Act was not major land-use reform in agriculture. Farming and its consequences were certainly at the forefront of reformer agendas, but Congress produced a modicum of policy development.[62] The ease with which it moved through the normally difficult legislative process supports this conclusion. It may have been a first step in that direction, however. The

1935 Act was noteworthy for the potential it contained and the opportunities for development that it could have set in motion. Within a year, whatever potential the 1935 Act may have held (or was intended by reformers to initiate) was eclipsed. On January 6, 1936, the Supreme Court upended the entire effort and sent matters on a different course.

The Final Piece of the Backstory: The Supreme Court Strikes Down Farm Policy

The year 1936 featured elections in which President Roosevelt and the New Deal would be on the ballot to be judged by the American voting public. Not content to permit voters the opportunity to render their judgment, a handful of New Deal policies were subject to very direct power plays in the federal court system. Conservative corporate and financial interests tried to stop and reverse the reforms being rapidly produced by the New Dealers. They found willing accomplices in the activist conservative faction of justices on the Supreme Court. The justices were known as the "Four Horsemen of Reaction" because they sought to nullify the New Deal by striking down key laws beginning in the spring of 1935. Each case presented different legal issues and reasoning, some more persuasive or valid than others; each case was an example of judicial activism. Combined, these cases were a shocking and arguably dangerous intrusion by the unelected justices of the Supreme Court on matters of policy and legislation enacted by duly elected officials as emergency responses to a global economic crisis.[63]

On January 6, 1936, the Supreme Court declared the 1933 Agricultural Adjustment Act an unconstitutional exercise of power by Congress. The cost of direct federal assistance to farmers was to be covered by taxes on the processors or first handlers of the commodities, such as grain elevators and cotton ginners, millers, or warehouses. Little surprise that such an obvious policy of redistribution drew strong ideological and partisan opposition. The case, *U.S. v. Butler*, involved Hoosac Mills, a bankrupt cotton mill in Massachusetts that initiated litigation to avoid paying the processing taxes.[64] The decision possessed monumental implications at the time but has long since been lost to history, buried beneath layers of American jurisprudential dust.

Ostensibly a dispute about the processing tax, the case was a legal Trojan horse for powerful political interests vehemently opposed to the New

Deal and all that it represented.[65] Litigation was initiated by the receivers of Hoosac Mills, who were prominent Republicans. The receivers were also backed by wealthy interests that included a large meat packer. It was a blatantly partisan strategy.[66] The case provided a clear example of powerful political allies converging to attack the New Deal and Roosevelt by attacking the 1933 AAA. Prominent among them was the American Liberty League, which had been formed in 1934 by members of the du Pont family and other wealthy conservatives.[67]

The majority's opinion was written by Justice Owen Roberts, and it managed to distort both the statute and the Constitution. Justice Roberts made for a rather conspicuous author. He had been appointed by President Herbert Hoover in 1930. One of the lawyers who argued the case for the mill's receivers was George Wharton Pepper. Scion of the Wharton and Pepper families in Philadelphia, Pepper had served in the U.S. Senate with Hoosac Mills receiver William Butler, and more notably he had mentored Owen Roberts, promoting his career at key moments.[68] Moreover, political rumor at the time had the American Liberty League pushing the Republican Party to nominate Justice Roberts to run against President Roosevelt in the 1936 election.[69] What was a highly questionable opinion legally and constitutionally appears much more unsavory with such an obvious conflict of interest by the author.

Substantively, the litigation was a legal long shot at best. It should have stood little chance in any court, let alone the Supreme Court. The Constitution explicitly provided Congress with the power to tax, and courts long held that taxpayers seeking to challenge expenditures by the government faced insurmountable legal hurdles.[70] Unable to confront the constitutional power to tax, the majority had to find another route. Justice Roberts's opinion pronounced that Congress lacked the power under the Constitution to regulate farming and agriculture.[71] To achieve this, the majority resorted to an absurd Tenth Amendment argument to concoct a limit on Congress that does not exist. Justice Roberts's opinion for the majority held that the regulation of farming was a matter reserved to the states under the Tenth Amendment and was therefore beyond the jurisdiction or power of Congress.[72]

The regulation of farming was overstated, if not misinterpreted. The 1933 Act paid farmers to plant less acreage to commodities that were oversupplied and very much part of interstate and foreign commerce. The program was voluntary, but Justice Roberts mischaracterized the support program as regulation of farming. Going further, Justice Roberts wrote, "Congress has no power to enforce its commands on the farmer to the ends sought by the Agricultural Adjustment Act," and therefore "it must follow that it may not indirectly accomplish those ends by taxing and spending to purchase compliance."[73] The *Butler* majority's retreat to the Tenth Amendment was more an exercise in ideology and politics than an application of law.[74]

The Supreme Court decision established no legal theory, nor did it provide any lasting jurisprudential contribution to American law. The dissenting justices had the better argument, which was written by Justice Harlan Stone. He pointed out the fallacy in the majority's logic, explaining that "a levy unquestionably within the taxing power of Congress may be treated as invalid because it is a step in a plan to regulate agricultural production and is thus a forbidden infringement of state power." This was, he added, "the pivot on which the decision of the Court is made to turn." The majority opinion amounted to "judicial fiat," according to Justice Stone, and the dissent warned against "the mind accustomed to believe that it is the business of courts to sit in judgment on the wisdom of legislative action."[75]

Voters delivered the ultimate verdict. In the 1936 elections, they voted overwhelmingly for Roosevelt and congressional Democrats, handing the Democrats a strong, clear mandate. With voters backing the New Deal over its opponents, the Supreme Court eventually reversed course and began upholding New Deal legislation, including farm support policy.[76] The *Butler* decision and its faulty reasoning, further shadowed by ideological pursuits and strong questions of conflicts of interest, has long been buried in the dustbins of history and law. The decision was not without consequences, however.

If Justice Owen Roberts had been right, the implications would have been far reaching, and his views would have drastically impacted the ability of the American people to govern themselves through elected representatives. The ideological fever dreams of those like the American Liberty League had been refuted by the reality of the Depression and the Dust Bowl. A farmer's actions in his fields damage not only his fields but those of his neighbor. The damage can be exported across state lines, possibly the length of the nation. Reality exposed critical fallacies in ideological purity. This worldview was the same one that prevented the development of incremental reforms that may have helped head off catastrophes. Adding irony to that reality, problems that were not addressed resulted in catastrophes and produced crises likely to catalyze more drastic responses.

As will be discussed in the next chapter, the decision provided the impetus for swift congressional action, but it also managed to limit the extent or scope of potential reforms. The faction of elites who had exercised immense power before the Great Depression were fighting back to hold onto as much of that waning power as they could, fearing they were being deposed in the arena of politics and elections. That faction fulfilled very clearly Hamilton's warning in *The Federalist*, no. 1, quoted at the end of chapter 2.

The backstory contains lessons for the political development of policy, beginning with the monumental challenges to achieving reform. The interconnections and feedback loops working through the Depression, the Dust Bowl, farming practices, and farm tenancy magnified the challenges confronting New Deal reformers. In the final analysis, a closer look at the 1935 law exposes the very limited nature of it. Those limits are particularly glaring juxtaposed against the Dust Bowl catastrophe and congressional recognition of the vast nature of soil erosion problems. What does it mean that Congress was unable to do better in the middle of the dust storms? Would Congress have done more to develop conservation policy that implemented the land-use reforms being worked on at USDA? There is no way to prove the counterfactual. It is not possible to chart the potential path of development in a Congress that was not controlled by southern factional interests, nor influenced by the tragic situation of that region's sharecroppers. It is also not possible to chart a different path had the Supreme Court's conservatives not intervened.[77]

4

Reform amid Ruin

LESSONS FROM THE 74TH CONGRESS

The 74th Congress of the United States (1935–36) was a Democratic juggernaut, surpassed in history by only the 75th Congress (1937–38). For the second half of his first term in office, President Roosevelt's party had increased its congressional majorities in the 1934 midterms. In 1936 Democrats had a 229-vote advantage in the House and a 46-vote advantage in the Senate.[1] This was, to put it simply, a congressional majority that was nearly unbeatable, with few structural limits on its power. Neither the double majority nor the complex legislative process presented real hurdles or barriers to any enactment.

While it may be far removed in time—and further removed in terms of politics—the early stages of the New Deal opened the path for an enormous amount of political and policy development in the United States. Agricultural conservation policy was one part of the much larger undertaking, and those first steps for conservation were critical. The course of conduct at the start guided future developments.[2] The interconnections and feedback loops working through the Depression, the Dust Bowl, farming practices, and farm tenancy were pronounced. These threads from the origin backstory worked their way into the efforts that unfolded in Congress. The complexities at work in American society left their mark on the policy just as the dust storms did on the landscape of the ravaged southern Great Plains. These combined disasters of the Great Depression and the Dust Bowl—one economic and the other environmental but overlapping and entwined—made the natural resource and environmental problems abundantly clear, more immediate, and pressing.

The single most important factor in the development of the 1936 Act, however, was a quirk in the federal system that permitted unelected, lifetime-appointed justices to shoehorn an ideological agenda on political development and the legislative process. The legislative record leaves no doubt that it was the Supreme Court decision that had an outsized importance in 1936. It catalyzed rapid action by USDA and in Congress for a second time in just ten months. Rather than dust storms, the 1936 Act was written in the dust from the Supreme Court's demolishing of the agricultural adjustment policy.

When that dust had settled, Congress amended the 1935 Act with the Soil Conservation and Domestic Allotment Act of 1936 and shifted farm policy to new programs built around soil conservation. Congress declared among its purposes the "preservation and improvement of soil fertility," the "promotion of the economic use and conservation of land," and the reduction of "exploitation and wasteful and unscientific use of national soil resources," while protecting rivers and harbors.[3] Doing so allowed it to work around the Supreme Court decision to deliver assistance to farmers but under the guise of conservation. Historical light, filtered by the Dust Bowl and diminished by time, has long been focused on the origins of agricultural conservation policy as an example of the political power of the dust storms. The work product of Congress complicates the story and provides perspectives on what limited that political power and the scope of intended reform in the policy.

Rapid Enactment in January and February 1936

The Supreme Court's decision was handed down on January 6, 1936. Within three days Senators Charles McNary (R-OR) and John H. Bankhead II (D-AL) attacked the decision on the Senate floor. Senator Bankhead's floor speech also laid key planks for the design of a new program. He stated that the goal was to rewrite the Agriculture Adjustment Act of 1933 (1933 AAA) but jettison the taxes on processors and the contracts with farmers. Congress should "eliminate the acreage contracts, which have so profoundly shocked the consciences of the Supreme Court majority," he emphasized. On this point, Senator McNary agreed. The Supreme Court had decided that "contracts for the reduction of acreage and control by production are outside the range of Federal power."[4] Neither senator mentioned soil conservation.

An effort to replace the 1933 AAA was already underway in the administration, led by Secretary Wallace and key appointees in the Agricultural

Adjustment Administration (AAA) such as Administrator Chester Davis. The intellectual force behind what would become the administration's proposal was Howard R. Tolley, who had led the AAA's Program Planning Division before returning to his faculty position at the University of California, Berkeley, in the fall of 1935. It is Tolley who has been credited with shifting the emphasis of agriculture policy toward conservation. Tolley had previously worked on efforts to reform farm policy as it was unfolding under the 1933 AAA. As one example, in 1935, Tolley led a study by USDA and academics that focused on what should be the proper methods for adjusting agriculture. The conclusions from the effort pointed toward conservation and better, more scientific land management, as well as adjustment based on regional needs and capacities. The search for a replacement for the 1933 AAA system provided a prime opportunity. Administrator Davis called him back into service in Washington and would eventually hand over the reins of the agency to him. Through Tolley, the eventual legislation connects to various New Deal reformers within USDA, including Hugh Hammond Bennett. And it was conservation-based policy that resonated strongest with President Roosevelt, who had highlighted the goal of moving toward such a policy for agriculture in October 1935.[5]

With a basic plan under development at USDA, Secretary Wallace called a conference of the farm organizations in Washington DC, just four days removed from the decision.[6] USDA's team had put together a proposal for the farm groups to endorse; the conference was not a chance for the farm groups to craft a replacement themselves. Chester Davis, for example, reportedly told the gathered farm leaders that there was only one constitutional path forward: using conditional appropriations (i.e., grants, payments, or rewards) to farmers based on "conserv[ing] their farm lands by scientific management in the planting of soil building legumes and the utilizing of other conservation practices."[7] The USDA plan was for conservation without a contract, providing a form of indirect production control without the constitutional concerns. Wallace's conference succeeded in getting farm leaders to coalesce around the administration's position. Coming out of the January 10 conference, the farm groups presented a consensus plan for Congress to enact a new law providing benefits to farmers by reducing acres for soil conservation.[8]

The well-orchestrated plan almost didn't survive first contact with Congress, however. The farm groups reported to the Senate Ag Committee on January 14, 1936, in the only public meeting of either committee regarding the legislation.[9] During the hearing, Senator George Norris (R-NE) confronted the key witness for the farm groups, Earl Smith of the Illinois Agricultural Association, with questions about the constitutionality of regulating agriculture. Senator Norris got Smith to acknowledge that the farm groups were effectively proposing another method for regulating farming, which the Supreme Court had ruled out. Seemingly a minor point, it was enough to send USDA scrambling. It also threatened the farm group consensus. It didn't take long for some of the groups to splinter off in their own preferred directions.[10] The effort to save the proposal led USDA back to the 1935 Act and further into the use of conservation. It was the path previously mapped by New Deal reformers but resisted by the insiders of the AAA.[11]

A little more than two weeks had elapsed since the Supreme Court decision when the negotiations returned to consensus. Tolley and USDA officials worked with American Farm Bureau lawyers in crafting a proposed legislative draft, which House Agriculture chairman Marvin Jones (D-TX) introduced in the House on January 22, 1936. A nearly identical bill was introduced the same day in the Senate by Senator John Bankhead (D-AL) instead of Agriculture Committee chairman Ellison D. "Cotton Ed" Smith (D-SC); authorship offered a hint of potential opposition by the chairman.[12] Both bills amended the 1935 Act and authorized broad powers to provide for soil conservation—paying farmers based on the acreage of soil-improving or erosion-preventing crops—but the program was to be administered by the Agricultural Adjustment Administration, not the Soil Conservation Service. Both bills prohibited contracts between USDA and producers.[13] Neither included taxes on processors, using appropriated funds instead.

The only noticeable difference was subtle but potentially important. The Senate bill initially did not include authority to make payments for changes in land use, but the Jones bill in the House did. The House bill also declared that promoting the economic uses of land was one of the five purposes of the act. Both provisions trace to the Tolley-led push for reforms.[14] While subtle, these differences add support to the view that the House bill was essentially the bill proposed by USDA and backed by the farm groups. In addition, if the New Deal reformers at USDA had intended for this to be the permanent program for a better adjustment of agriculture, then including changes in land use was a key element.

The difference signaled that the Senate Ag Committee, or at least key players on it, may not have been on board. Further demonstrating the existence and source of resistance, the initial challenges to the proposed conservation-based farm program arose in the Senate Agriculture Committee and with Chairman Smith. Sorting out his opposition required two days of closed-door negotiations with USDA. The compromise with the Senate Chairman modified the Bankhead bill by adding a second program designed to begin in 1938. This program would be operated through the states with USDA in a cooperating role. Before the month was over, the solicitor general defended the bill's constitutionality, and the Senate Ag Committee voted 15 to 2 to report the bill.[15]

The Senate Ag Committee appears to have taken the lead, and the House Agriculture Committee worked to revise the Jones bill to match. The new soil conservation program was estimated to cost $500 million each year with USDA indicating a goal to remove between 30 million and 36 million acres from production.[16] The Agricultural Committees reported to the House and Senate nearly verbatim what had been reported to them by USDA. Nearly 50 million acres of farmland had been destroyed by erosion, and an additional 50 million acres "were in almost equally bad condition." Moreover, erosion problems were noticeable on another 100 million acres, and "deterioration threatens the great bulk of 360,000,000 acres of cultivated lands in the United States," with potentially perilous consequences for the nation's farmers and consumers.[17] This information, important in its own right for a report to the respective chambers, also made clear the close coordination between the committees and the administration. This could have signaled the extent to which the president and congressional leadership were unified behind the bill as the policy preference for the Democratic Party in an election year. Such a display of unified support would likely dampen the enthusiasm that any Democrats might have had for demanding changes or offering amendments. Republicans, with such a vast vote deficit, would have little chance of stopping the bill's progress or making any changes. The juggernaut was moving, and the only options were to get on board or get out of the way.

As the legislation moved from the Senate Ag Committee to the Senate floor, committee leaders were still working out the design for the state-led program that was increasingly proclaimed to be the permanent program.[18] The bill was undergoing an important change from what had been proposed by USDA and introduced by Senator Bankhead. The bill introduced in the Senate was a straightforward federal program to pay farmers based on soil-conserving cropping practices or a percentage equal to the crop's normal production for domestic consumption, known as the domestic allotment. The reported bill, however, now split the program in two: the federal payment program prior to January 1, 1938, "and thereafter by assistance to and cooperation with the States in State action calculated to effectuate" the purposes of the act.[19] The reported text also included promoting the economic use of land as one of the purposes. Incorporating that component of the reform push gives the reported bill the look of a compromise between Chairman Smith (the state program) and USDA (economic use of land).

One month removed from the Supreme Court decision, on February 6, the Senate began considering the AAA replacement bill. Opening debate on the bill in the Senate, Senator Carl A. Hatch (D-NM), author of the 1935 Act, stated that "if we live long enough, we shall see, to our sorrow, that ... the dangers from depletion of the soil and the improper use of the soil resources of the Nation are not imaginary dangers; they are destroying forces which are already at work on millions of acres of farm lands in America today."[20] Leadership's backing helped Chairman Smith defeat any changes to the bill by amendment on the floor. The chairman was able to easily overcome what little opposition existed. The Senate passed the bill by a large margin on February 15, 1936.[21]

While the Senate Ag Committee had the full support of Senate leadership, the effort did not have unanimous support. For example, Senator Charles McNary (R-OR) led opposition to the bill, and he questioned its constitutionality throughout the process. He argued, "We are not trying to conserve soil, but we are trying to avoid the decision of the Supreme Court."[22] His opposition was notable because he was the ranking Republican on the Agriculture Committee, as well as a longtime champion of farm policy; it was McNary who had unsuccessfully pushed for farm support legislation throughout the 1920s against two presidents from his own party. Senator McNary also had alternative policy preferences from those earlier legislative efforts. More importantly, he was the leader on an amendment backed by dairy interests who were concerned that conservation-based payments would drive expansion of competition in their sector. Dairy interests wanted Congress to prohibit using acres in conservation for pasture. It was a key dissent from within the farm coalition, but the amendment stood little chance because it faced an overwhelming Democratic majority and the strong interest in getting the AAA replaced. The amendment's defeat, however, was blamed on the Supreme Court's decision, as even Republicans like Senator Norris argued it would require regulation of farming.[23]

The legislative text agreed to by the Senate and sent to the House for its consideration featured a permanent program that was more fully designed. The bill required USDA to cooperate with the states "in the execution of State plans to effectuate the purposes" of the act and by "making grants under this section to enable them to carry out such plans."[24] The new state program permitted the states to submit a plan to the secretary, and if the secretary approved the plan, the state was entitled to funds. As before, the states had until January 1, 1938, to have plans submitted and approved. The policy declarations and purposes of the act had also been revised to "(1) preservation and improvement of soil fertility; (2) promotion of the economic use and conservation of land; (3) diminution of exploitation and wasteful and unscientific use of national soil resources; (4) the protection of rivers and harbors against results of soil erosion . . . and (5) reestablishment and maintenance of farmers' purchasing power" while also maintaining a stable supply of agricultural commodities.[25] No small task for states to accomplish.

Wasting little time, the House debate on the modified bill began on February 19, 1936. It being the House, the debate reads as more contentious, the partisan edges sharper.[26] In general, arguments in favor of conservation and the general desire to replace what the Supreme Court had torn down carried the day; with a 229-vote margin, passage was not in doubt. House Agriculture Committee chairman Marvin Jones (D-TX) opened dramatically. He emphasized, "We have swept over this country like a swarm of devastating locusts . . . destroying our capital and letting it flow into the creeks and rivers and off to the sea."[27] After three days of intense debate, the House voted overwhelmingly to pass the bill on February 21, 1936.[28] The House had made some notable changes to the bill that had come over from the Senate, but the general policy of two programs—a federal payment program to 1938 and a state-based program through plans approved by the secretary thereafter—had remained intact.[29]

With calendars turning to March, planting season was soon to begin in the South. For members of Congress, the election loomed on the horizon. Working under pressure, conference between the House and Senate took less than a week, and the conference report faced no opposition in either the House or the Senate.[30] The effort to replace the 1933 AAA was accomplished on February 27, 1936, and President Roosevelt signed the bill into law on February 29, 1936. According to President Roosevelt, when he signed the 1936 bill, the federal payments to farmers were "conditioned upon actual evidence of good land use" and the conservation of a natural resource vital to the health and well-being of the nation.[31]

Only eight weeks separated the Supreme Court decision and the new law. It was no small accomplishment for $500 million in annual assistance to farmers through policy burdened by constitutional questions and concerns. If nothing else, enactment was a pronounced demonstration of political power by large elected majorities in both chambers of Congress for the president's party in an election year. The two branches worked in tandem to move forward along lines that the third branch had sought to impede. The Supreme Court's opinion was effectively moot two months after it had been handed down.

Digging Deeper: Soil Conservation, Domestic Allotment, and the Legislative Text

The speed at which Congress worked obscured important developments. The 1936 Act was hybrid policy within a hybrid bill, beginning with the title: soil conservation and domestic allotment. Congress authorized two programs to help farmers. One was intended to be permanent policy that operated through the states; the other was temporary, providing federal payments directly to allow states time to act. Even the temporary federal program was hybrid policy. Farmers could receive payments based on any combination of soil conservation and land-use change, or the domestic allotment. The domestic allotment was for payments based on the percentage of the farmer's "normal production of any one or more agricultural commodities" and the crop's normal domestic consumption.[32] These options, at the very least, provide an indication of competing, sometimes conflicting, goals.

The temporary program was also a stopgap, intended to operate only long enough to allow states time to implement their plans. If operationalized, this design would result in the reform plans being temporary, unless the states elected to incorporate those elements in their own plans. It offered cold comfort for reform. It was also a considerable gamble because it could well have resulted in sending federal funding to the states to use largely as they wanted with great difficulty for USDA or congressional oversight. Of course, two years can be a long time. Part of the gamble could have been that it would be replaced and not go into operation; a smart bet, as it would turn out. Adding the state program held, at the very least, the potential for an extraordinary redirection of federal agricultural policy.

The state-operated program provided arguably the strongest evidence of the Supreme Court's impact on political development. Digging deeper, further proof of the court's impact can be found in a provision toward the end of the temporary program. It was a relatively peculiar and puzzling restriction for the temporary federal payments. Congress instructed that "in carrying out the provisions of this section, the Secretary shall not have power to enter into any contract binding upon any producer or to acquire any land or any right or interest therein."[33]

This limit on the secretary's contracting power was a feature of the 1936 legislation from the start, and both the House and Senate bills, as introduced, contained a version of it.[34] Both committee reports made certain to emphasize that the "Secretary is expressly denied the power to enter into contracts binding any producer to any course of action or to acquire any land or right or interest in land under the bill."[35] Senator John Bankhead argued from the beginning that the "Court laid considerable stress upon the requirements in the rental contracts that farmers should comply with the Government regulations."[36] During the Senate Ag Committee hearing on the proposal from farm groups, he again emphasized that the Supreme Court decision provided "a line of demarcation, well recognized, between control by contracts under which the farmers agree to abide by regulations of the Government on one side, and a mere conditional appropriation giving to them benefits or bounties upon compliance, preceded by a contract to do so."[37]

This interpretation of the Supreme Court decision was widely accepted in Congress and in the administration.[38] When he got to the floor, Senate Agriculture Committee chairman Smith (D-SC) explained to senators that "the proposed bill specifically prohibits the Secretary from entering into contracts binding upon producers or acquiring any land or right or interest therein."[39] And in the House, for example, Representative Wall Doxey (D-MS) stressed that under the temporary program the farmer can receive direct payments, but "there is no contract" and "nothing is binding on anyone . . . no contractual relations between anyone" for the payments authorized by the bill.[40] Even President Roosevelt in his signing statement noted that "there will be no contracts with farmers."[41]

Puzzling because it complicated the operation of a program that paid farmers in return for conservation practice adoption, it would also have questionable legal force. Any arrangement in which a farmer was paid based on having undertaken actions on the farm would effectively be a contract, and the argument was not accepted by everyone. Senator James Pope (D-ID), for example, argued that the prohibition on contracts replaced by conditional grants to farmers was an "illogical distinction" that would not fool the court.[42] Under a close reading of the *Butler* decision, Senator Pope appears to have had the better argument, at least technically. The Supreme Court's conservatives had concluded that the regulation of farming and farm production was a state matter reserved to them by the Tenth Amendment. It is difficult to conclude that they would have had a different view of direct federal payments to farmers to divert acres from soil-depleting to soil-conserving crops. The inclusion of a provision that prohibited the use of contracts with farmers provided an argument but likely would not offer much legal protection.[43]

Unexplored were the implications for the program and any of the reforms intended for it. At its most basic, the power to make a contract is also the power to enforce its provisions. Would USDA merely pay farmers for undertaking practices in good faith but with little actual enforcement or consequences for not complying? This placed a high level of trust in the farmers but also by the farmers in USDA. It created a high degree of risk for achieving a soil conservation program, let alone broader and more far-reaching reforms. Whatever the risk, the provision carried through from the beginning of the process to the end and was administered accordingly.[44] Its inclusion was the result of a questionable interpretation of a suspect Supreme Court decision that was more convenient than accurate. The greatest significance of its inclusion was as a clear reminder of how much that decision drove development of the 1936 Act.

These twenty-six words were evidence of the interplay of power between the branches. They highlighted the ability of the Supreme Court to reach into the legislative process but also the limits of that reach. The Supreme Court can strike down a law, but it cannot write a new one to replace what it struck down. Congress still possessed the legislative power—the power to initiate the making of law—and can always write new laws if it can muster the votes to do so. It can also respond in any number of creative ways, including working around a court's decision. But this is not to say that the court cannot do damage. It has been a rare situation in American history that a president's party—or any party—has a majority of the magnitude that was enjoyed by Roosevelt and the Democrats in the 74th Congress.

Digging even deeper uncovers two other provisions that are notable for offering perspectives on the politics at work in this early stage of development. These two provisions reveal the impact that the issues of the day—explored in the backstory in chapter 3—had on the final legislation. These provisions also stand out because they were added by amendments in the House. First, Congress directed the secretary of agriculture to administer the temporary program "in every practicable way" to "encourage and provide for soil conserving and soil rebuilding practices rather than the growing of soil depleting commercial crops."[45] The soil-depleting crops were, unsurprisingly, the bulk commodity crops that had been the focus of previous programs (e.g., corn, cotton, and wheat), including the 1933 AAA. The provision was a compromise introduced into the House debate by Chairman Jones. The compromise was the result of a heated debate over an amendment that sought to protect the northern dairy industry.

In the House, Representative Gerald Boileau (R-WI) was the author; as noted above, Senator McNary led the unsuccessful effort in the other chamber. Boileau was concerned about the impact on the dairy industry of the proposed thirty million acres put into soil conservation. If a number of those acres were used as pasture or to feed more dairy cows, the additional production that would result would harm dairy farmers. Therefore, his amendment sought to preclude the sale or harvest of crops on the acres enrolled in conservation, including grazed or pastured livestock and products. His concerns were magnified by the provision prohibiting contracts.[46] Facing resistance and certain defeat, Boileau modified his amendment to get the support of Chairman Marvin Jones (D-TX). The problem for Boileau was that the modification of his text did not modify his intentions. When he interpreted the modified amendment as a mandatory requirement on the secretary, he lost the chairman's support rather quickly in the debate and complicated matters further.[47]

His amendment exposed another impact of the Supreme Court's dictate. Pursuant to the accepted reading, any mandatory requirement on farmers was not acceptable. Because Boileau and his allies sought to restrict what farmers could do with the acres diverted to soil-conserving crops, they triggered the constitutionality and Supreme Court concerns. Opponents of the amendment argued it regulated agricultural production. Chairman Jones explained that the "Supreme Court says we cannot regulate production" and that the amendment would "make one of the primary purposes of this soil-building measure regulation of production which the Supreme Court says is a right reserved to the States. . . . We have stayed within the limits of the Court's opinion."[48] Representative William Bankhead (D-AL) added that the amendment "undertakes in definite, direct, and specific terms to regulate agriculture or some phase of agriculture," which is forbidden under the court's decision. He emphasized to members that "if you want an agricultural bill, you certainly cannot afford to vote for this amendment."[49] Adding any restrictions into the bill on the use of the land would push it into the territory forbidden by the Supreme Court's decision.

Left unexplained was why the conditional grant design could not include a condition that protected the dairy industry. If payments were conditioned upon action by the farmer, would it matter that the conditions included one that did not allow those soil-conserving crops to be used for commercial purposes? It was a fine line at best, visible to those who wanted most to see it.[50] In the end, Chairman Jones won. His modified amendment that the secretary "shall in every practical way encourage and provide for soil-conserving and soil-rebuilding practices rather than the growing of soil-depleting commercial crops" was accepted by the House without a vote. It was included in the final text signed into law.[51]

The Boileau amendment most clearly highlighted the regional and partisan nature of the disputes, clear evidence of the factional competition at work among those within the farm coalition. The southern cotton faction was in a direct dispute with the northeastern and midwestern dairy faction. Midwestern corn farmers who mostly produced feed for livestock and dairy animals were also in conflict with the South. Representative Francis Culkin (R-NY) attacked "the Members from the South" for being willing to harm the dairy farmers to "rectify the economic condition of your people" and wanting "to have their cake and eat it too" out of "selfishness, influence of section" and the "misuse of political power."[52] Representative Gilchrist (R-IA) argued that "under the A.A.A. we paid the Iowa farmers for taking corn land out of production, and then went into the South and paid the farmers to put cotton lands into corn production."[53] Representative Boileau (R-WI) attacked southerners for seeking to diversify cotton farms with federal payments, arguing, "Do not ask us to give you a subsidy . . . to pay you money to compete with us" and spread surplus.[54]

The dispute reinforced theoretical concerns about factions, that small groups with relatively homogenous interests can exercise outsized power, helping themselves and harming others.[55] The dairy amendment serves as a reminder that the framers' design did not account for political parties or Supreme Court decisions. The dairy districts were, at the time, represented mostly by Republicans, while the cotton South was a Democratic stronghold. Given the size of Democratic majorities, the dairy faction stood little chance if its only supporters were Republicans. That the amendment was contested would seem to indicate that it possessed some political strength. More importantly, the debate offers an example of a faction in action through the dynamics of the process. With no control over how its decisions were used by factional interests, the Supreme Court's opinion was an effective justification to override the dairy industry's concerns. The court's intrusion may have altered the factional competition, contributing to existing asymmetries.

Over time and through iterations, powerful factions can exhibit the self-destructive tendencies of closing off other interests, in this case northeastern dairy farmers. The faction with the power, the southern cotton interests and Democrats, demonstrated little interest in helping the dairy faction. Rather than compromising to expand or maintain the coalition, they drove them away. The reasons for doing so are not explained in the debate nor discernible from the record, but it was a harbinger of the problems for farm policy in the years to come, a preview of the regional feuding that nearly destroyed it and the farm coalition. From just the perspective of maintaining a coalition in Congress, the dairy issue highlighted problems with the policy as it was developing in its early stages and should have been a warning to supporters. Instead, the faction with power sought only its own advantage, without much concern about the implications.

The second and final issue was a striking mark on political development that was left by the problems roiling the South over sharecroppers and the cotton programs. Congress instructed that "in carrying out the provisions of this section, the Secretary shall, as far as practicable, protect the interests of tenants and sharecroppers."[56] It is safe to presume that this issue was not a high priority of the controlling factional interests and that the southern faction would have wanted to avoid public debates and any federal interference. The strong push for reforms coming out of USDA presented real hazards for the southern planters, whereas success for reformers could have opened opportunities for more changes. The amendments and debate differed notably from that of the dairy amendment. The South wanted neither, but the sharecropper issue presented the greater threat. And yet, the provision was included due to the efforts of a few southerners in the House, and it faced little on-the-record opposition. The debate did not produce the same level of confrontation as the dairy amendment. The issue was not addressed in the Senate.

Representative Malcolm Tarver (D-GA) was the lead author of an amendment to add protections for tenants and sharecroppers. He argued that the bill would not do "anything for the sharecropper.... No reference is made to the payment of any benefits except to the landowner."[57] He was not alone among southerners; Representatives Frank Whelchel (D-GA) and Maury Maverick (D-TX) also pushed for similar protections. Representative Maverick argued that "the sharecroppers, the workers, and the tenants were not adequately protected.... The least we could do in this bill is to recognize such classes as existing like other human beings." Representative Whelchel added, "It has been a matter of great grief to me, the hardship that has been brought about on the croppers and tenants in my district, and it can be termed in no other manner than unfortunate."[58] In their arguments, the least the bill could do was to require some fairness in, and sharing of, the federal benefits. It was hard to argue against, and there was little open opposition.

Chairman Jones claimed that everyone wanted to protect tenants and sharecroppers.[59] Whether true or not, the problems caused by the Supreme Court's confused command presented a convenient obstacle for those who wanted to use it. For example, Representative Wall Doxey (D-MS) lamented, "I am convinced that the language of the majority decision of the Supreme Court makes it exceedingly difficult for Congress now to enact practical and beneficial legislation in behalf of the farmers, especially the small farmer—the tenant and the sharecropper."[60] The court decision proved effective enough. Representative Tarver's original amendment was deemed too prescriptive, and he negotiated a compromise. Requirements on, and any consequences for, farmers were considered unacceptable risks even if, as in the case of fair treatment for sharecroppers and tenants, they did not regulate production. On this topic, the reading of the Supreme Court's conservatives was rather convenient for southern cotton planters.

As modified, the Tarver amendment added tenants and sharecroppers to the definition of "producer." It also required apportionment of the federal payments to tenants and sharecroppers on fair terms based on their contribution to the land put into soil conservation and to offset the loss of income to them from acres taken out of cash crop production.[61] Although it had been watered down, Representative Tarver argued that it would "demonstrate clearly to the Secretary of Agriculture the purpose and intent of Congress that in carrying out the provisions of this legislation the tenant and the sharecropper shall receive a fair share of the benefits."[62] The amendment was accepted without a recorded vote.

The Tarver amendment served up a process lesson as well. It created a difference with the Senate bill that would have to be resolved in conference. It was a significant vulnerability for the protections. Conference resolved the issue in the Senate's favor and did not accept the House provisions. The meager protections were watered down further, delegated to the secretary's discretion if deemed practicable to do so.[63] Senate Agriculture chairman "Cotton Ed" Smith took credit for weakening the provision, explaining that the House conferees "agreed to an amendment which very considerably modifies the provision as adopted by the House."[64] Senator Smith had not been at all shy about his opposition to the provision. After the House passed its version of the bill, for example, he was quoted as responding with "an angry blast," calling it a "fool thing" for the House to have adopted it because it was not "fair" to the landowner.[65] His opposition reflected his troubled, racist views. The outcome also represented the depths of the problem and the extent of the planter power in Congress. House Agriculture Chairman Jones tried to put the best face on the outcome: "I think that is the best that we can have, and I think it will assure that they will be fairly treated."[66]

By 1936 there was little to support Chairman Jones's statement. It was impossible to ignore how the New Deal farm policies were making matters worse for tenants and sharecroppers in the South.[67] Members of Congress needed to look no further than the Southern Tenant Farmers Union strike going on in Arkansas as they debated the 1936 Act.[68] Watering down the protections and leaving them to the discretion of Secretary Wallace and the Cotton Section of the AAA was not going to change things. The Cotton Section of the Agricultural Adjustment Administration had been clear that they did not consider tenants and sharecroppers to be farmers, and as such, they were not eligible for federal benefits.[69] Leaving this matter to the discretion of the offenders was a guarantee the problems would continue. True to form, the Cotton Section did not change its position after the 1936 Act, and the horrendous treatment continued.[70]

Speculating through the Thick of Dust and Time: Concluding Thoughts

The contemporary mind exhibits myopic vision when it views the New Deal era as ancient history, too far removed in time and politics to be relevant. It can seem too unique, with its Great Depression, Dust Bowl, and more archaic society in the times before World War II and the onset of the modern world. But to see our modern situation and ignore the New Deal era is to see the plant but ignore its roots. Much of our modern system of government traces there, and our politics continue to feed from the changes that the Great Depression wrought. To cast it aside as anachronistic fodder for historians is also to ignore the path-dependent nature of political development. The vast shifts in the federal policy apparatus during the Depression and the first two terms of the FDR administration included the origins for agricultural conservation policy in the farm bill. Reviewing the 1935 and 1936 Acts informs an understanding of those policies and the path opened for future development.

Given the importance of conserving soil, there is some value in noting that the Soil Erosion Act of 1935 and the Soil Conservation and Domestic Allotment Act of 1936 were not a part of the legendary first one hundred days of Roosevelt's presidency. Agricultural conservation policy was not included in the first New Deal.[71] Conservation of the soil could have been a secondary priority to addressing the economic concerns of the Great Depression. If correct, this reality highlights a fundamental challenge for conservation policy that was raised in chapter 1. Soil and society operate on very different time horizons, and economic problems are typically more immediate and pressing. Legislators facing frequent reelections are going to be more responsive to the short-term priorities of their constituents, which by their nature do not often include natural resource concerns. This demonstrates Madison's argument that the mild voice of reason is too often drowned out by impatient avidity—a consuming greed for which time is too short.[72]

Arguments for soil conservation policy were made by the dust storms of the drought-devastated Great Plains in the middle of the Great Depression; a broken economy and broken lands demonstrated to a broken people that many things had gone terribly wrong. The Dust Bowl brought the natural resource and environmental problems to the fore and imprinted itself on the public mind. How much that public mind connected dust falling on the Eastern Seaboard to decisions made thousands of miles away and over decades is unknown to us in our time. Likewise unknowable were the depths of any public understanding that policymakers had failed to deal with these issues in a manageable, strategic manner on the public's behalf. In 1935 and 1936, the message had broken through, at least temporarily and to an important degree.

In 1937 President Roosevelt wrote to governors that the "nation that destroys its soil destroys itself," and his administration reported that "erosion of our soil has its counterpart in erosion of our society," echoes of America's first commissioner of agriculture.[73] Hugh Hammond Bennett also continued the crusade: "And not mere soil is going down the slopes, down the rivers, down to the wastes of the oceans," he wrote in 1939. "Opportunity, security, the chance . . . to make a living from the land—these are going too," and "to sustain a rewarding rural life as a bulwark of this nation . . . we must defend the soil."[74] Looking back on it in 1953, however, Murray R. Benedict wrote of the 1936 Act, "If soil conservation had been the primary objective, it would have seemed logical to expand and modify the work of the Soil Conservation Service," but doing so "did not contemplate a gentle shower of federal checks evenly and undiscriminatingly distributed over eroding and noneroding areas . . . paying farmers . . . to take the kinds of action dictated by their own best interests and those of society as a whole."[75] In 1999 Chris Rasmussen wrote, "Desperate for some means to keep the nation's agricultural economy afloat, the Roosevelt administration was only too willing to jettison soil conservation to preserve its broader agricultural policy, which was predicated on curbing production and keeping checks flowing to the nation's farmers," and the "federal government thus had a strong inclination to treat soil erosion primarily as an unfortunate consequence of the malfunctioning farm economy, rather than as a distinctly environmental concern."[76] It is difficult to square these perspectives.

In the U.S. Code, beginning in 1936, the American taxpayer could provide $500 million per year to encourage farmers to conserve the vital soils of the nation. The 1936 Act was short-lived, however. Less than two years after it was enacted, on February 16, 1938, President Roosevelt signed the Agricultural Adjustment Act of 1938 into law.[77] The state-led program that was intended to be a permanent policy for agriculture never became operational, discarded in 1938. Under the 1936 Act, farmers increased planted acres despite hundreds of millions of dollars in payments and reportedly tens of millions of acres reduced. According to USDA data, in two years of operation the acres planted to both cotton and wheat increased.[78] These were the crops for which erosion was the most significant—water erosion in the cotton South and wind erosion in the Great Plains wheat region.

The Agricultural Adjustment Administration (AAA) blamed drought for the "belief that the permanent solution to the agricultural problem lay in a conservation approach" and laid blame on the Supreme Court for having "hastened the acceptance of this approach almost to the exclusion of all others."[79] A return of good weather in 1937 led to big crops and lower prices on increased acres; corn, cotton, and wheat all exceeded USDA's goals for reduced acres, with farmers planting far above the goals for wheat. According to the AAA, the experience with the 1937 crop "revealed the inadequacy of the conservation approach" and the need for broader controls. In 1938, AAA further explained, Congress implemented a compromise between the 1933 and 1936 approaches with "a legislative strengthening and continuation of the conservation programs" coupled with "marketing and storage provisions for control of surpluses" in the good years that Secretary Wallace termed the Ever-Normal Granary.[80]

The AAA reports presented a more anodyne version than the Congressional Record. With pressure from the countryside, led by the American Farm Bureau, President Roosevelt called a special session of Congress that began in November 1937, with a direction from the president to fix farm policy. The congressional debate continued into February of the new year, resulting in the 1938 AAA.[81] The House continued some emphasis on soil conservation, and the final law amended the 1936 Act. Congress also declared its intention to continue the Soil Conservation and Domestic Allotment Act with amendments that included replacing the state-based program with a continuation of direct federal payments. The focus of the 1937 and 1938 debates, as well as the primary policies in the statute, were not conservation, however. Congress implemented Secretary Wallace's call for an "ever-normal granary" through the parity system of price-supporting, nonrecourse loans, acreage controls through allotments, and marketing quotas.[82] With bumper crops and depressed prices, it was a return to economic concerns as the priority, not conservation.

History has not been kind to the 1936 Act, and part of the harsh judgment has to be based on our perspective of the dust storms in the southern Great Plains as a driver of politics and policy. From this perspective, not even the moving mountains of dust were enough to move Congress to do more for soil conservation. We may be overreading the power of the dust storms, however. A counterintuitive, maybe controversial, view is that all the catastrophic consequences of the dust storms may not have proven as effective or prominent a political catalyst as we might think. Hindsight and the perspective of studied history may well produce a vision of the event in the imaginations of the present that does not match the actual lived experience of those at the time.

This is not to say that the dust storms were not catastrophic and historic; it is to acknowledge that they were relatively limited in time and space. Concentrated far from most of the population and political and media centers, the dust storms delivered their worst in the southern Great Plains on the outer rim of the old frontier. The political result may have been a form of political myopia that is familiar. Falling dust on the Eastern Seaboard may not have been enough. If the Dust Bowl was a problem mostly for poor farmers in a remote region of the nation, and a slight oddity or inconvenience elsewhere, then it did not pack the political power necessary for more substantive changes in policy. The Great Depression was by far the more immediate and dominant catastrophe, affecting a wide swath of the American public and economy. In agriculture, depressed crop prices and farm incomes were the more pressing problems to be addressed. Concerns about soil erosion could be tied to the larger problem, blamed on the inability of struggling farmers to better care for the soil. In that view, improving prices and incomes would lead to better soil conservation; erosion was a symptom of the greater disease, even if it may have been the other way around. Given the rather short political memories prevalent in this country, moreover, the dust storms of 1934 and 1935 could well have fallen into the collective memory hole, buried beneath the more immediate issues of the Depression and the Supreme Court's decision at the starting line of a major election year.

Ultimately, much of the blame for any shortcomings in developing more far-reaching, consequential conservation policies for agriculture in 1936 must be laid at the feet of the Supreme Court. Farm policies implicated redistributionist concepts and centralized planning anathema to the conservative elite who had held power going into the economy's collapse. Adjusting farming, reducing erosion on private property, and conserving natural resources were extraordinarily complicated. Centralized control across millions of independent farmers implicated everything from property rights, which had traditionally been the purview of state and local authorities, to the constitutional issues conjured up by the Supreme Court in its efforts to protect private industry. Forced to retreat by the voting public, the last redoubt of this factional interest in American politics was the court. To the court they turned, and the court delivered. The conservative justices were the biggest reason more was not done the moment the catastrophe of the Dust Bowl collided with the Great Depression.

Neither the theoretical conception nor the constitutional design of Congress accounts for the Supreme Court. The nation's highest court presents the potential for a powerful stronghold in the highest levels of government, and from that privileged space, the power to exert a near-complete veto over policy if it so chooses. The decision in *U.S. v. Butler* demonstrated that clearly. The Supreme Court's application of an ideological agenda was more than legal error at a time of depression and dust storms. A small faction, unelected and with lifetime appointments, deployed unusual legal theories in an attempt to thwart the work of the branches elected in response to the catastrophes. The efforts to institute limits on the federal government amounted to little more than an attempt to displace the voice of the voters who had rejected that governing ideology at the ballot box. For the development of policy, it also left a vacuum and caused a crisis.

Congress—neither its members nor its designers—can escape blame for any shortcomings in the 1936 Act, the development of soil conservation policy, or a reform of farm policy and land use. The drafting pens and gavels were firmly in southern hands, and the southern reputation for achieving its narrowest, factional ends in Congress has been well documented.[83] There are clear hints of that influence on the 1936 process, from protecting the narrowest interests against those of dairy farmers in other regions, to the more sinister efforts against sharecroppers and tenants that perpetuated the Jim Crow system of white supremacy and oppression. The latter was an egregious and unjustifiable lack of adequate action as the Southern Tenant Farmers Union suffered brutal, violent retaliation for a strike against cotton planters. Senate Agriculture Committee chairman Cotton Ed Smith appears to have been the primary source of resistance, pushing back on the minor protections in the House bill. Aside from the obvious problems with shielding reprehensible actions in the South, the resistance from Chairman Smith would have impacted any efforts by New Dealers to push further reforms in agriculture, including for conservation. It is thus notable that he also worked to include the state program.

Production control, long-term land-use planning, and social reform have been identified as the three competing directions for New Deal agricultural policy. Conservation policy was flexible enough to fit in all three, or at least to implicate issues for each. This was both a strength and a weakness. Social reform was anathema to the South and struggled to gain any traction in Congress despite herculean efforts at USDA; it would be killed outright during the war. While land-use planning was initially reinforced by the drought and the dust storms, it was also viewed skeptically as a threat to property rights and did not develop in Congress (or by Congress) beyond the limited versions in 1935 and 1936.[84] The outcome in 1936, coupled with that of 1938, demonstrated that even amid the ruin of the dust storms, the driving forces behind moving farm bills in Congress remained the economic concerns of the most-connected farmers. Crop prices and farm incomes were prioritized above all else.[85] By seeking to control production to boost crop prices, those policies substantially narrowed the scope of beneficiaries. One need look no further than the simple fact that this was the priority at the same time that many Americans were hungry, malnourished, underfed, and underclothed.

Efforts at reform suffered. Tolley had "feared that Triple-A programs would become rigid and freeze existing patterns of production," and his "greatest fear was that the farm program would serve only the interests of established commercial farmers."[86] In 1935 Secretary Wallace had accepted that "some of the worst characteristics of the American tenancy-system are found in the South" and that the "operation of the cotton programs has probably added to the immediate difficulties just as relief policies have injected additional complications into the usual tenant and farm-labor relationships."[87] A vast understatement, the secretary's comment was at least evidence that the New Deal reformers grappled with the problematic consequences of the narrowed scope of farm policy, beginning with the treatment of tenants and sharecroppers. They also understood the extent to which that treatment held relevance for soil conservation.[88]

If they were not careful, however, soil conservation policy could feed a tendency to blame the most ill-treated of the farmers for the problems of soil erosion and further displace many of them, while consolidating farmland and farming with the largest and most sophisticated farmers. A policy that paid to shift acreage into soil conservation could do more harm. Acres no longer producing a commercial crop had little need for labor, and the least productive acres would be the acres most suitable for soil conservation. It is reasonable to conclude that those acres most suitable for soil conservation, or most in need of efforts to address erosion, would be those most likely worked by the worst-off tenants and sharecroppers—poor people and poor land went hand in hand.[89]

In addition to the amendment for tenants and sharecroppers, the House debate further demonstrated an uncomfortable honesty about the early developments of farm policy. If a program didn't work for the large farmers, they would refuse to participate; if these large farmers refused to cooperate, the entire program would fail because there could be no realistic controls on production without the largest producers. To add another example, the House had briefly debated limiting total payments to any farmer; the effort faced opposition from both within USDA and on the floor. Representative Hampton Fulmer (D-SC) supported limiting payments. He proclaimed that such reforms were "opposed by the Department of Agriculture," and he was clearly frustrated that "this class of farmers [was] not getting a square deal" because they were being punished for cooperating while large planters were able to increase their operations and get more federal payments.[90] Representative Harold Cooley (D-NC), who would go on to hold the House Agriculture Committee gavel for many years after World War II, argued that limiting payments "might ultimately force the large landowners and producers of the Nation to resort to subterfuge and fraud."[91]

The debate on payment limits, coupled with those on the amendments for dairy farmers, sharecroppers, and tenants, provides compelling evidence that Tolley's fears were well founded. Farm policy had been captured by this faction within agriculture, and their interests were being hard-wired into the policy. Policy reforms were under the control of the very faction that benefited from the status quo and likely would lose out from reforms. Seemingly proving the point, the final statutory text contained no limit on payments but did include watered-down protections for small farmers in nearly identical language as that for sharecroppers and tenants.[92] The end result for any attempt that ran afoul of the narrow interests was always the same, right down to the words in the statute.

Nowhere were these layers of complication more evident than in the program that never went into effect. Arguably, it was the most potentially problematic development for conservation policy. Congress would have delegated to the states the pressing challenges of addressing soil erosion and conservation and improving soil fertility—which included protecting rivers, harbors, and navigable waters in addition to protecting "national soil resources"—as well as reestablishing the purchasing power of farmers based on the 1909 to 1914 era.[93] Representative Cooley acknowledged that the dust storms and erosion "know no State lines and respect no State rights."[94] These were problems national in scope, but USDA would have assumed a secondary role, passing through federal funding. The program would have been voluntary for the states, and once the state plan was approved, any assistance to farmers was also voluntary for those electing to participate.

The idea of deploying state governments in the efforts may have been part of the evolution in thinking among many of the New Dealers. For much of the year prior to the Supreme Court decision, for example, Hugh Hammond Bennett had been working on the concept of soil conservation districts and a model state law to implement them.[95] It was an idea that Secretary Wallace reportedly considered a form of economic democracy that could be a counterweight to entrenched political power.[96] Unstated and certainly unexplored, at least publicly, this idea conveniently aligned with southern views on states' rights. With the Supreme Court deploying the Tenth Amendment and a states' rights argument to strike down the 1933 AAA, it may have seemed only logical that the long-term solution was to delegate much of the effort down to the states.

For the southerners, it would have been a solution nearly too good to be true. Devolution of federal power to the states was the preferred method of the Jim Crow South.[97] With the state standing between the feds and the farmers, the program would further insulate their private property; for the southern cotton planters, this meant strong protection for their system of oppression and control. Senate Chairman Cotton Ed Smith apparently latched onto it early, and it was added at his behest. Little doubt he and others in the southern faction grasped that it would send federal money to the states to distribute as they saw fit but with very limited federal interference; USDA could not enforce any obligations on the states beyond approving the plan. The state program was unlikely to achieve soil conservation or reduce soil erosion and would have little chance of helping farmers selling in national and international markets. The best aspect of this program was that it never went into operation.

The final irony is that the only permanent outcome of the 1936 Act was the system of local conservation districts known today as the Soil and Water Conservation Districts (SWCD) and organized nationally as the National Association of Conservation Districts (NACD). USDA produced a model law for states to enact that authorized soil conservation districts and land-use regulation. This was the most lasting and arguably one of the most important products of the 1936 Act. By 1945 all states had laws in place authorizing soil conservation districts and had created 1,900 districts covering more than a billion acres.

State-authorizing laws vary, but most provide districts with general powers, both administrative and legislative in function, as governmental subdivisions of the state. They generally derive power from the state's police powers to promote public health, safety, and morals; very few, however, authorized land-use regulation powers for the districts. In general, districts focus on controlling or reducing soil erosion and protecting water quality through voluntary conservation efforts with farmers and landowners, and in cooperation with federal and state authorities. Organized on the basis of either watersheds or counties, upon a petition of landowners in the proposed district, they can include towns or municipalities. Mostly they provide technical assistance, advice, education, and demonstration, as well as promote improved soil and water quality by seeking funds for improvement projects within the district, particularly federal conservation program dollars. The most permanent, lasting achievement of the 1936 Act, SWCDs are often also included on the long list of shortcomings and frustrations in this policy area. Those frustrations are often exemplified by the lack of actual land-use regulatory authority and by close relationships between districts and farmers.[98]

Any attempt at an assessment of New Deal farm policy, including the origins and early developments of conservation policy, must be based in political reality. When the Supreme Court intervened, Secretary Wallace and USDA officials converted their long-term planning into an immediate replacement for the 1933 AAA. They turned to the major farm organizations, led by the Farm Bureau, for consultation and endorsement. There were no conservation or environmental factions or interest groups available, at least none operating in agricultural policy. It was only farm interests at the negotiating table; when the only tool was a hammer, every solution had to be a nail. Moreover, the only competition in the legislative arena was among farm interest factions. That the outcome was anything other than the status quo was due to the Supreme Court's decision and the ability to put to creative use the obvious problems of soil erosion. For land-use reform and conservation, the 1936 Act was a half measure at best. This half measure, however, was an accomplishment given the very limited political competition.

As such, soil conservation policy originated as a dual-purpose, hybrid undertaking. It was not targeted at controlling erosion on those lands that were most problematic. What it offered in the moment was a convenient way around the Supreme Court. The assistance was for improvements in soil fertility and land use rather than preventing or decreasing deterioration, conserving soils while also providing acreage reductions and supply controls. The temporary program presented a tenuous balancing act of competing priorities and contradictory demands. The more the policy moved in the direction of soil conservation, the firmer the constitutional footing, and the stronger the political and public support. Diverting acres into conservation, however, was less likely to deliver effectively for the farm interests backing political development, especially in the South. Worse still for that region, it might bring New Deal reformers closer to interfering with their property and cause them problems. The more Congress shifted the bill toward domestic allotments, the stronger the benefits for the farm interests, but the constitutional footing fell away precipitously.

Seeking reform amid ruin, the entire New Deal was a vast governmental undertaking. It was expansive, changed frequently, and produced a series of emergency programs in the middle of a worldwide economic depression. It helped millions of Americans. It may well have helped stave off existential threats to self-government and helped America avoid the worst outcomes experienced in Europe and elsewhere. But at the same time, New Deal policies harmed millions. Worse still, the most harmed were too often those who were most in need. The New Deal farm policies demonstrated this point clearly, working mostly to assuage the moderately afflicted while devastating the truly destitute, such as Black farmers in the Jim Crow South.[99] Ira Katznelson has written arguably one of the best concluding observations about the New Deal: "If history plays tricks, southern congressional power in the last era of Jim Crow was a big one" when the "ability of the New Deal to confront the era's most heinous dictatorships by reshaping liberal democracy required accommodating the most violent and illiberal part of the political system, keeping the South inside the game of democracy.... The triumph, in short, cannot be severed from the sorrow."[100]

Where does this leave the origins of conservation policy in the farm bill? Any conclusion is complicated and often mixed and muddled. The origin and early development of agricultural conservation policy in 1936 confirm Katznelson's observation while adding to it. Between the Supreme Court and southern factional power at its apex, the realpolitik conclusion is that major reforms could not overcome the political obstacles of the time. While southern factional interests fought against New Deal reformers, a conflicted Justice Owen Roberts played Sinon and delivered the Trojan horse legal theory on behalf of a narrow elite.[101] The decision upended policy development and left a vacuum. New Deal reformers made the first moves but were hemmed in by narrow factional interests and the preferences driven by the Great Depression for continuing the economic policy status quo. In Congress, reformers were limited further by the protectorate for the Jim Crow system, especially those on the Senate Agriculture Committee. Some of the methods provide lessons relevant today, and the outcome offers warnings.

The reform agenda for land use and conservation in Congress reached its peak potential in 1935 and 1936 but largely dissipated thereafter; it nearly disappeared during World War II.[102] The intervening years were mostly interregnum, but that may have contained some blessing in disguise. At its origin, conservation policy risked inheriting many of the problematic consequences of the farm support policies. Congress, USDA, and farm groups would spend decades locked into a counterproductive, internecine fight over those farm policies as American agriculture underwent a technological revolution after World War II, and the American body politic changed significantly. Conservation policy would largely lie dormant, reemerging in the middle of the 1950s as another potentially creative solution to problems in farm policy, some of which were concerns about the consequences of our treatment of natural resources.

PART 3 The Soil
Bank Saga

5

Out of Surplus and Policy Failure

THE RISE OF THE SOIL BANK

On January 9, 1956, President Dwight D. Eisenhower delivered a special message to Congress in which he proposed the enactment of a new policy for American agriculture. He proposed that Congress authorize paying farmers to remove from cultivation those lands "poorly suited to tillage, now producing unneeded crops and subject to excessive wind and water erosion."[1] The name for his proposed new policy was the Soil Bank. On May 28, 1956, he signed into law the Agricultural Act of 1956 and proclaimed that the "heart of the bill is the soil bank," a policy that was "rich with promise for improving our agricultural situation" by helping to "bring production of certain crops into balance with their markets" and reduce "market-destroying surplus stocks of farm products."[2]

Twenty years had passed since enactment of the Soil Conservation and Domestic Allotment Act. As if recycling from that history, President Eisenhower's promising new direction would barely survive three years. In 1958 his budget proposal to Congress recommended terminating part of the Soil Bank early. Congress complied, ending part of the program for the 1959 crop, and allowed the remaining authorities to expire in 1960.[3] This chapter and the next chart the origins of the Soil Bank and its relatively swift demise.[4] The conclusion attempts to understand what happened to this second attempt at developing conservation policy for American agriculture, as well as the lessons to be drawn from it.

War, Technology, Drought, and Factional Feuds

The political pressures of the Depression and the Dust Bowl catastrophes gave way to those of war soon after the Agricultural Adjustment Act of 1938 replaced the 1936 Act. The 1938 Act instituted a prescriptive support policy combining flexible loan rates with supply controls through acreage allotments and marketing quotas. Under the demands of war, Congress moved the price-supporting loan program to high, fixed loan rates that encouraged production for the war efforts. Between World War II (1941–46) and the Korean War (1950–53), Congress enacted the Agricultural Act of 1949, which continued the wartime high, fixed price-supporting loan rates coupled with the acreage allotment system. Total acres in cropland, cotton, and wheat all hit post–New Deal highs in 1949. With the onset of the Korean War in 1950, the push for production continued, but when the Korean War ended abruptly in summer 1953 with crops already in the ground, surplus problems began almost immediately.[5]

Planted acres were only part of the problem. Coming out of World War II, American agriculture entered a technological revolution that had a direct result on the performance of the parity system's acreage allotments. The mechanical cotton harvester was introduced in 1942 and further advanced mechanization of farming, which replaced hand labor. Arguably the most consequential innovation was the ability to commercially produce synthetic fertilizers, especially plant-available nitrogen. It was an advancement that was an allegory unto itself. Science synthesized ammonia from elements in the air to produce explosives and ammunition for destroying life on the battlefield; it could also be used to produce fertilizers to feed life in farm fields. Similar innovations from war advanced herbicides and insecticides that helped cut losses from weeds and insects. In addition, plant breeders and researchers produced better seeds.

The result was more intense production on each acre farmed and greater yields. Technological advancements rendered futile any attempt to control production by reducing acres because the changes drastically increased production per acre farmed. With each acre planted yielding more of the crop, the only way to have controlled production through acres would have been sharp reductions in acres planted. Such acreage allotments were not politically feasible. For example, by the middle 1950s, yield gains in wheat should have resulted in an allotment roughly one quarter of what farmers were planting.[6]

Drought returned to the southern Great Plains in the winter of 1949 and 1950, becoming increasingly worse through the fall of 1956. By one measure, the drought was as bad, if not worse, in Kansas and Oklahoma than during the Dust Bowl. The drought was relatively short-lived, however;

rains returned in 1957. The return of drought could have provided painful reminders because millions of acres of submarginal land had been put back under the plow for wheat during wartime. Policymakers and farmers effectively repeated at least some of the misguided decisions from World War I that had helped bring about the Dust Bowl in the 1930s. Once again, millions of acres of farmland were being damaged by wind erosion. Massive dust storms plagued the region, but a full-blown Dust Bowl was avoided. Or, at the very least, the dust storms did not capture the national media or political attention the way they had in the 1930s.

According to Douglas Hurt, there were likely three reasons the plains avoided another Dust Bowl. First, farmers had learned some lessons from the earlier catastrophe. Many had adopted conservation practices in its wake, practices that helped mitigate the worst outcomes. Second, many farmers in the region had adopted and implemented irrigation to provide water when the weather did not. Improved irrigation technology was another critical byproduct of the war effort, as wartime equipment was converted and repurposed to deliver water to the farmers in the region. Third, and arguably most important, the drought had hit after years of good rains, bumper crops, and stronger prices thanks to wartime demand. Without the Great Depression, and riding stronger prices from two wars, farmers were in much stronger financial positions and could afford to undertake conservation and especially irrigation. The federal government also provided assistance, such as emergency loans and emergency feed programs for livestock producers.[7]

The actual foundation of the Soil Bank was laid in central Illinois in 1953, and it was the result of midwestern frustration with acreage diversion by southern cotton farmers under the parity system's acreage allotments. The surplus problems were just beginning with the first harvests after the Korean War. Melvin P. Gehlbach of Lincoln, Illinois, first proposed the creation of a soil bank to the House Agriculture Committee in a field hearing in Bloomington, Illinois, on October 17, 1953. He argued that parity price support policy was a production incentive for row crop farmers that hurt livestock producers and diversified (crop and livestock) farmers. It also perpetuated poor farming practices that damaged the soil. Parity policies "helped most those farmers needing the least help," and the soil was among the victims. Farm policies, he asserted, should prioritize helping "an acre of soil-building crop . . . be more profitable than an acre of corn or wheat or cotton" on lower quality soils and should assist farmers in shifting their "land-use pattern" for better efficiency and higher income.[8] Seemingly innocuous, Gehlbach's soil bank proposal in October was part of a brewing political maelstrom over farm policy.

Under the parity system at the time, farmers could take out price-supporting loans on the supported bulk commodities (e.g., corn, cotton, and wheat). If at market time the prices were below this loan rate, the farmer simply forfeited the grain to USDA without penalty. USDA was effectively the buyer of last resort for the supported commodities at the loan rate, and the loan rate served as a floor on market prices. In order to be eligible for these beneficial loans, however, farmers had to comply with acreage restrictions in the form of allotments. Prior to each crop year, USDA would estimate expected demand, translate it into production estimates, and then allot acres out to farmers with a history of planting the crop. The allotment equaled the acres they could plant to the crop in the coming year and still be eligible for the loan. Because the loan rate was a production incentive to the farmer, the acreage allotment system was considered necessary for keeping price-supported crops from being produced in excess of needs or demand. This system largely failed to control production and supplies, however. It was a policy failure made abundantly clear to all.

Failure to control supplies resulted in massive stocks of the supported commodities that were owned and held by the USDA through the Commodity Credit Corporation (CCC). In 1953, for example, the new Eisenhower administration reported to Congress that the CCC held over $62 million worth of cotton, over $5 million worth of the other feed grains, more than $330 million worth of wheat, and nearly $420 million of corn. A year later, USDA reported inventories approaching $5 billion: $2.1 billion in wheat, $1.2 billion in cotton, $1 billion in corn, and nearly $400 million in dairy products as it requested an increase in borrowing authority from Congress. Seeking another increase in CCC borrowing authority in July 1955, USDA reported $361 million in cotton, $1.2 billion in corn and other feed grains, and nearly $2.5 billion in wheat held in storage.[9] It was a failure obvious to farm policy experts and the general public alike.

To the farm coalition, the failure to control supplies depressed crop prices, while the acreage allotments encouraged acreage diversions. Acreage allotments required reductions in the acres planted to crops USDA estimated to be oversupplied, mostly cotton and wheat. Acres not planted to the allotment crop could be planted to another cash crop. Many acres were diverted to corn or other feed grains, such as barley, oats, rye, and sorghum. Diverting acres to corn and feed grains caused or contributed to surpluses in those commodities. Among other problems, spreading the surplus problems to feed grains angered midwestern farmers who felt they had to compete with government-subsidized production. Given cotton's chronically oversupplied situation, the acreage reductions were consistently applied to that crop. Southern farmers were the most frequent offenders in planting cross-subsidized feed grains that competed with corn for the domestic feed market.

An increasingly intense political dispute arose along the traditional regional fault line where pressure frequently built between the South and the Midwest. The initial soil bank proposal was born of this frustration. For example, Gehlbach told Representative Harold Cooley (D-NC), "I do not want you cotton boys in the South to put your diverted acres into crops in competition with certain grains in our area," and that "we will leave you fellows alone, if you will promise to put your diverted acres to soil-building crops, and not in competition with our basic crops."[10] This put the soil bank proposal squarely on the side of midwestern corn farmers and directly in opposition to southern cotton interests. Ultimately, understanding the soil bank proposal depends on grasping the increasingly problematic political, regional feud over acreage policy in the parity system.

At the time, the vast majority of corn was fed to livestock domestically, and these other feed grains could be a substitute source of feed. Planting those crops directly competed with corn. In 1954 American farmers planted more acres to the other feed grains than they did to corn. In 1955 they planted nearly thirteen million acres more to the other feed grains than to corn. The years 1954 to 1958, in fact, represent an anomaly in the entire history of American agriculture. They were the only years corn lost its acreage position relative to the other feed grains. From Texas to the Carolinas, USDA data make clear that southern farmers significantly increased acres planted to the other feed grains.[11]

As conceived by Illinois farmers, the soil bank concept was an alternative or supplement to the parity system's acreage policies. It may have offered a way forward for farm policy and a new path for the development of conservation policy. It was saddled with political problems from the start because it emerged from the Midwest rather than the drought-stressed Great Plains or the South. Those problems would increase the more it was viewed as an alternative to the preferred policy of the southern cotton planters and in response to the impacts of that policy on midwestern farmers. Directly confronting southerners in the 1950s was loaded with political risks and challenges; proposals by midwesterners for diverting cotton acres were not well received.

At roughly the same time, an environmental movement was also forming. It was built upon increasing concerns with the technological developments taking place post–World War II. The nascent movement raised concerns about the increasing use of pesticides and reliance on other technological advancements in farming. The collision of technology and nature in agriculture has consistently produced challenges full of consequences, intended and unintended. In the 1950s those consequences were increasing in prominence as citizens were better informed about the impacts on soil, water, wildlife, and more. At its most profound, this movement called for a land ethic.[12] As the problems and consequences built, so did political pressures for policy responses. Environmental issues are bound by neither local nor state lines; they are not limited by regional boundaries or partisan constructs.

A year after Gehlbach first proposed the soil bank concept, there was a noticeable uptick in the focus on conservation policy. President Eisenhower promoted his administration's conservation plan in his State of the Union address. He recommended that Congress "strengthen agriculture conservation and upstream flood prevention work."[13] A Republican-led Congress agreed, enacting in August the Watershed and Flood Prevention Act of 1954. This law primarily focused on flood prevention efforts but also included goals for "furthering the conservation, development, utilization, and disposal of water and thereby of preserving and protecting the Nation's land and water resources."[14] Senator Edward Thye (R-MN) introduced a bill in February that would have created a program to pay farmers to divert acres out of production in favor of "soil-conserving practices normally used in such area."[15]

The Agricultural Act of 1954 did not include any new conservation policies, however. Instead, the debate in Congress featured a fight between the president's allies in the Midwest and against the South over price support policy. The president won a modicum of flexibility in the loan rates that would permit them to be lowered. That victory in Congress was followed by midterm electoral losses for the Republicans. Democrats regained control of Congress, and southerners again took hold of the gavels in the Agricultural Committees.[16]

In 1955 multiple bills were introduced that would have created an acreage reserve program to divert acres out of production and into conservation.[17] Senator Hubert H. Humphrey (D-MN) raised the issue on the Senate floor, stating that the Ag Committee had designated a subcommittee to hold hearings on bills to retire or divert "submarginal lands from the production of surplus agricultural commodities."[18] Also in 1955 the American Farm Bureau Federation adopted a resolution in support of a revised version of the soil bank policy as the organization's recommendation for Congress.[19] The Farm Bureau backed conservation as an alternative to acreage allotments and production controls. It proposed that the federal government rent acres out of production to reduce surpluses, with the acreage in conservation rather than in competing production. The Farm Bureau's endorsement did not signal broad support in the farm coalition. At the time, the organization was closely aligned politically with midwestern corn farmers, President Eisenhower, and many Republicans in Congress. The divisions over farm policy were deepening and breaking along lines that were both regional and partisan. The political friction among the factions within the farm coalition was getting worse.

Surveying the Political Terrain

John Mark Hansen concluded that "the Soil Bank precipitated the destruction of the bipartisan farm bloc." He added that, to Democrats on the House Agriculture Committee, "the political motives behind the Soil Bank were transparent."[20] This conclusion overlooks that the bipartisan farm bloc had been unraveling for much of the postwar period. Regional differences became increasingly partisan and worsened as the parity policy system failed. The 1954 Act had featured a win for the president. Worse still, in partisan minds, it was a win for Secretary of Agriculture Ezra Taft Benson, who was loathed by southerners in Congress. By the time President Eisenhower proposed the Soil Bank to Congress, it was a thoroughly midwestern policy and not supported by the cotton interests who held power in Congress. These were early warning signs that conservation policy was unlikely to cool political friction. The bipartisan farm bloc deteriorated much more during the Soil Bank's brief existence. It hit rock bottom a few years after Congress terminated the programs. The farm bloc's deterioration cannot be blamed on the Soil Bank, but the program contributed to it. Just as significant, fuel was added to the regional, partisan feuding by larger political and societal events.

In 1956, when he proposed the Soil Bank to Congress, President Eisenhower was up for reelection. Politically, Eisenhower was significant. Elected president in 1952, he was the first Republican to win the White House since Herbert Hoover in 1928, prior to the Great Depression. Democrats had won every presidential election for twenty years, five straight victories. Eisenhower had brought with him a Republican congressional majority in 1952 but lost it in the 1954 midterms. It was a congressional majority in the House that the Democrats would hold for forty years.

Changing partisan control of the White House and Congress was joined by multiple societal and political firestorms during the Eisenhower years. While peripheral to farm policy, these matters were important because of the implications for the southern faction. During the mid-1950s the Jim Crow system of segregation and oppression in the South was mortally wounded. It was also a time in which American society was rapidly changing during the postwar era. No single event was more notable or consequential than the Supreme Court decision in *Brown v. Board of Education*, handed down on May 17, 1954. It overturned the notorious 1896 decision of *Plessy v. Ferguson*. The Supreme Court struck down the separate but equal doctrine and ruled that segregation in public education violated the Constitution's equal protection provisions.[21] The implications in the South were clear and momentous, exemplified by the violent reactions of segregationists and white supremacists. A prominent, tragic example was the murder of Emmett Till in Tallahatchie County, Mississippi, in August 1955.[22]

One intersection of the societal issues around race and farm policy can be found in the Citizens' Councils, which were formed by white elites mostly before the Supreme Court's decision. These white supremacist organizations, as James Cobb has explained, were spawned in the Delta's "planter-dominated society and economy" among the "agriculture-dependent business and professional classes." The councils provided funding and leadership in the South and "drew considerable support from planters and farmers in the surrounding countryside."[23] The councils effectively stoked white grievance, drawing support and power from long-standing insecurities. Pete Daniel explained that they melded "segregationists, right-to-work partisans, and businesspeople" in the South to deploy intimidation and other "intricate racial tactics used to divide black and white workers, to encourage right-to-work legislation, and to discourage unions." The Citizens' Councils were involved with federal farm policy on the local level and were known to deploy "intimidation and economic power to attack African Americans who pushed for the vote or school integration." Among the levers of power and intimidation they pulled were those of local banks and county USDA offices, from which "local whites administered federal agricultural programs" and through which "retribution could be disguised as policy" decisions.[24]

For cotton planters, federal interference in their operations became a much greater threat to their power and control in the wake of the Supreme Court's *Brown* decision. Consider the obvious issues in rural areas where the prospect of integrated schools could mean that the children of planters would have to sit in the same class as the children of sharecroppers. The threats only grew as the civil rights movement began to take hold and gain political clout. At the very least, the convergence of these matters would have exacerbated the perceived threats to southern planter power from reformers and the federal government. This, in turn, added hostility toward any farm policy proposal from outside the region and one to be implemented by a president from the opposing party. John Mark Hansen noted that "under the pressure of the civil rights movement, southern congressmen had increasingly hardened into sullen intransigence, precisely at the moment that northern liberals had reinstated social reform on the congressional agenda."[25]

These developments in Eisenhower's first term add important context to the situation into which the Soil Bank was proposed, fought over, enacted, and implemented. Even as it failed, farm policy benefited the southern cotton interests by effectively underwriting the modernization and transition they wanted. Federal assistance and diverting acres to other commercial crops helped the planters consolidate acres, invest in bigger and better machinery, and purchase more fertilizer and chemical inputs each crop year. This, in turn, narrowed the faction and furthered its enclosure. Whereas the 1936 debate had pushed out dairy interests, the policy fights in the fifties were driving away corn interests and some wheat interests.

From the Dust Bowl forward, the political development of conservation and natural resource policy was saddled with significant conflicts and complications. Conservation policy better served the public interest but not necessarily the factional interests in the farm coalition. Those challenges grew with the technological revolution in agriculture. Acres diverted to other commercial crops would still require equipment and chemical inputs, but acres diverted into conserving cover would not. Diverting acres to conservation not only would have been less lucrative but would have cut against the modernization effort. Created and funded at the federal level, it involved actions on individual private property requiring implementation at the local level. The planters held power at all levels but were facing political problems from a parity system that was unraveling at the same time as the Jim Crow system. If conservation policy was found to be helping smaller, poorer farmers—or, worse still, if it helped in any way tenants, sharecroppers, or Black farmers—it would have committed the gravest of political sins in the South; politically toxic before the Supreme Court required desegregation of the schools but even more explosive after.[26]

The 84th Congress and the Fight to Enact the Soil Bank

With Democrats back in control of Congress in 1955, southerners in the House wasted little time trying to undo what had been achieved the year before. They ran headlong into unyielding opposition from President Eisenhower, Secretary Benson, and the American Farm Bureau. The House Agriculture Committee reported out a bill in March, and House Agriculture Committee chairman Harold Cooley (D-NC) set the tone from the start. Antagonistic and highly partisan, he laid blame for agriculture's problems at the feet of Secretary Benson for mismanagement in the wake of the Korean War. Chairman Cooley's fast start was not a demonstration of strength. The bill barely passed the House. That close brush with failure, coupled with a presidential veto threat, convinced southern senators to put the effort on ice. It was possible that they were hoping for better political terrain in 1956, when the president was on the ballot for reelection.[27]

President Eisenhower responded in January 1956. He blamed acreage reduction policies for having harmed efficient farm management while spreading surpluses because diverted acres were used to produce other competing commodities. In other words, he sided with the Midwest against the South. On the growing surplus he laid blame for depressed prices, lost export markets, increased foreign production, declining exports, and even foreign imports. He was taking direct aim at the preferred parity policy system created and long defended by southerners in Congress. In no uncertain terms, he pronounced it a failure and proposed the Soil Bank as an alternative.

In place of this failed policy, the president proposed two programs under the name of the Soil Bank. The first was an annual acreage reserve that would divert excess acres into conservation, while the use of payments-in-kind would decrease stored commodities that had been forfeited by farmers. It was a unique feature of his proposal. Forfeited commodities would be used as part of the compensation package for farmers reducing acres through the acreage reserve. A farmer could receive a certificate for surplus commodities in storage in lieu of planting and harvesting more of the oversupplied crop. President Eisenhower argued that it would help "work off surpluses" and relieve "excessive production from acreage diverted from surplus crops." The second program was in addition to the short-term diversion program. The president proposed a conservation reserve into which landowners could retire acres "poorly suited to tillage" and that should not be under intense row crop cultivation. These fields were causing a "wastage of soil and water resources" and were estimated to equal nearly twenty-five million acres.[28]

Three days after President Eisenhower proposed the Soil Bank to Congress, Secretary Benson made it clear to Chairman Allen Ellender (D-LA) and the Senate Agriculture Committee that time was of the essence. The administration felt it was necessary to get at least the acreage reserve in place prior to the 1956 planting season if there was any chance to address surpluses. Additionally, the administration was clear with Congress that the intent was for the entire Soil Bank to be an aggressive effort for both acreage reductions and conservation, enrolling as many as forty-five million acres (twenty-five million for the conservation reserve), but that the acreage reserve was to be short-lived and of an emergency nature. Secretary Benson took an aggressive political stance against the parity system, continuing the direct attack launched by the president. Southerners would not have failed to understand.[29]

The payment-in-kind feature of the acreage reserve was also a potential political liability. It amounted to a gamble that "as production is pulled down by this underplanting, a vacuum is created in the market into which the [CCC] would move a part of its present stock" and reduce its surplus holdings. Commodities stored by USDA were forfeited to the CCC under the price-supporting loan programs, which meant that the commodities used as payments-in-kind had been purchased from farmers by USDA. Purchased from farmers with taxpayer funds, which also paid for storage and handling costs, these commodities would be used to pay many of the same farmers to reduce acres. Or, in Secretary Benson's words, USDA "would use the surplus to use up the surplus."[30] If it didn't work, more commodities would be forfeited back to USDA, and costs would skyrocket. Even if it did work, it would be incredibly expensive for taxpayers but generous for those farmers in the program. It was clearly policy born of desperation and an indicator of the degree of problem that already existed, which was itself expensive. But the Soil Bank was a unique development because it provided an alternative. To its supporters, the program provided a "place to put the diverted acres" in return for "some just compensation" on those acres, and an "opportunity to reduce his plantings and at the same time not go bankrupt."[31]

The committee got off to a quick start, but USDA opposed the initial draft, providing another sign of the challenges in the politics of an election year with planting season just around the corner. As the policy developed, the committee authorized a billion-dollar-per-year Soil Bank, 75 percent of it for the acreage reserve and farmers of just five crops: wheat, cotton, corn, rice, and tobacco. It was also in addition to the existing price support programs for these same crops. The Senate Ag Committee bill offered an illusion of bipartisanship, with Democrats moving a key priority of a Republican president in the election year. Beneath the billion-dollar price tag were real problems. Southerners on the committee decided to use the president's proposal as leverage against his opposition to their preferences on loan rates. They also escalated the regional feuding over acreage policy because the committee's version of the acreage reserve was viewed as harming midwestern corn interests; partisan and regional conflicts played out in the esoteric details of the program's design.[32]

Title I of the Senate bill was the most glaring challenge because it restored high, fixed price supports for two crop years. Proposed by Chairman Ellender, this policy was opposed by the president and his allies on the committee. The issue of price supports had complicated the committee's work and was agreed to by a single vote during markup.[33] The argument for high price supports was generally that farmers were being squeezed between increasing costs and stagnant or depressed prices, held down by surpluses. The cost side of the squeeze, however, appears to have been due to adoption of the increasingly expensive products of the technological revolution (e.g., machines, seeds, pesticides, and fertilizers). It was a double bind: more expensive inputs were producing larger crops, causing surpluses that made those crops less valuable or lower priced. To the traditionalists, most notably the southern senators such as Chairman Ellender and their allies like Senator Milton Young (R-ND), the only acceptable prescription was high price supports that at least gave the appearance of boosting farm income. To opponents of the policy, such as former secretary of agriculture Clinton Anderson (D-NM), it was akin to trying to "help the farmer by giving him another dose of the medicine that has made him sick."[34]

President Eisenhower had made it clear that he would veto any bill that included the price support provision. If members of Congress wanted to get a farm bill enacted ahead of planting season, they were taking a big chance that the president either was bluffing or would fold. A veto held the potential that the Soil Bank would not be enacted if it left Congress and the administration at a stalemate. In an election year this was no small risk. But there were risks on the other side as well. If the Soil Bank succeeded where the parity system failed, it would serve as a direct refutation of two decades of policy efforts. The Soil Bank presented a direct challenge to the southern policy priorities and, thus, their political power. The Senate debate further ratcheted up the political pressure when senators voted to reject Chairman Ellender's price support provision on the Senate floor.[35]

Compared to the price support debate, the acreage issue was an odd, confusing mess that had been long in the making. The committee design for the Soil Bank created a substantive problem for corn farmers in the

Midwest because it linked participation in the acreage reserve to the acreage allotment system designed in the Agricultural Adjustment Act of 1938. Since 1938, corn production had spread beyond the Corn Belt—some of this was the result of acreage diversions to corn and feed grains under the parity system. More areas and farmers were added outside the traditional Corn Belt of the Midwest, which required more fields, farmers, and acres to be squeezed under any allotment. The technological revolution was also driving significant increases in corn yields at the same time the production area increased. Combined, these developments contributed to a surplus of corn and feed grains, which, in turn, would require lower acreage allotments. Stuck in 1938, the acreage allotment system exacerbated the regional and political disputes.[36]

The Senate bill required compliance with the problematic acreage allotments to be eligible for the Soil Bank (acreage and conservation reserves). Complicating matters further, USDA had announced an allotment for 1956 that was considered unreasonably low. If the allotment were required for participation in the Soil Bank, it became effectively punitive on midwestern corn farmers. Not unreasonably, southerners accused Secretary Benson of sabotaging the acreage allotment policy. He, along with midwestern farmers, Republicans, and the Farm Bureau, strongly opposed acreage allotments and blamed them for causing or helping contribute to the feed grain surplus, benefiting southern farmers at the expense of midwestern farmers. Many corn farmers had, in fact, simply not complied with allotments for years. They were effectively opting out of the parity system, which made the situation worse for midwestern farmers when southern senators used that very system for determining participation in the Soil Bank.

Further complicating matters, the bill made the new acreage system for corn subject to approval by a two-thirds vote of corn farmers in a referendum after 1956. The referendum was a source of confusion from the start, even with its authors. If implemented, it represented the very real possibility that the new program would be damaged early in its operation if farmers rejected the referendum. Corn farmers had never been through a referendum and had historically opposed any acreage reductions or supply controls. Midwestern opposition to these provisions had become more intense, and it would only take a minority of corn farmers to vote down the compromise a year into the program.[37]

On the Senate floor, Senator Bourke Hickenlooper (R-IA) led the midwestern fight to change the committee's design. Senator Hickenlooper's amendment, however, went too far in the other direction. This not only undercut Hickenlooper's position but also escalated the regional feuding. The result muddied the argument that existing policies were unfair to corn farmers. Senator Hickenlooper conceded to political reality and compromised, which the Senate accepted. The entire debate accomplished very little. At most, it served only to increase the political heat of the already overheated partisan and regional feuds.[38]

Southern senators were all too willing to increase the acrimony and vent frustrations to inflict punishment on corn farmers and the administration. At one point in the debate, Senator Richard Russell (D-GA) proclaimed, "I shall vote against the pending amendment, on the theory that if agriculture must go down the drain, all components of it ought to go down together, and let the corn farmers feel what has been done to the producers of other commodities in other sections of the country."[39] Southerners also clearly understood that success for the Soil Bank might translate as success for the increasingly unpopular Secretary Benson. The vitriol exposed the deteriorating political state of farm policy. It did not, however, prevent the Senate from passing the bill by a wide margin without the price support provisions.[40]

The Senate bill included Eisenhower's Soil Bank but not the price supports. The 1955 House bill contained the price supports but not the Soil Bank. Not surprisingly, the final version of the first 1956 Farm Bill combined Eisenhower's Soil Bank program with a restoration of the high price supports.[41] Conferees explained that the "acreage-reserve program is not a soil-conservation measure but a short-term program designed to curtail production of specific commodities." The program would reduce the surplus by "induc[ing] farmers to reduce their acreages . . . and to make no other use of the land so retired."[42] All told, the final agreement allocated specific levels of acreage reserve funding among the commodities, with $375 million designated for wheat and $300 million for corn. The bill also provided $300 million for cotton, a comparative windfall at a time when cotton acres were about one-fourth of corn acres and one-third of wheat acres.

The final version of the first 1956 Farm Bill passed in the House and Senate on April 11, 1956. As promised, President Eisenhower vetoed the farm bill on April 16, 1956—reportedly calling it a "private relief bill for politicians" rather than a farm bill—and the House failed to override the president's veto on April 18, 1956.[43] The House quickly became "realistic" and moved a new bill within two weeks that removed the provisions opposed by the president. Thereafter, the Senate moved expeditiously on a bill revised to agree to the president's position, passing it without a recorded vote one week after reported out of committee. Much of the fight had gone out of the combatants. Price support advocates conceded to reality, passing the new bill just over one month after President Eisenhower vetoed the first one.

The president signed the second one and the Soil Bank into law on May 28, 1956, but with concern that the delay would harm the effectiveness of the acreage reserve. It was therefore notable that Congress required USDA to operate the Soil Bank beginning with the 1956 crop. It was a decision made knowing "that the larger part of this year's plantings have already taken place." The requirement was inserted because southerners in the House demanded it.[44] The provision was a rather loud warning that problems for the Soil Bank did not end with enactment.

With the benefit of hindsight, the 1956 Act contained multiple such warnings. It had taken Congress nearly five months and two attempts to move President Eisenhower's proposal for a new(er) direction in farm and conservation policy from concept to law. The Soil Bank had changed little in the process. The problems for the policy were not based on any substantive concerns but were purely political, regional, and partisan. It had been held hostage as those matters were fought out against a rapidly advancing calendar. The final bill also contained further potential political land mines buried in the text.

First, a farmer—or at least a small farmer—could enroll his entire farm in the Soil Bank using both the acreage reserve and conservation reserve. Conferees also removed payment limits. Third, the final bill authorized that "the funds of the Commodity Credit Corporation may be used for carrying out the soil-bank program" in the first year of the program, "but that after that date such funds must come from regular appropriations."[45] It was one year of mandatory funding similar to price support programs, but after that year Soil Bank funding was completely subject to congressional appropriators. This would turn out to be central to the policy's demise.

Finally, receiving "special mention" were changes that once again purported to provide protection of tenants and sharecroppers. According to the conference report, "No single problem connected with the proposed soil bank has caused the committee more concern than that of guaranteeing adequate protection of tenants and sharecroppers under the program." Conferees claimed that the protections in the Senate bill were "inadequate to afford the protection desired." On the issue of dividing payments between landlords and tenants, however, conference left it to "each landlord applying for participation in the program to stipulate in detail how he proposes to share the benefits with his tenants or sharecroppers." They further required that the "county committees made up of farmers who are thoroughly familiar with local conditions to approve the proposed division of benefits before the farmer will be permitted to participate in the program."[46]

The historic record presented overwhelming evidence that this would directly work against the interests of sharecroppers. Important discretion was left to the secretary, and he could strengthen the statutory protections but would do so with the knowledge that he risked blowback from southern planters and their congressional allies. A cynical read of these provisions suggests that southerners meant them to have no teeth but, in effect, dared the secretary to do more through implementation and risk southern wrath.

As enacted, the Soil Bank was a billion-dollar-per-year policy (approximately $10 billion, adjusted for inflation) that consisted of two separate programs, an acreage reserve and a conservation reserve. The acreage reserve program compensated farmers who reduced their crop acreage and diverted those acres into conservation, agreeing not to "harvest any crop from, or graze" the acres. The Conservation Reserve Program authorized the secretary to enroll acres for up to ten years that were to be maintained in "protective vegetative cover" for the entire contract. Congress declared concerns with "improper land use" in farming, "soil erosion, depletion of soil fertility, and too rapid release of water from lands where it falls." Congress also returned to the national, general welfare concerns "that the soil and water resources of the Nation not be wasted and depleted," which justified policy "to protect the national soil, water, and forest and wildlife resources from waste and depletion . . . and to provide for the conservation of such resources." Congress emphasized that "the production of excessive supplies of agricultural commodities depresses the prices and income of farm families." Natural resources and farmers were "depleted in the production of such burdensome surpluses," and therefore new policies were additionally justified "to protect and increase farm income . . . an adequate, balanced, and orderly flow of such agricultural commodities in interstate and foreign commerce."[47]

Assessing the Origin and Rise of the Soil Bank

The first thing that stands out with the developments in 1956 was how they signified the diminished power of the South. The political situation contrasts rather drastically with that of the 1936 debate. Any threats about southern power were ineffective in deterring midwestern interests from pushing for their preferred policy alternatives. Chairman Ellender and southerners lost on the Senate floor over loan rates. The president vetoed the bill, and they were unable to deter him in a reelection year. Moreover, while they tried to call his bluff, the president called theirs and forced southerners to concede. All of this and more signaled the extent to which the political situation for the southern faction had deteriorated. The farm bill fight served as a microcosm of the larger changes roiling American society and politics in the postwar era—within a decade after the Soil Bank, Congress enacted the Civil Rights Act of 1964 and the Voting Rights Act of 1965.

The fight over the Soil Bank further demonstrated the factional competition in Madison's design. The midwestern corn faction fought it out with the southern cotton faction in the process, which was *some* increase in competition over the New Deal era. It took the power of the presidency as exercised in the veto to shift the balance against the South, which demonstrated the power of that feature in the constitutional design but not without consequences for the policy. As in 1936, however, this factional competition was limited to the farm interests. The nascent environmental movement did not possess the strength to compete, and no faction representing the conservation interests was at the negotiating tables or pushing for policy changes in the process. Designed by farm interests, conservation policy remained primarily designed for them. Accordingly, conservation policy in 1956 was heavily focused—although not exclusively—on production controls and related farm economic concerns. The political competition had yet to be opened to interests outside the farm faction, and the policy demonstrated as much.

In the Soil Bank was further proof of path dependence, which endured across two decades of interregnum. Agricultural conservation policy as represented by the acreage reserve was a continuation of the 1936 concepts in which conservation served merely as an alternative method for diverting acres out of production. By 1956 the parity system may well have failed, but the concept of centralized control over farming retained its hold on policy development. The acreage reserve remained within the basic New Deal policy framework and continued the idea that production could be controlled by reducing acres despite what the decade had demonstrated. Neither Congress nor the administration addressed the fundamental fallacy of that policy as exposed by the technological revolution. As in 1936, conservation policy in 1956 had far more to do with the control of supplies and helping farm income than with the conservation of natural resources. Conservation was once again the lesser priority, secondary to the perceived needs of farmers and commerce burdened by surplus in the postwar technological revolution.

The best argument for the acreage reserve was as an attempt to resolve the political fighting among farm interests over policies. It provided the potential for a solution. Diverting acres into conservation instead of competing crop production, especially feed grains, might have cooled the political friction generated by the parity system. It was a partisan alternative by midwestern corn interests and the Farm Bureau, proposed by a Republican president. The parity system may have failed, but it did help the southern planters who held outsized political power on the Agriculture Committees. Acres diverted were still producing commercial crops for sale, and the planters also received high subsidies on their cotton crops. If these partisan problems could have been diminished or set aside, the Soil Bank may have offered Congress and farmers a way out of the self-inflicted problems they had created. For conservation to serve as an effective alternative, however, it would have to be very lucrative for these farmers, offsetting not only the cost of the land but also the potential dual benefits from marketing feed grains combined with cotton subsidies. It was a high bar for any federal policy to clear.

By comparison, the conservation reserve held out the most promise of a new direction for farm policy, one that invested taxpayer funds into conservation efforts that would benefit them more directly. Because it was policy designed in the direction of conservation purposes, the conservation reserve was also the more important development. It provided an annual rental in return for the establishment and maintenance of conservation cover (i.e., grasses, trees, etc.) for up to a decade, and it offered a solution to the handling of those fields that should not have been in the production of row crops. It may have been the result of drought problems in the Great Plains and concerns over another round of dust storms, but some have speculated that the main purpose of this program was to help transition smaller, inefficient farmers out of farming in order to reduce overall production. The conservation reserve, while limited, could have offered a different path for future development, an adjustment toward wise land use over the long term, and a policy shift in favor of the public good. At its peak, more than twenty-two million acres of (presumably marginal) cropland were enrolled.[48]

By any measure, President Eisenhower had won a significant victory. He had pushed through substantive changes to federal farm policy that held real potential for conserving natural resources. It could have changed the course of farm policy going forward. His victory tallied an investment of more than $1 billion each year, a significant portion of which was expected to deliver real conservation benefits for soil and water. This would seem to have been a matter of no small political value given the rapidly changing dynamics in postwar America and the near-complete failure of parity policy. The Soil Bank could have represented a triumph but had inherited the interregional farm factional feuds over acreage and price support policy that it was intended to resolve. It could have launched development of conservation policy anew, achieved against the myriad challenges of the legislative process. Any progress made in 1956 was potential that would go unrealized. Instead, the legacy of the Soil Bank was to be another brief episode that further demonstrated the power of the southern faction in Congress and that left behind questions about what might have been—about the path not taken.

6

Southern Sabotage

THE SWIFT DEMISE OF THE SOIL BANK

On May 28, 1956, President Dwight D. Eisenhower signed the Agricultural Act of 1956 into law after vetoing an earlier version. His signature put into law a version of his proposal for a soil bank. According to the president, the new Soil Bank was "rich with promise," offering the "most significant advance in the conservation field in many years," a development in policy that would "result in improved use of our soil and water resources for the benefit of this and future generations." The promise at enactment fits uncomfortably with the Eisenhower administration's request that Congress terminate the Soil Bank's acreage reserve program early. The administration also did not fight expiration of the conservation reserve authority in 1960.[1]

Somewhere between the promise and the relatively swift demise of the Soil Bank, specific, concerted actions were taken that effectively sabotaged the program. These began with the final version of the legislation, in which the House conferees demanded that the acreage reserve operate in 1956 despite the late start. Sabotage proceeded deliberately through appropriations. That much is relatively clear in the public record. The record also provides strong evidence—some of it circumstantial—as to the culprits and their motives for destroying the Soil Bank.

The culprits were from the powerful bloc of southern Democrats who held the gavels in key committees, as well as leadership posts in Congress; among them were the House Agriculture Committee chairman, Harold D. Cooley (D-NC), and vice chairman, W. R. Poage (D-TX), as well as the Senate Agriculture Committee chairman, Allen J. Ellender (D-LA). More than anyone else, however, sabotage of the Soil Bank was the handiwork of Representative Jamie Whitten (D-MS), the powerful chairman of a subcommittee of the House Appropriations Committee. Chairman Whitten's subcommittee had jurisdiction over the United States Department of Agriculture (USDA), and he was often referred to as the permanent secretary of agriculture.[2] Senator Richard B. Russell (D-GA), one of the most powerful senators and the chairman of the subcommittee of the Senate Committee on Appropriations with jurisdiction over USDA, was a key accomplice. To be opposed by Whitten and Russell in the late 1950s was a death sentence for any program or policy; the Soil Bank proved no exception.[3]

Background on the Methods for Sabotage

Untangling a political sabotage effort on a set of enacted policies through public congressional records presents multiple challenges. Undoubtedly, much more went on behind the scenes among congressional appropriators, their staffs, and the appointees and employees at USDA and in the administration. The public record delivers sufficient evidence of sabotage, and reviewing it provides valuable lessons in political development. Piecing the matter together requires some background on the obscure and esoteric operation of federal budgeting and appropriations. As such, this discussion begins with an explanation of the funding mechanism and the appropriations process as background for how it was deployed against the Soil Bank.

The first year of operation of the Soil Bank (1956) was funded with mandatory funds through the Commodity Credit Corporation (CCC), but the subsequent years of the program were at the discretion of appropriators in Congress.[4] This distills to the difference between mandatory funds and those funds that were discretionary or appropriated. In one sense, mandatory funding operates like a credit card, whereas discretionary funding operates more like a gift card; a credit card permits expenditures up to a certain limit at the discretion of the spender and then is paid off, but a gift card is only for a fixed amount that has been preloaded before being spent. In farm policy, mandatory funding uses the authorities of the CCC, which can borrow directly from the U.S. Treasury and is reimbursed by appropriators in Congress. With mandatory funds, USDA can offer programs that are not limited to a specified funding level, and CCC authority meant that funds were available to be spent based on the acres enrolled and the rental rates at which those acres are enrolled. This was important because the Soil Bank operated by paying farmers to rent some portion of their fields out of production for either one year (acreage reserve) or up to ten years (conservation reserve).

Discretionary funding, by comparison, had to be appropriated in advance by Congress; operation of the programs was dependent on the decisions and discretion of congressional appropriators like chairmen Whitten and Russell. If appropriators limited the amount of funding for the program, that, in turn, limited the program's operation because USDA cannot spend more than the appropriated amount. If Congress appropriated too little for the program to meet the demand, USDA must either limit who can enroll or reduce the rental rates, or some combination of both. Farmers who want to enroll will be left out, or they receive less than expected in rental payments. For USDA, the farmer was unlikely to be aware that any problems with his or her ability to enroll in the program were due to decisions by congressional appropriators. Sign-up and payments were with USDA, and problems were easily blamed on USDA. Those problems would also result in the farmers souring quickly on the program and policy. Whitten used this method effectively over his many years holding power. He deployed it on the Soil Bank at the first available opportunity.[5]

Timing on federal calendars was also critical. At that time, the federal fiscal year ran from July 1 to June 30, meaning the Soil Bank was enacted near the end of fiscal year 1956 (ending June 30, 1956). Appropriations for fiscal year 1956 were enacted on May 23, 1955, and more than a year prior to the Agricultural Act of 1956. This would have been a primary reason for funding the first year of the Soil Bank through the CCC because in addition to planting season being near complete, so was the federal fiscal year. Moreover, the appropriations for USDA for fiscal year 1957 that began on July 1, 1956, were just days away from being enacted.[6] Congress authorized CCC funds through June 30, 1957, or the end of fiscal year 1957, and required USDA to seek appropriations beginning July 1, 1957, or the beginning of fiscal year 1958.

Adding confusion and complexity, crop years do not align with federal fiscal years. The acreage reserve was intended to be an option to reduce planting, and it operated across federal fiscal years.[7] Enrollment for the acreage reserve after July 1, 1957, would be subject to the fiscal year 1958 appropriations funding, which would have coincided with the 1957 corn and cotton crop years (planted spring, harvested fall 1957). Winter wheat added further complexities, with harvest in summer 1957 of a crop planted in fall 1956. Any impacts from appropriations decisions would likely begin in summer 1957, around wheat harvest but prior to fall wheat planting for the 1958 harvested crop, while the 1957 cotton and corn crops would have been planted earlier in the spring. This background and general timeline will prove important.

Sabotage, Step One: Politics and an Assist from the Weather

The late May enactment of the Soil Bank contributed substantially to the poor start for the new acreage reserve program. By May 28, 1956, most of the planting for spring wheat, corn, and cotton would have been completed and the harvest for winter wheat (planted fall 1955) would be beginning in the southern areas of the country. As noted in the previous chapter, House conferees from the South required USDA operate the Soil Bank in the 1956 crop year. Requiring implementation of the acreage reserve for the 1956 crop year meant that USDA would have to figure out how to divert acres after farmers had already decided to farm them. This was no small matter to reduce acres planted to surplus crops if those crops had already been planted. Participating farmers would need to plow under a growing crop, the controversy of which would not be lost on southerners after the experiences with the 1933 Agricultural Adjustment Act. By forcing implementation for the 1956 crop, opponents of the program provided themselves plenty of opportunities to attack the administration's implementation and policy in an election year; they received a well-timed assist from the always unpredictable weather.

Opponents of the program did not wait long. Secretary Benson also seemed almost incapable of avoiding controversy. By early July 1956 USDA was already under attack in the House Ag Committee for its operation of the acreage reserve. This was less than two months after enactment. Drought conditions were continuing in Texas, and Vice Chairman Poage accused Secretary Benson of "discrimination" against cotton. He claimed the program created "conflict . . . only in the South."[8] From the hearing record a more complete story emerges. Some USDA field employees had reportedly informed certain cotton farmers in Texas that there would be no Soil Bank in 1956 because the president had vetoed the first bill. This turned out to be incorrect because conferees from the House Agriculture Committee, including Vice Chairman Poage, required implementation in 1956.

Problems due to unpredictable weather were entirely predictable at enactment. Not only were weather problems predictable, they were known. Drought had returned to the southern Great Plains in the 1950s and was worsening in 1956. Because the acreage reserve was intended to take acres out of production in order to reduce the surplus, the drought posed a difficult operational issue regarding how to treat those acres for which nature was doing the work of reducing the crop. If USDA paid farmers for taking out a crop that wasn't going to produce, the program operated more like disaster assistance than surplus reduction. A disaster program in an election year added political problems. On the other side of the equation, if the Soil Bank paid some farmers less because they enrolled acres after it was known that the crop was damaged, the agency stayed true to the purposes expressed by Congress. Doing so, however, would result in paying farmers different rates and, more pointedly, paying less to those farmers who were also experiencing damage from drought.[9]

To be fair to USDA, this was precisely the situation into which southerner conferees in the House put the administration. USDA chose to play it close to the statutory purpose despite the risk of adding problems for farmers hit by drought and increasing political risk. USDA paid a lower rate to farmers who plowed under a crop damaged by drought than it paid to those who didn't plant or those who enrolled before damage was known. While justifiable, USDA's decision provoked political blowback. It appeared that USDA was treating farmers differently, that those who voluntarily cut back acreage in anticipation of the Soil Bank were being treated better than those that planted into the drought and then plowed under the withered crop. These decisions allowed southerners on the Ag Committees to protest that the 1956 program had been too generous to midwestern corn farmers and unfair to southern cotton and peanut farmers. The 1956 drought and USDA's decisions for the acreage reserve played into these perceived regional grievances.

There is likely more to this story as well. Conveniently overlooked, the farmer who plowed up a crop likely did so only once it became clear that drought was going to damage the crop enough to make the acreage reserve payment seem the better option. It is very probable that some farmers who received a lesser payment attempted to game the program. Either way, USDA's decision hit hardest in the South and for cotton. Cotton farmers had expected as much as forty-five dollars per acre. Those cotton farmers who plowed under cotton damaged by drought received only six dollars per acre. For their farmers it was a problem, but for southerners it was a political opportunity and another chance to attack the secretary and the administration.

When Representative Poage attacked Secretary Benson, accusing him of having discriminated against cotton farmers, he presented an interesting argument. Logically, those cotton planters who held the right political connections in Congress were probably the ones most impacted by USDA's decision. They were mostly likely to have received initial intel from USDA field employees that the Soil Bank would not be operational in 1956—intel that turned out to be inaccurate. They were also far more likely to have had the resources to take chances on planting a crop into a drought. And the planters would have been most likely to game the acreage reserve once drought proved a problem and then to take their complaints to members of Congress. Maybe most striking in this episode was that Representative Poage went so far as to tell the secretary that he (Poage) was going to personally instruct local USDA officials in Texas to countermand the directives from the secretary of agriculture, rig the acreage reserve payments, and deliver the maximum payments to the complaining cotton planters.[10] The episode speaks volumes about southern perspectives on farm policy in this era and the methods of operation. It also supports the argument that southern members were more interested in sabotaging the program than in fixing it.

To complete the problems, the 1956 legislation had included a requirement that corn farmers vote in a referendum about whether to continue the Soil Bank's acreage program. Rejecting that acreage program would return the policy to the much tighter and more problematic acreage allotments with which most corn farmers had refused to comply. As designed, the referendum vote permitted a minority of farmers to defeat the referendum; little surprise, then, that a minority of farmers defeated it in December 1956.[11] The result was taken as a rejection of the acreage program by the very corn farmers for whom Congress created the policy. For southerners in Congress, the acreage program was considered a gift, and rejection added to their grievances. Corn farmers, moreover, would be subject to an allotment requiring a reduction of nearly one-third of the corn acres in the Corn Belt.

Setting aside the various oddities in operation, the acreage reserve had been designed in part to address the failures and complications of the parity system's acreage policy. When farmers rejected the corn base referendum in 1956, the result piled further complications on the Soil Bank in its first year of operation. Fallout from the referendum's failure meant that farmers who failed (or refused) to comply with drastically reduced allotments would be eliminated from participating in the Soil Bank. Instead of being a solution to its problems, the Soil Bank became entangled in them. President Eisenhower's alternative was at risk of repeating the very failures it was meant to address. The increasingly absurd system of farm policy at this point left the impression that it was only capable of producing self-inflicted problems.[12]

Sabotage, Step Two: Appropriators Launch Their Attack

The Soil Bank acreage reserve program received mandatory funding from the Commodity Credit Corporation (CCC) through June 30, 1957, but would need appropriations for fiscal year 1958, beginning July 1, 1957. This is notable in its own right because it differed completely from how Congress designed commodity support programs. Those support programs did not suffer this change in funding sources; all nonrecourse parity loans received mandatory funding from the CCC. In February 1957 Chairman Whitten opened a hearing by exclaiming, "I think the soil bank, and I am saying it from my heart, as it is now administered, is the worst thing that ever happened to the American farmer."[13] This was quite a statement after a single year of operation. That any federal payment program could be the worst thing that had happened to farmers exceeded exaggeration and hyperbole.

Not to be overlooked, Chairman Whitten clearly opposed all acreage reduction or production control policy, not just the acreage reserve. He argued that reducing acres planted only encouraged farmers to increase production on the remaining acres. He also claimed that reducing acres hurt small towns and rural businesses, particularly the ginning and other cotton industries. Finally, he understood that for an export-dependent crop like cotton, reducing U.S. acres generally resulted in an increase in the acres planted to cotton in other countries. Reducing cotton acres therefore effectively pushed production overseas while also opening market room for synthetic materials.

Be that as it may, Whitten's remark exposed his clear animosity for the program, which was important given that he possessed the power to act upon it. And act upon it, he did. Whitten understood that reducing the funds for the acreage reserve would harm the program and effectively sabotage the policy. A tried-and-true strategy for appropriators, it was one for which he was particularly adept. At the first opportunity, Chairman Whitten sought to eliminate nearly half of the original budget request for the acreage reserve.[14]

On the shortcomings of acreage policies, Secretary Benson largely agreed with Chairman Whitten. The acreage reserve is where they parted ways. Whitten viewed the acreage reserve as a continuation of this failed policy. Secretary Benson stressed the need to get the surplus under control, and he initially considered the acreage reserve the best option. He argued that the Soil Bank was an "emergency tool" for an "emergency situation" due to "unwise Government policy" and necessary for the postwar economy.[15] Chairman Whitten could not get past partisanship and considered it a political stunt by the Eisenhower administration to pay farmers in the election year. Along with his southern allies, he was unwilling to give the new program a chance to prove whether it could work.[16]

These matters were asymmetrical. Opposed to acreage reductions, Whitten showed little concern with the surplus spreading to other crops; his concern was only for cotton and the South. Whitten's position was not available to Secretary Benson or the Eisenhower administration. Surpluses were piling up in federal storage, and they were stocks owned by USDA, by the administration politically. The surpluses also continued to depress crop prices. These problems were national in scope and involved multiple commodities. They were problems for which the administration had few good options.

The real puzzle was the administration's tepid response in the face of withering attacks. Either unwilling or unable to engage in a political fight to save the Soil Bank, the administration communicated relatively weak support and little political backing for the program it had requested. Worse still, the administration dug a deep hole for the program. In April 1957 the president sent an amendment to its earlier budget request that suggested a reduction for the acreage reserve based on recomputed cost expectations. The amendment proposed a substantial reduction from the original two-year budget estimate but does not appear to have been based on expectations for a reduction in the acres enrolled. To reach the reduction in spending on the enrolled acres that were expected, USDA would need to significantly reduce the payments per acre. It presented an option unlikely to win the program or the department much support from farmers.[17]

Conceding cuts to an appropriator opposed to the program was almost too easy. Chairman Whitten took credit for the administration's reduction and sought to cut it further. His subcommittee reported to the House that "the acreage reserve program was established under dire necessity to get some funds in the hands of the farmer last year" but that the program had failed to reduce production. The report added that it had become "apparent that, if this program were to be pushed, it would have a damaging effect on local business."[18] When the appropriations bill was on the House floor, southerners sought to eliminate the acreage reserve entirely. They claimed concern for the budget implications, but those concerns were hard to justify, given a $1.2 billion expenditure to reimburse the CCC for price support programs—half was just to store and handle the forfeited commodities. Adding intrigue, southerners also sought to attach a payment limitation on Soil Bank payments. If it survived, a tight payment limit would damage the program further. It was a curious move given the long-standing and strong southern opposition to payment limits on price supports. Charges of hypocrisy proved ineffective against the South.[19]

As if to prove the charges of hypocrisy, House Ag chairman Cooley and vice chairman Poage were accused of trying to kill a program they authorized. Representative August H. Andresen (R-MN) attacked them for using their majority on the 1956 conference committee to rig the game and then turning around to destroy it on the sly through appropriations rather than by addressing it in the authorizing committee they led. They wrote the Soil Bank's provisions, he argued, and "now before the law has had a fair chance to operate, they want to destroy it" with dishonest claims.[20] He also made it clear that any problems with the Soil Bank were mostly in the South and were likely the result of decisions at the local level in that region. Chairman Whitten and his allies were undeterred by the Republican counterattacks, and more importantly, they had the votes. The House agreed to the southern amendment to defund the acreage reserve in May 1957 and then passed the appropriations bill without a recorded vote.[21]

In August 1957 Congress finished its work on the appropriations bill for the 1958 fiscal year that had begun in July. Senators saved the acreage reserve in conference. According to Senator Milton Young (R-ND), one of the "big accomplishments on the Senate side in the conference was the restoration of the acreage reserve program."[22] The bill also included a cap on USDA's ability to enter conservation reserve contracts in any one year, and a limit on spending for the acreage reserve. In its first year operating under appropriations rather than mandatory funding, the Soil Bank's programs were severely limited by Congress (with an assist from the administration) and these limitations would have real consequences for the policies, as well as the farmers those policies were intended to help. Chairman Whitten very nearly bragged that the "House voted to cut out the soil bank after this year on a very close vote." He also continued to argue for the payment limit that was ultimately rejected. He added that it "appears to me to be a reasonable solution to the problem, at least as reasonable as we could agree upon," but southerners were unable to get an agreement on including a tight limit.[23]

Looking back on it, USDA's first encounter with Whitten and appropriators had been confounding. In April 1957 Benson stated that the "real test of the soil bank will, I think, be this year."[24] This statement conflicted with the administration's recalculation and the revised budget proposal to reduce spending for the acreage reserve, which had been sent to Capitol Hill that same month. Not spending the funds to make the program work was a guarantee of its failure. One guess on the reasoning for the budget revision is that it may have reflected Secretary Benson's own conservative ideology, as well as that of the administration and its budget staff. Uncomfortable with a federal program of the Soil Bank's scope and costs, Benson and the administration may have started to back away.

After the program's troubled first year (1956), the acreage reserve looked increasingly unlikely to achieve much of a reduction in supplies. Farmers increased fertilizer and intensity on the acres that remained. The technological revolution was taking over, and there was no good way to plan for the impacts of weather. Worse still, USDA had to admit that some farmers had enrolled fallow land in the acreage reserve and were able to receive payments. It also did not seem to matter that this problem was most prevalent in the South.[25] These matters could also have factored into the administration's decision to recommend a funding reduction. With these realities, and the political attacks by powerful members of Congress, the administration may have become reluctant about the program and uncomfortable with little to show for the spending as early as spring 1957. The Soil Bank may have seemed like a good idea until the cost was being incurred without the reduction in supplies. If so, this was shortsighted. USDA and the administration had to have known that it would take time for the program to show progress with the surplus and that operation in 1956 was not indicative of the program's potential.

Political problems compounded the lack of progress with the surplus because USDA and the administration were under unceasing attack from southerners in Congress. If they were trying to buy peace with southerners and time for the program, it should have been little surprise that appeasement failed. The conflict was very partisan and regional. Republicans from the Midwest defended the Soil Bank and attacked southerners for the underhanded attempts to kill it after using the price support system to divert their acres into competing crops. The fight over an amendment to an appropriations bill further demonstrated the deterioration in farm policy politics. On the political score, the Soil Bank had not helped ease the tensions but rather became a source for further inflaming them.

Sabotage Succeeds: Eisenhower Surrenders in 1958

Southern opponents of the Soil Bank did not let up. In 1958 they added anger with the 1957 sign-up to their attacks. They conveniently ignored the fact that they had cut funding in the appropriations bill enacted in August 1957, impacting farmers in the subsequent sign-up months after the funding reductions. Some farmers had been unable to enroll in the program because USDA had to limit acreage due to the reduced funding. Unable to enroll in the program, farmers complained to members of Congress. In February 1958 Senator Lister Hill (D-AL) testified to the Senate Agriculture Committee about these matters. He stated, "One problem which demands our immediate attention is that raised by the present administration of the cotton acreage reserve features of the soil-bank program," a situation that he saw as "one of confusion, inequity and injustice."[26] To his colleague's concerns, Senator Aiken responded that "the administration requested the Congress to give them money to carry out the soil bank," but "Congress refused to give them money enough to carry it out," and that was "the reason the cotton farmers are not able to put land into the soil bank at this time."[27]

The full testimony of Senator Hill is also instructive. He complained on behalf of cotton farmers who had expected to be able to enroll in the acreage reserve program but were not able to enroll. He claimed that these cotton farmers had made no preparations for planting because they expected to enroll in the acreage reserve. Maybe exposing more than they intended, Senator Hill also claimed that some of these planters had even dismissed their laborers based on the expectation. When they were not able to enroll, they were caught in a real bind, if Senator Hill's account can be believed. The situation also exemplified the entitlement mentality of the southern cotton planter toward federal programs. It was a further reminder, if one was needed, of the ease with which their laborers bore the brunt of farm policy. The planters refused to take any responsibility for their own decisions. At the same time, the southerners in Congress who cut funding for the program were able to avoid blame for the problems they had helped cause. All anger and frustration were directed at USDA.

The weather complicated matters once again. The fall 1957 harvest had been plagued by wet conditions that harmed a significant portion of the cotton crop; the crop was harvested, but wet conditions lowered its quality. This brought the cotton industry into the discussion, adding their complaints against the program in 1958. The cotton industry claimed that they were facing a major emergency due to a shortage of high-quality cotton to meet their needs. The wet weather had damaged quality but not overall production. Cotton farmers in 1957 harvested a large crop of low-quality cotton that the industry claimed it couldn't use. As problems mounted, blame was cast on the Soil Bank and the administration, not weather or appropriators.

The situation bordered on comical, adding a dose of irony to the mess. A loss of high-quality cotton had driven up its price—something that cotton farmers had long claimed they wanted, and that southerners in Congress had long claimed to seek for them—and bad weather reduced the quality of the cotton that was harvested. The Soil Bank had also reduced acres planted to cotton, with over three million cotton acres enrolled in the program in 1957.[28] The higher prices, however, made cotton less competitive with synthetic fibers, according to the industry. The only solution was to demand an increase in cotton acres. One bad weather year for cotton, so the argument went, was more than the Soil Bank could handle, and the program was accused of "culling people in a bad year." Closer to the truth, it was a convenient excuse to do what the industry had wanted all along, but it clashed with a quarter century of stated policy goals for cotton.[29] There may have been more to this request; among the demands to change the program was one to allow acreage reductions for some farmers while increasing acres for other farmers.[30] Given the history, it is not difficult to imagine who would win and who would lose in cotton country. USDA did not grant the cotton industry's demand for additional cotton acres. The department noted that cotton farmers had already signed contracts to put over three million acres into the acreage reserve for 1958.

In the end, the southern saboteurs won. It was a victory assisted by the weather, the costs of the program (especially to a conservative administration), and an odd piling-on from the cotton industry. In January 1958 President Eisenhower and USDA requested that Congress terminate the acreage reserve so that it would not be used for the 1959 crops. President Eisenhower informed Congress in his annual budget message that he had concluded "after careful consideration" that the acreage reserve wasn't worth it. He added that "more material and lasting benefits per dollar spent, both in reducing production of surplus crops and in obtaining enduring conservation of the Nation's agricultural resources, will be achieved under the conservation reserve program."[31] The push for more funding for the conservation reserve was, at least, some level of continued support for conservation policy.

USDA officials explained further the decision to Congress a month after the president's budget message. The timing is also notable because it coincided with continued attacks from southerners, such as those by Senator Lister Hill mentioned above; surrendering to the South did not stop the efforts to denigrate the policy or the administration. Congress complied in the appropriations bill for fiscal year 1959 (July 1, 1958, to June 30, 1959), which was enacted in June 1958. The final year of the acreage reserve experienced "unexpectedly heavy sign-ups" and higher costs, however. Whitten and Russell continued to lead the fight for further reductions in Soil Bank spending. Declaring an end to the program, Congress made clear that "no part of this appropriation shall be used to formulate and administer an acreage reserve program with respect to the 1959 crops."[32] The fiscal year 1959 appropriations bill effectively closed the book on the Soil Bank. USDA could enroll acres in the conservation reserve through 1960, but appropriators made certain to keep the funding tight.

Sabotage Supported: Whitten's Investigative Reports

As noted above, proving a sabotage effort from the public record is difficult. Arguably, there is sufficient evidence of it in the public records, without any explicit admissions. The records of the appropriations committees and leaders, especially Chairman Whitten, demonstrate a concerted effort to limit the program, harming its administration and effectiveness, while attacking it and USDA at every opportunity. The contrast with how Congress treated commodity price support programs is critical. Even as they were rapidly unraveling in the postwar technological revolution in agriculture, Congress placed no limits on the funding or the policy; even as costs escalated and the failures of the policy quite literally piled up in USDA storage, congressional leaders like Whitten did little to rein in those programs or to subject USDA operations to tighter controls. It is a stunning contrast in political treatment of policies for the same factional interests. And those are not the only sources of evidence, moreover. Chairman Whitten provided important support for the sabotage effort he led. He ordered two investigations of the Soil Bank by his committee staff. These investigative efforts, which undercut his arguments against the Soil Bank, were never used against him or the southern opponents and saboteurs. USDA did not use the findings to defend itself or the Soil Bank. The findings provide important insight into operation of the programs. The findings and conclusions also provide strong circumstantial evidence to support a conclusion that the outcome for the Soil Bank was the result of sabotage.

Investigative staff produced their first report in March 1957, reviewing the Soil Bank's operation in the 1956 crop year. The investigation unearthed plenty of problems. For one, production of corn in 1956 exceeded production in 1955, with nearly 10 percent of corn base acres enrolled in the acreage reserve. Favorable weather in the eastern Corn Belt led to record yields, which more than offset production lost to drought in the western Corn Belt; those yields also made up any production lost to enrollment in the acreage reserve. For supporters of the Soil Bank, it was a painful reminder of how little control an acreage program can have as compared to the overwhelming impacts of weather. The acreage reserve "failed to curtail production by any significant amount" because the "farmer has rented the poorer part of his land to the Government" and "increased his per-acre yields on the land not placed in the acreage reserve, to offset the loss of production on the land taken out of use."[33] On this point, Chairman Whitten and Secretary Benson were in agreement.

Reality was not a surprise; farmers wanted to produce crops on their land. In response to acreage reductions, the "farmer just spreads more fertilizer, cultivates fewer acres more intensively and in many cases gets about as much production than he did on a greater number" of acres.[34] Farmers increased the use of fertilizer on the acres that remained in production, including some fertilizer purchased with USDA assistance. Some farmers also gamed the acreage reduction program by plowing under oats, which were not covered by the program, rather than corn. That USDA regulations allowed many midwestern farmers to work around the acreage reductions only added fuel to the fires of southern anger. Adding more, the investigators found that nearly 90 percent of the funds were not paid to cotton planters. Due in part to the drought, Iowa was the largest participant in the acreage reserve for 1956. This bolstered claims that USDA had used the program for drought relief in key midwestern areas for President Eisenhower's reelection prospects.[35]

The 1957 investigative report also provided important findings that were conveniently overlooked by Whitten and his southern allies. Most telling, large cotton farmers had turned down the program because the payment rate was not high enough. Most of the participation in Whitten's home state of Mississippi was from small hill farmers, many of whom had off-farm employment. Investigators reached similar conclusions for Arkansas, Georgia, and Louisiana, where they also found connections between the acreage reserve and the "loss of tenants and sharecroppers to industry and other States."[36] In fact, investigators found that some large cotton planters did not participate because they were concerned they would lose tenants and that they would have a difficult time finding new tenants in future years if those tenants moved to towns and cities. Those most critical to the decision to participate in the Soil Bank, however, were the cotton ginning industry and bankers. Their opposition to farmer participation was based on concerns that if cotton farmers participated, the ginners and bankers would lose income.[37]

Whitten's investigators reached similar conclusions with respect to participation in the Conservation Reserve Program of the Soil Bank. In those areas where land values were high because they were productive, such as in the Corn Belt and the Mississippi Delta, farmers were not interested in the CRP. Instead, participation was largely driven by farmers with less productive land, as well as interest in Texas, where drought had been a recent problem. Some exceptions were found in the southeastern states where land could be leased to paper companies or if the farmer was nearing retirement or considering off-farm employment. Many farmers were not enticed by the program, given the relatively low rental rates and the long-term contracts that locked up land in the program.[38] Large farmers and farmers with good, productive land were simply not interested in the Soil Bank for the most part, according to Whitten's investigation.

Placing this first investigative report in context has value. The report was produced in March 1957, about a month after Whitten opened a hearing proclaiming the Soil Bank the worst thing that had happened to farmers. A month after the report, the administration submitted its amended budget request that proposed reducing the appropriation based on a recalculation by USDA of the costs. At the time, the program still operated under the mandatory funding of the CCC but would shift to an appropriation, which was enacted in August 1957 (for fiscal year 1958). Of course, planting for the corn and cotton crops would have taken place in spring 1957 and for the wheat harvest that summer, much of it before the fiscal year 1958 appropriations bill in August. Additionally, the Soil Bank had only operated in a single year (1956), a year in which it had been enacted too late to impact planting decisions while also being implemented in the middle of a severe drought.

The investigative staff produced a second report in February 1958 on the 1957 operation of the acreage reserve program. The focus was again on the impact of the acreage reserve on local economies and businesses. The report generally concluded that the program helped in areas suffering from drought, with potential adverse impacts on local communities where weather was normal. Once again there was little participation by the largest cotton producers in the Mississippi delta regions of Arkansas and

Mississippi in the 1957 program. The acreage reserve program remained attractive only to the smaller farmers. In Arkansas, investigators found that "participation was the smallest in those areas where the most cotton is produced and highest where the least cotton is grown" and that "the greatest participation was from farms with allotments of 15 acres or less." Again, "the owners of large farms who are the heavy producers did not participate" in the acreage reserve because "these farms are highly mechanized and equipped for production and the producers are encouraged to produce by the ginners, lending agencies, and other persons interested in cotton production." Investigators added that "the farms with tenants are not likely to participate . . . where the owners were apprehensive about losing tenants whom they might need at a later date."[39] Also of note, Arkansas ginners had complained about adverse impacts on their businesses but "three ginners themselves participated" in the acreage reserve.

Importantly, the investigators concluded in 1958 that problems in the 1957 program were not due to the Soil Bank. The "weather probably reduced cotton production as much as, or more than, participation" in the acreage reserve. Among Mississippi cotton farmers, Whitten's investigation found that "the heaviest participation was in the small-production areas of the State, while there was light participation in the large-production areas" and that participation was concentrated among "farmers with small allotments of 10 acres or less" in the 1957 program. The largest producers in the Delta region stayed out of the acreage reserve "because the cotton factors, cotton associations, and lending agencies were opposed" to participation and "their influence contributed to the continuation of cotton production."[40] The Soil Bank was not a program for large farmers or those with good, productive fields.

The investigations eroded one of the main arguments wielded against the Soil Bank but provided clues to the motivations behind its sabotage. The largest, most sophisticated cotton planters were still producing cotton, and there was little decrease in total production. If that were the case, there could have been little actual economic harm to industry or communities. Just as in Arkansas, some of the ginners complaining about the acreage reserve were themselves participants in the program. The investigators were clear that whatever problems had impacted cotton, they were caused by the weather, but the ginners blamed the Soil Bank.

Texas cotton farmers were the outliers. Among them, the program was more attractive to large farmers who could enroll some acres and use the guaranteed income from the program "to gamble with the rest of their land to make a good crop, depending on weather conditions." In a drought-stressed state, the income from the acreage reserve was considered by some farmers to be a "lifesaver" that allowed them to cover costs or debts and provided a stabilizing economic effect. But for the smaller or family-sized farms in Texas, Whitten's investigators found that the 1957 program provided too little income to entice participation. This was noticeably different from what had been found in the other southern cotton states.[41]

Instead of the partisan, political complaints lodged at it, the real problems for the acreage reserve were the same problems that existed for other attempts to control production by acreage. Regardless of the program, reducing acres had little real impact on production because farmers simply increased the intensification of their production practices on the remaining land, or found ways to game the policy. Not one single aspect of this was new or unique to the Soil Bank. The same issues and problems had plagued all acreage-based programs since the New Deal. What was unique to the Soil Bank, particularly the acreage reserve, was that diversion to conservation was not as remunerative as diversion to other cash crops. On this point, Whitten's 1958 report was clear. Investigators wrote that "diversion to other crops lessens overall economic impact because by substituting one crop for another," a farmer would expect to continue at least some of the "demand for labor, equipment, and other services" on those acres. Any conclusions about the economic impact, however, were very speculative and acknowledged as likely overstated.[42] That farmers could not make as much from conservation and payments as from switching to other cash crops was not exactly a shocking conclusion.

In fact, most of the problems for the cotton sector that were not related to weather were caused by the parity system, not the Soil Bank. The failures included holding cotton prices too high to be competitive on world markets and against synthetic fibers. Reducing acres of cotton in the United States also helped increase acres and production overseas by foreign competitors.[43] Loss of income, ginning business, and foreign markets were a large part of the problem with the allotment system. These problems could not be blamed on the Soil Bank.[44] And yet they were, and the attacks on the Soil

Bank were effective; Whitten's second investigative report was released approximately a month after President Eisenhower surrendered in the budget request for fiscal year 1959 in January 1958.

Were reason to carry the day, any concerns about hurting cotton ginners and others in industry would be most pronounced by acreage diversions to other cash crops. Planters made far less from the government payments than a cash crop. Two investigative reports by Whitten's own staff had found that planters were more likely to divert acreage than to enroll it in the Soil Bank. And yet, southerners in Congress opposed the Soil Bank but defended the parity allotment policy. The facts reported by Whitten's staff belied, if not outright contradicted, the arguments used to attack the Soil Bank and the Eisenhower administration. In those reports is damning evidence in support of the accusation that southerners sabotaged the program, while also begging questions about their real motivations and reasons.

Searching for Answers in the Rubble of the Soil Bank

Representative Whitten and his cosaboteurs consistently and persistently raised two reasons for their opposition—damage to local economies and a failure to control production because of increased intensity on the acres not diverted—but those reasons do not hold up to scrutiny. Either could have provided a substantive challenge to agricultural conservation policy. Logically, however, they could not both be true; the second contradicted the first. More fundamentally, both were refuted by Whitten's own investigations.

Concerns by the ginners and others in the industry were often based on little more than rumors and anecdotes. Whitten's investigative efforts uncovered little actual damage to local businesses in cotton country. His investigation concluded that local businesses and banks effectively steered the large planters away from the acreage reserve. In addition, production data supported the accusation that the programs failed to control production, and there was solid understanding that acreage reductions led many farmers to increase production on the acres that remained in farming, usually through purchasing more inputs and equipment.[45] Failure to control production made it less likely that the Soil Bank caused problems for local businesses. If the Soil Bank led to increased intensity of production on the larger farms and better acres, it should have helped agribusinesses and the communities in which they operated through purchasing more fertilizer and chemicals, possibly additional equipment. It defies logic and common sense that the Soil Bank could have both led to an increased intensity of production and hurt local businesses.

Moreover, the available program data indicates that the Soil Bank's acreage reserve may have worked best for cotton. There was not a reduction in the grain surplus (corn, wheat, feed grains) during the three crop years it was in operation. Farmers did reduce acres of cropland used for crops and those planted to corn. They substantially reduced acres planted to wheat, but feed grain acres initially increased before falling in 1958. In those three years, acres planted to cotton were reduced, and total cotton production was lower. Based on Whitten's investigative reports, however, these reductions were accomplished mostly by small farmers and not by the planters (except in Texas). The cotton surplus fell in 1957 and 1958, and cotton prices reached their highest levels since 1950 in those years. If accurate, these investigative findings would establish that the Soil Bank was a significant benefit to the cotton planters: it cut acres and production by renting land from small farmers and poor acreage, allowing the planters to continue to produce on their best acres and receive better prices from the market. For the 1957 and 1958 crop years, furthermore, cotton farmers also received significantly more per acre from the acreage reserve than did wheat or corn farmers. Only in 1956 did corn farmers receive higher per-acre payments.[46] Arguably designed for corn, the acreage reserve achieved the most success for cotton. Southerners in Congress, however, barely permitted it to operate, let alone succeed.

Whitten's reports indicated a possible reason. The Soil Bank did not work to the advantage of the planters. This may have provided some reason for attacking it. Enrolling in the Soil Bank was no "sure thing," and payments may not have been as favorable to southern crops because the program lacked the "gimmicks" that benefited cotton planters more than others.[47] If they did enroll ground in either program, the payments were unlikely to be as financially rewarding as having federal policy subsidize their diversion of acres to other cash crops. But such technical matters could have been easily fixed by southern committee chairs and other experienced legislators. Ultimately, such matters do not provide good reason to sabotage the program rather than fix it.

Another explanation for the Soil Bank's demise is the political one, in both partisan and regional senses. The Soil Bank was proposed by midwestern corn farmers, pushed by the Republican Eisenhower administration (including use of the veto pen), and defended by Republicans in Congress. An unpopular secretary of agriculture was its face. The strongest opponents of the program were southern Democrats in Congress fighting on behalf of the powerful faction of cotton planters in their states and districts. As such, the Soil Bank was yet another victim of the long-running regional dispute between these factions in the farm coalition. By the late 1950s this regional dispute had taken on a very partisan flavor.

It is also possible that the Soil Bank presented a threat to the preferred policy of the cotton interests. To the extent that the Soil Bank or even just the acreage reserve offered a better alternative to the failed policies of the parity system, it would have jeopardized the outsized benefits that the system provided to the southern cotton planters. As noted previously, the acreage allotments effectively subsidized the diversion of acres for one factional interest (southern cotton) above the others, especially in relation to midwestern corn. If the Soil Bank was an attack on the parity system, it would have been perceived as an attack on southern agriculture by partisan and regional political adversaries or enemies.[48] Having stood guard in protection of the parity system for the better part of two decades—the only two decades of actual experience with direct farm assistance—southerners would be expected to respond in kind. Any such threat would be magnified if the Soil Bank held strong political appeal because decreasing soil erosion or other harmful externalities of farming would have delivered benefits to the general public as well.

That conservation policy could have posed a particular threat to southern interests helps better explain the reliance on arguments that the policy damaged businesses and communities. The consequences of conservation for local economies provided a powerful political argument because it would directly undercut any other benefits the policy might provide the general public. It is no surprise that the attack arose again and again. That it presented, or was perceived as, a threat to southern planter interests offers understandable reasons for the sabotage effort. That the Soil Bank threatened the parity system seems less convincing, however, given the serious problems plaguing that policy by 1956.

The Soil Bank was agricultural conservation policy as it had been in 1936. It was narrowly defined as paying farmers to remove acres from production. In cotton country especially, Whitten's investigators found that the Soil Bank enrolled mostly small farmers and paid for poor-quality ground. From these conclusions in the investigatory reports, an important reason emerges. The Soil Bank likely committed fundamental sins in the eyes of the planters that could not be fixed and would have necessitated a severe response. The Soil Bank spent more of its federal funds on small farmers and poor ground than it did on the planters. This was particularly true for the conservation reserve, designed to remove land from row crop production that was ill-suited to it. Both were in direct contrast to the parity system, from which the benefits went almost exclusively to the planters. The Soil Bank was a problem not because it hurt local economies but rather because it angered large, politically connected cotton planters and their allies in Congress.

If the program was most popular with smaller farmers, it would be expected that they were the farmers who might have had jobs in town, farmed on the side, or even worked acres for the large planters. The Soil Bank likely provided these smaller farmers with options that they otherwise would not have had. The Soil Bank may have provided these farmers with options that were contrary to the aspirations of the large planters looking to consolidate landholdings. A small farmer could enroll acres in the program, for example, rather than sell out to a large operator, pocket the minor additional income, and work a job in town or walk away from sharecropping or tenancy. If it prevented them from acquiring more lands and farms, or at least complicated their efforts to do so, then the Soil Bank presented a problem for the planters, which might also have caused some bankers consternation if they lost business for land purchases.

In addition, southerners like W. R. Poage (D-TX) disapproved of the "whole philosophy" of the Soil Bank. To Chairman Whitten, the Soil Bank was a form of welfare policy. It paid farmers "not to farm" their land. The program was guilty of "giving the farmer dollars for nothing," and "he cannot ever live it down" because it "would ruin his standing with other Americans." This conveniently distinguished conservation policy from the parity system, which subsidized diversion to other cash crops. With acreage diversion to other cash crops, the parity system was not paying farmers not to farm. Making matters worse from this perspective was the fact that the Soil Bank also paid for substandard land owned by smaller, poor farmers. Whitten complained that USDA had "no control over what land the farmer may put in the soil bank" and that funds were spent on the "thinnest, sorriest land."[49]

Relegating conservation policy to the status of a welfare program put it in direct conflict with the ideological views of conservatives, including southerners. Poor people were too often (and too easily) deemed not worthy recipients of federal public assistance. This could be applied to payments for poor lands and small or poor farmers. Whitten's expressed views on conservation exposed an ideology under which the spoils of federal policy are the rightful entitlement of the strong, the well-off, and those viewed as productive contributors to society. In effect, the attack on the Soil Bank was a statement that federal support should only go toward what can be considered good for businesses and the economy. Such a view reduces the definitional scope of public benefits in order to limit the pool of recipients to smaller, discrete factional interests; it encloses the beneficiaries of public funds, helping the businessman/farmer to largesse from the taxpayer but does not accept helping the taxpayer with the consequences from farming. It is a perspective that too easily sees fraud and abuse in all policies that help those outside the factional interests deemed worthy of support.[50] From here conservation policy not only pays too little, not only lacks inherent value, but begins to tap into a darker vein in American society and politics.

Consider also the timing: The Soil Bank came about during the twilight of the Jim Crow system, enacted just two short years after the Supreme Court's landmark decision in *Brown v. Board of Education* against segregated public schooling. It was preceded by, for example, the brutal murder of fourteen-year-old Emmett Till (August 1955), the Montgomery Bus Boycott following Rosa Parks's arrest (December 5, 1955, to December 20, 1956), and Autherine Lucy enrolling at the University of Alabama, followed by riots (February 1956). On March 12, 1956, nineteen southern senators and eighty-one southern representatives published the "Southern Manifesto" against the Supreme Court's decision. A month later President Eisenhower vetoed the first version of the 1956 farm legislation. In 1957 a comfortably reelected President Eisenhower sent federal troops and the Arkansas National Guard to protect nine Black students trying to attend Little Rock High School and signed the Civil Rights Act. More broadly, the 1950s also witnessed McCarthyism and anticommunism—Senator Joseph McCarthy (R-WI) was discredited in March 1954 and died in March 1957. It was an era in which conservative, reactionary responses to the New Deal system were being born anew and advanced with academic theories, all fueled by the Supreme Court's decision.[51]

Caution counsels against connecting these matters too closely, and against attributing causality completely, however. What was critical was that the Eisenhower administration's attempted pivot toward conservation and away from the self-defeating, self-destructing parity policy coincided with a very acute and difficult moment for southern political power. Concern and anger about federal interference with matters in the segregated South were at a fever pitch. It would be folly to think that agricultural conservation policy could escape or be immune to this reality, especially considering the various implications for private property. If conservation policy had been perceived as another avenue for federal meddling, the reaction would have been intense and disproportionate. At that particular crossroads in American politics and history—where conservative anticommunism and antigovernment reactionaries intersected with southern segregation and states' rights revanchists—being cast into the realm of redistributionist welfare policy would have been extraordinarily detrimental for conservation policy and the death knell for development.

In context, Whitten's investigative findings that the Soil Bank programs were favored by small farmers and owners of relatively poor land take on a different light. The federal benefits were often rejected by the large cotton planters. At times, those planters claimed they were unable to enroll. This would mean that federal funds were not flowing to the planters. It also meant more direct interaction between these poorer, smaller farmers and local USDA employees. In the South, many of the smallest and poorest farmers were Black, or they were in the category of poor whites the planters had long manipulated against African Americans.[52] Gilbert Fite has made this point clearly, writing that "help[ing] poor farmers throughout the [South] meant assisting black producers" and that "by the 1950s any program for small, poverty-ridden farmers in the South became entangled with the civil rights movement."[53] It would not have taken many instances of helping poor Black farmers to the detriment of a local planter to spark a political backlash.

In more ways than one, the Soil Bank arrived at a particularly fraught moment in history. Barely a decade into the mechanical harvesting of cotton—a prime example of the technological revolution in agriculture—the southern effort to displace tenants and sharecroppers had been underway for nearly a quarter of a century. For example, USDA reported that 55,348 tenant and sharecropper families were "forced off farms due to 1955 reduction in cotton allotments." While very limited, the data on the displacement of Black farmers in the South indicates about half as many Black farm families at the end of the 1950s than at the beginning of the decade. Pete Daniel has explained this effort as an "enclosure movement" that used the "forces of government intrusion and mechanization" to achieve "whitening" of the South and cotton production.[54] It would have hit a critical, precarious moment in the wake of desegregation, as the civil rights movement gained momentum. Possibly the only thing that could make the situation worse for the Soil Bank in the South was if the planters who did enroll in the program had been required to share their payments with and protect the interests of their tenants or sharecroppers.

In fact, that may well have been part of what happened. Protection of tenants and sharecroppers had received "special mention" in the conference report for the 1956 Act, but it was clear in the report that conferees had weakened any protections. When it implemented the new program, USDA instructed its local offices that farmers could not enter Soil Bank agreements unless they fairly shared the payments with tenants and sharecroppers. In addition, the offices were instructed that a farmer should not receive any benefits if they reduced tenants and sharecroppers. USDA went further, providing explicit instructions to the county committees to determine whether "duress has been used to secure signature of tenants or sharecroppers to contracts and agreements" as well as whether the "division of any payment between landlord and tenant is in accord with the respective contributions" to the crop production. Here was direct and potentially inflammatory federal interference in the southern planter system. It would have threatened their control over farming, sharecroppers, and tenants.[55]

If USDA went further in protecting tenants and sharecroppers than southerners in Congress found acceptable, then that was likely the final, critical reason for sabotage. The congressional records do not provide definitive proof that the Soil Bank implicated segregation or the white supremacist power structure in the South. These records do, however, provide enough evidence that the policies could have gotten close enough to have been burned as the South began scorching the political earth. It is not outlandish speculation that the Soil Bank may have crossed this peculiar southern red line. If it helped Black farmers, sharecroppers, and tenants in the South, then this was more likely than not the final nail in the Soil Bank's political coffin.[56]

Sabotage and the Damage Done: The Consequences for Development

If the process for enacting the 1956 Act demonstrated the diminished power of the southern faction, the sabotage campaign against the Soil Bank served a reminder that diminishing power is not necessarily the same as power lost; it certainly is not powerlessness. The more accurate assessment is that southern political power was only beginning to wane in a changing postwar America but that the South retained real power in Congress. Jamie Whitten was the exemplar.[57] The 1956 enactment process proved that southerners lacked the votes to override a popular president's objections and that their unpopular policies were losing support in the Senate. The Soil Bank's entire brief existence, more accurately, was a case study of political power in the negative sense. President Eisenhower deployed it with his veto pen in order to force the southerners to back down on loan rates; Whitten and his allies deployed it to destroy the president's policy preference. The outcome resembled little that could be considered any ideal of legislating or policymaking. Going into the 1960s, the state of farm policy was tattered and torn with few good options to repair it. The trend and the momentum, in fact, were in the direction of further degeneration through escalating factional retaliation and the settling of scores. Therein lie some of the self-destructive tendencies of factions that earned them Madison's moniker as the mortal diseases of self-government.[58]

Sabotaging the president's preferred policy and forcing him to abandon it before it had a chance to work was a victory of sorts, but not much of one. It may have demonstrated some power—and it certainly reinforced Whitten's reputation, position, and strength—but it also demonstrated significant weakness. Sabotage took time and multiple moves. It appeared underhanded, especially by those on the agricultural authorizing committees who had worked to enact the program only to help destroy it.

Each tactical move in the larger strategy served up costly reminders of the deteriorating situation and held the potential to further drain power as opponents absorbed lessons and reacted or adapted.

The political, factional fighting did little for the farmers for whom the policy was supposed to help, let alone the general public. In many ways, destroying the Soil Bank was a Pyrrhic victory because it delivered another loss for farm policy. The acreage reserve was an alternative necessary because previous policies had failed (except, maybe, for cotton planters). Failure of the alternative magnified those failures. Farmer behavior—specialization, intensification of production, and the rapid adoption of technological advancements, especially synthetic fertilizer—and the weather were simply insurmountable barriers to successfully controlling production. In the middle of the technological revolution, diverting acres into conservation was not going to overcome these fundamental shortcomings. Conservation payments were not going to ever make up for lost markets, nor were they going to match the subsidized diversion into other cash crops.[59]

The Soil Bank saga damaged agricultural conservation policy, arresting or stunting its development. It had not developed much from the 1936 version. Conservation in 1956 remained anchored to commodity support policy as little more than another method for controlling production by paying farmers to reduce acres, although the conservation reserve represented some developmental progress. Southern saboteurs didn't just kill a couple of programs; they cast conservation policy in very negative lights, including as akin to welfare policy. They ginned up concerns that it would harm rural businesses, economies, and communities. If conservation policy was viewed as ineffective, or worse, caused damage or unintended consequences, further development would become infinitely more difficult.

The entire episode also could have left an impression that conservation was partisan policy. Conceived by midwestern corn farmers, endorsed by the American Farm Bureau Federation, and proposed to Congress by President Eisenhower, the Soil Bank stood as an alternative to the parity system southerners used to divert acres into other cash crops and to modernize cotton production by displacing tenants and sharecroppers. For southern farmers accustomed to these benefits of diverting acreage to other cash crops, conservation offered far less as an alternative, and according to Whitten's reports, the larger planters generally rejected the program.

If conservation became a partisan, regional policy, the consequences for further development would become unpredictable. Development could become overly subject to changing partisan winds or political fortunes and limited by matters beyond its purview.

In addition, conservation policy raised issues with respect to private property. Conservation paid for the least productive acres; by removing them from production, the policy implicated matters of land-use regulation and private property rights. Property is generally the province of state law but not exclusively, the Southern Manifesto aside.[60] Conservation policy could deliver benefits for the public interest. It did so through actions of individuals undertaken on private farmland, compensated by public funds. Conservation policies deemed too strict or overly burdensome could intrude on property prerogatives even if they effectively served the public's interests. If they were too lax, the policies would be little better than subsidies in terms of delivering public goods. These presented substantial challenges for the policy, weighing it down in the difficult political arena. Partisan and regional political baggage, or other unrelated matters, added burdens that could not easily be removed or reduced.

The final lesson of the Soil Bank may have been the most important because it implicated all others, as well as the fundamentals of the constitutional design. Madison's design was based on a theory of factional competition, with the ambition of each helping keep the others in check in order to end up as close as possible to the greater public interest. For conservation policy, something was missing once again. There was no argument on the other side, no factional interest competing with the farm interests. There were no interest groups speaking up for the environment or conservation, and none represented the public interest beyond the president's concern with the costs of commodity surpluses. Whitten's distorted arguments and misinformation went unchallenged in the public debate. Without a voice or a seat at the table, conservation was not able to refute private property arguments, nor could the interests build bipartisan and cross-regional coalitions. Conservation interests, to the extent they existed, were not in the competition. They did not defend the programs against attacks and appropriators, nor did they push back against funding reductions; conservation interests were not informing farmers or other interests about what had happened with the program.

The end result was more than a terminated program or another failure for farm policy. The consequences extended beyond the political, regional infighting among factions in the farm coalition. The Soil Bank risked teaching that the conservation of natural resources and addressing soil erosion simply didn't pay, or worse. The episode jeopardized conservation as public policy worthy of the expenditure of public funds, blaming it for harming rural economies and violating private property rights, all for partisan purposes. No future existed for any policy under that scenario, and Madison's competition was of little use when it was completely one-sided. The lack of competition produced consequences for the Soil Bank worse than it had in 1936, and from its demise, conservation policy would go back into a form of developmental hibernation for nearly a quarter of a century. It would be awakened by another dual catastrophe with severe consequences for the natural resources upon which the production of food is dependent and for many of those who produce it.

PART 4   The Food Security Act of 1985

7

From Dust to Dust

THE SEVENTIES INTERLUDE AND BACKSTORY

From the dust of the thirties to the dust of a demolished Soil Bank, development of agricultural conservation policy had been separated by twenty years and two wars. Congress would not substantively revisit the policy in a farm bill for another twenty-five years. When it returned, however, conservation finally attained the privileged position of the status quo for farm bills. The landmark Food Security Act of 1985 created a new foundation for agricultural conservation policy. It went beyond the traditional policy of paying farmers and landowners to retire land from production to include conservation efforts on active or working farmlands. As of 1960 and expiration of the conservation reserve authority, the new foundation was a long way off. The achievement would require another twin crisis (economic and erosion) for agriculture. This chapter reviews the interlude between the Soil Bank's demise and the return of conservation policy in the 1980s.

Southern sabotage of the Soil Bank did not save the parity system, if that had been the intent. The acreage policies unraveled further and quickly. Corn farmers voted themselves out of the system in 1959. The surplus problems grew worse and were inherited by the incoming Kennedy administration. By 1961 massive stocks of federally owned commodities, especially feed grains and wheat, were causing skyrocketing Commodity Credit Corporation (CCC) costs.[1] Acres in production and planted to the major commodities also increased after the acreage reserve was terminated.

The puzzle that was the Soil Bank deepened further in 1961, when President Kennedy proposed a "special agricultural conservation program under which acreage previously planted to feed grains would be diverted from production for 1961."[2] Barely a month later, the Democratic Congress enacted legislation to effectively resurrect the acreage reserve program for the 1961 crops of corn and grain sorghum. At the time, House Agriculture

Committee chairman Harold Cooley (D-NC) reported to the House that the federal government was sitting on an investment of $9.5 billion in agricultural commodities. In response to the compounding problems of surplus, Congress created the proposed "special agricultural conservation program" to pay farmers for diverting acres out of those two crops and into conservation. Over the opposition of midwestern Republicans, Congress pushed through the feed grains program. Congress renewed the program throughout the remainder of the decade, expanding it to other feed grains. Between 1961 and 1969, the program would divert from the production of feed grains to conservation an average of thirty-seven million acres, topping fifty million acres in 1969.[3]

The Soil Bank acreage reserve and the emergency feed grains program were very similar programs. Both compensated producers for diverting acres into conservation. Both programs included a payment-in-kind mechanism but with minor differences in calculating acreage reductions. There was one major difference. The Soil Bank also applied to cotton; the feed grains program did not.

The acreage reserve had been proposed as a solution to the growing surplus problem for all supported commodities. If measured by acres diverted to conservation and a reduction in the surplus of feed grains, the 1960s programs would tend to suggest that, if it had been continued, the acreage reserve of the Soil Bank would have been effective policy. Those in Congress representing the cotton-growing states and districts attacked the Soil Bank program most fiercely and worked to sabotage it. They were also the ones who pushed the feed grains program into existence, continuing the basic policy on crops and farmers not in the South. The 1960s feed grains programs support a conclusion that the South opposed the use of conservation to reduce acres and the surplus only if the policy applied to cotton.[4]

Aside from the feed grains programs, there was little development for agricultural conservation policy throughout the Congresses of the 1960s. But for the farm bill itself, the early 1960s were instrumental. Midwestern corn interest and Republican frustration with the feed grains programs and the actions of southerners in Congress led to an escalation of the regional feuding. In 1962 midwestern Republicans used civil rights issues to help defeat the 1962 Farm Bill effort. After the bill finally made it through Congress, the wheat program was rejected by wheat farmers in the 1963 referendum. Traditional farm policy was at its lowest point politically, at risk of self-destructing; 1964 provided an important development that continues to impact farm bill debates. Attempting to move legislation to fix problems with the cotton and wheat programs, House leadership paired votes on the bill with votes for the Food Stamp Act of 1964. Both passed, marking the first effort at building a coalition across the regional divide in the Democratic caucus. In doing so, food stamps pried open the political coalition for the farm bill, which had been exclusive to farm interests. The Food Stamp Act of 1964 arguably saved farm policy from its self-inflicted demise. It changed the politics of farm bills going forward and eventually helped conservation policy.[5]

The Agricultural Act of 1970 created a system of set-aside acres that were a requirement for farmers to be eligible for loans and other federal assistance. Proposed by President Richard Nixon's USDA, the set-aside policy made it through Congress with little change or resistance. The policy meant that if USDA concluded that supplies exceeded needs, farmers could be required to set aside acres and place them into approved conserving uses as a condition of eligibility for farm assistance. They generally did not receive any payment on the acres set aside, however. Under the 1970 Act, set-aside acres would fluctuate from a low of 37 million in 1971 to a high of 61.5 million acres in 1972. That year represented the high-water mark for acreage set aside before farm policy changed significantly in 1973. The 1970 Act's most consequential contribution to farm policy may have been that its unpopularity likely helped drive Agriculture Secretary Clifford Hardin into retirement, replaced by Earl Butz.[6]

Nixon and the Environmental Movement of the Early Seventies

A key part of the backstory to the new foundation for conservation policy had nothing to do with farming directly. On January 28, 1969—just days after Richard M. Nixon had been sworn in as the thirty-seventh president of the United States—inadequate safety measures on an oil rig off the coast of Santa Barbara, California, resulted in an explosion that "cracked the sea floor in five places" releasing more than a thousand gallons of oil each hour for a month.[7] It was, at the time, the worst oil spill in U.S. history. Fast forward to June 22, 1969, when sparks from a train ignited an oil slick and driftwood in the polluted Cuyahoga River in Cleveland, Ohio. The relatively small fire burned two railroad trestles but sparked a rather extensive legend. These two environmental events are credited with inspiring the first Earth Day on April 22, 1970, which, in turn, is credited with launching the modern environmental movement.[8] The modern environmental movement would quickly achieve a trifecta of legislative, policy, and political success. It was an achievement that has not been repeated since.

On January 1, 1970, President Nixon signed into law the National Environmental Policy Act (NEPA), creating the Council on Environmental Quality in the Executive Office of the President and a system for assessing the environmental impact of federal agency actions. The legislation had been introduced in the Senate by Senator Henry M. "Scoop" Jackson (D-WA) on February 18, 1969, and it sailed through Congress with little opposition in the last half of the year. In his New Year's Day 1970 signing, President Nixon emphasized the need for the United States to address pollution. The first Earth Day followed four months later. It was, however, nearly trampled under by the troops and tragedies of the Vietnam War. Nixon's announcement of the invasion of Cambodia on April 30, 1970, and the tragic shooting of students at Kent State University on May 4, 1970, followed closely behind the first Earth Day.[9]

For the environmental movement and policy, President Nixon rounded out the tumultuous year by creating the Environmental Protection Agency (EPA) and signing into law the Clean Air Act Amendments of 1970 (CAA). President Nixon created the EPA with congressional acquiescence as he competed with Democrats over claiming the environmental political mantle. The Senate approved the nomination of William Ruckelshaus to be the first administrator on December 2, 1970. Next, President Nixon signed the CAA into law on December 31, 1970. It had also sailed through Congress with almost unanimous support and little opposition. NEPA, CAA, and EPA were major policy and political victories, breathtaking achievements for the nascent environmental movement in a single year, but they also obscured the politics coursing beneath the statutory surface.[10]

Nixon's embrace of the environmental movement and its policy priorities was tenuous at best, mostly a political gamble to harvest partisan gains for Republicans. Nixon, the consummate politician, could not resist making a play for the growing public support for addressing environmental problems, but President Nixon could only go so far on environmental issues in a competition with a Democratic Congress. His problematic relationships with Congress complicated his ability to negotiate and take credit. For example, Nixon attempted to jam Democrats in Congress with his proposal to create EPA and sought to gain the initiative over Democrats in the midterms with the CAA.[11]

Environmental interests had little motivation to compromise or to take the weaker policies pushed by the president. Voters for whom the environment was a priority were unlikely to be (or become) Republican voters. Neither the president nor his party was credible on the issue, and while the environmental issue was a key driver in the midterm elections, it did not help lift Republican candidates against the headwinds from the economy (inflation), the Vietnam War, and the usual midterm challenges for a sitting president. His political gamble failed for its purpose but did advance major environmental legislation, which raised major concerns among conservatives and corporate interests. Focused on Senator Edmund Muskie (D-ME) in the 1972 election, the political lessons were not lost on candidate Nixon. President Nixon fell back in line with business and conservative interests, walking away from his big bet on the environment.[12]

The clearest indications that Nixon's political calculations flipped were on matters near and dear to farming. First was legislation amending federal pesticides law; second was water pollution law. Both had been signed by President Truman in his first term, and both were subject to increasing scrutiny throughout the 1960s as the environmental movement began to take shape and demand reforms. Amendments to both laws were enacted in October 1972, but President Nixon signed the pesticides amendments into law while he vetoed the Clean Water Act, and Congress overrode his veto. The different paths to becoming law in 1972 marked the end of Nixon's gamble on competing for political support over environmental regulation. He was easily reelected in 1972. Environmental issues captured the political spotlight, but federal policy to protect the environment had proved unfavorable ground for Republicans and Nixon.[13]

The Clean Water Act amendments were driven in part by "burning rivers, massive fish kills, declining shellfish populations, and closed beaches" that "captured public attention" and media coverage. Under the jurisdiction of the Committees on Public Works in the House and Senate, the Clean

Water Act moved easily through Congress and with little opposition. Introduced by Senator Edmund Muskie (D-ME) in October 1971, it cleared the Senate a month later and the House in the spring of 1972; both chambers passed the conference report overwhelmingly in early October 1972, and no senator voted against the bill at either stage. Despite the large votes in Congress, President Nixon vetoed the Clean Water Act on October 17, 1972, emphasizing concerns about the costs of addressing water pollution and the budgetary implications, such as inflation and taxes. Congress voted to override his veto almost immediately.[14]

Amending and updating pesticides law came about after Rachel Carson's book *Silent Spring* was published in 1961 and also in response to the various eradication campaigns by USDA during the decade that resulted in backlash from citizens and environmental interests. The path for pesticides legislation was more problematic, burdened by narrower political interests and a complicated history with chemicals like DDT. The nearly insurmountable hurdle for major reform involved farm interests and southerners in Congress. Southerners at this point held control through a conservative coalition with Republicans. They did face more of a fight in the Senate, where the Agriculture Committee had to share the work with the Commerce Committee's strong advocates for the environment and real reform of pesticides regulations.[15]

Rachel Carson's book and Cleveland's burning river sounded alarms over the "rapidly rising costs of manmade destruction of the natural environment."[16] The environmental movement that emerged in the early seventies diverged from a focus on the more traditional issues over public lands and a general support for nature. It focused on protecting the public and public resources such as air and water from the damage caused by industry or other actors. It was also a movement increasingly aligned with people's rights and the critical role of the federal government in protecting or ensuring them. This was not a shift comfortable for Republicans nor for conservatives. It was especially ill-suited to southern Democrats in Congress less than a decade removed from major voting rights and civil rights legislation.[17]

The environmental movement's trifecta of legislative achievement (NEPA, CAA, and the Clean Water Act) also indicated just how insulated the farm faction was from those alarms; the 1972 Clean Water Act exempted agricultural nonpoint sources of pollution from the regulatory framework. The movement's hopes for vast pesticides reforms were dashed in the House Ag Committee. Whether ignored by environmentalists as a political backwater or recognized as a concession to political realities in Congress, the environmental movement stopped at the farm field's edge. This hands-off approach to farming had real consequences. It would take more than a decade of damage before Congress took up changes to federal farm policy.

Nixon and Butz Put Acres on a Collision Course with the Environment

The roots of the next major development in agricultural conservation policy can also be found in the year 1972. Events of that year provided the major catalyst for another round of massive soil erosion problems and set farm policy on a collision course with the environmental movement. In short, the seventies brought about the "first major interruption of several decades of U.S. agricultural surpluses" and resulted in a substantial change to the farm economy.[18] The total acres of cropland devoted to the production of crops as reported by USDA provided one measure of the drastically altered situation: after more than a decade in which production acreage remained within the range of 330 million acres, total acres used for crops began a sharp increase after 1972. That spike would not hit its peak until 1981, when it reached 387 million acres of cropland—a 55-million-acre increase in barely a decade, outpacing even the increase from World War I. The increase in acres presented a graphical Matterhorn upon which agricultural conservation policy finally found its modern political foothold, but that development came at a severe cost to the nation's soils and waters. The consequences of decisions by policymakers in 1972 and 1973 reverberate to this day.[19]

A spike in crop prices provides only a partial explanation for the increased acres under planter and plow, in part because the price movements were without precedent at the time. Average crop prices from 1973 to 1975 more than doubled for corn and wheat as compared to 1970 to 1972, and nearly doubled for soybeans and cotton. There were two main causes of the spike. On July 8, 1972, President Richard M. Nixon announced the largest grain deal in American history. The Soviet Union would purchase at least $750 million in American grains—including seven hundred million bushels of wheat or one quarter of the entire American wheat crop—over three years. The Russian purchases were financed by low-interest loans issued from the Commodity Credit Corporation.[20]

During winter 1972–73, an El Nino event off the coast of Peru drastically harmed the anchovy harvest. Anchovies were an important source of protein for animal feed; the loss of the source contributed to a soybean price spike at the same time corn and wheat prices were spiking from the Russian grain deal. The Nixon administration was trying to control inflation while battling congressional investigations. The reelected president began his second term under the growing clouds of the Watergate scandal. In what can only be described as political panic, the president on the hot seat ordered a temporary embargo of soybeans, setting in motion market and other forces still felt today.[21]

Put the price spike in perspective: for the twenty years after the Korean War ended in July 1953, farmers had been pinched between crop prices held down by surpluses and costs increasing as farmers purchased expensive inputs (fertilizer, pesticides, and mechanized equipment) that boosted yields and contributed to the surpluses. From parity acreage allotments to the acreage reserve of the Soil Bank and the feed grains programs of the 1960s, federal policies had cycled through versions of centralized efforts to reduce the acres farmers planted in failed attempts to control production. The events of the early seventies provided a moment for new export exuberance, and Secretary of Agriculture Earl Butz was the chief salesman and cheerleader. He infamously encouraged farmers to "plant fence row to fence row" and to either "get big or get out" of farming, but those were not his most consequential comments. In 1973 Secretary Butz proclaimed that agriculture had reached "the promised land" after forty years in the wilderness of low prices and failed, outdated policies. Under the illusion of this new promised land, Congress enacted the Agriculture and Consumer Protection Act of 1973, signed into law by President Nixon on August 10, 1973.[22]

The 1973 Act established a new direction in farm support policy, prioritizing income-supporting payments when prices were low over price-supporting commodity loans. Congressional authors viewed the policy as a production incentive. The new policy unleashed American farmers to chase expected export market demand while having the federal government share the price risks. Pent-up demand in the farm sector, magnified by the spike in crop prices, was released by the 1973 Farm Bill's switch to production incentives. At all levels of the federal policy apparatus, from Congress to USDA, farmers were being encouraged to expand and put more acres into farming. It worked, proving yet again the ease with which acres can be brought into production.

One year later, however, President Nixon had resigned from office, and crop prices were either at their peak or already in decline. The monthly average of wheat prices had fallen more than a dollar per bushel since January 1974. And yet, total acres in production continued to increase, approaching 380 million acres by 1977 on the way to the 387-million-acre peak in 1981.

The millions of acres put back into production during this time frame included many acres that had previously been removed from production by federal policies. Most of these acres would have long been understood as ill-suited to intensive row crop cultivation and crop production. Those acres that did not fit this category were likely newly broken native sod or drained wetlands. It should not have been a surprise when massive soil erosion problems followed. In 1974 and 1975, for example, Iowa experienced its worst erosion in twenty-five years, with some fields losing as many as fifty tons of topsoil per acre. More than "one-third of the U.S. cropland is suffering annual soil losses in excess of the limit at which soil productivity can be sustained over time." Farmers had already increased cropland acres by more than thirty million acres; they would add another twenty million.[23]

American agriculture was repeating the mistakes of the past. When "farmers plowed under more land to take advantage of rising prices," the "old conservation practices were lost. . . . Partly as a result of these trends, the quality and quantity of America's croplands became a national concern."[24] In almost one fell swoop, Secretary Butz and President Nixon wiped out decades of efforts at combatting soil erosion from wind and water. The mirage of export demand and the temporary euphoria from spiking crop prices triggered "a great 'plow-out' of marginal U.S. farmland, followed by a great 'wash-away.' . . . [The] assault was not just on grain stocks; it was on the American land." American croplands were losing two billion tons of soil per year to erosion, with as much as 141 million acres experiencing moderate to severe erosion by 1981.[25] These were among the consequences of the shortsighted land rush that unfolded under the mirage of Butz's promised land; few appeared to consider the implications for natural resources or soil erosion.

If you've made it this far, you can be forgiven for wondering what the hell happened in the 1970s. American farmers were within living memory of the Dust Bowl. Agricultural policy was barely four decades into the halting half measures for addressing soil erosion. By 1972 the environmental movement had achieved a legislative trifecta and EPA was in operation; the Russian grain deal took place as Congress was working on the Clean Water Act. As farmers plowed under more ground, the environmental movement continued to notch legislative and political successes. For example, President Nixon signed the Endangered Species Act into law on December 28, 1973, and a year later his successor, Gerald Ford, signed the Safe Drinking Water Act into law on December 17, 1974. Farmers moved in the opposite direction, away from public opinion and into direct conflict with the environmental movement. Deep into the technological revolution in agriculture, the problem was more than just losing soil to wind and water. Farmers were applying increasing levels of chemical inputs and using heavy equipment to work the lands. Every ton of eroded soil carried with it chemicals and fertilizers, likely at levels not before experienced. As farmers lost topsoil, they increased applications to make up for the productivity lost with the eroded soil. It did not go unnoticed even if it went unaddressed. Decisions made under the influence of greed and bad policy threatened to turn significant swaths of American farmland into wasteland.[26]

Things Unravel in the Seventies

Explanations for what happened in the seventies likely begin with the combination of an energy-oil crisis and what economists called "stagflation" as unemployment and inflation remained stubbornly high. The energy and oil crisis hit in October 1973, when Egypt and Syria attacked Israel. Saudi Arabia and some of the other major oil exporting nations in the Middle East cut production and exports while also threatening to reduce supplies further. Congress quickly pivoted away from environmental priorities to address the oil embargo and energy crisis. The crisis would have also impacted the post–1973 Act acreage response by farmers. Row crop farming under the technological revolution was drastically impacted. From the diesel to run tractors and combines, to the production of pesticides and synthetic fertilizers, farming relied on oil. Increased oil prices squeezed farm incomes by increasing costs while also challenging supplies. Farmers have a long history of chasing bad economic situations by trying to increase production, and under the 1973 Act they had federal policy incentives to do so. If the oil crisis helped take the wheels off the environmental movement, agricultural conservation policy, having not broken through with the rest of the environmental policies, found no opening. Times of resource shortages have unfortunately not been times of wise conservation of natural resources.[27]

The political environment was exceedingly volatile, from the end of the Vietnam War to social unrest and the swiftly imploding Nixon presidency; within a year of signing the 1973 Farm Bill, Nixon would resign the presidency in disgrace. Congress was also in the throes of procedural and political reforms. On July 12, 1974, President Nixon—just weeks away from resigning—signed into law the Congressional Budget and Impoundment Control Act of 1974. The federal budget and cutting spending were Nixon's favorite weapons against Democratic policies, and he singled out environmental policy as well as the meager conservation policies at the time. As if it were an omen, one of Nixon's last official acts as president of the United States was to reject the budget for the EPA he helped bring into existence.[28] In time, budget reforms would come to have profound consequences for policymaking and legislating, but they did little to control the federal budget or reduce debts and deficits. Picking up what Nixon fumbled, the budget would become central to the Republican and conservative ideological playbook, an effective way to attack or limit Democratic and liberal policies, programs, and priorities without having to attack them directly.

American agriculture appeared to be unraveling in tandem. In June 1974 the Soil Conservation Service reported that Iowa had its worst spring soil erosion in twenty-five years, with some fields eroding at rates above forty to fifty tons per acre. Senator Dick Clark (D-IA) laid bare the extent of the problems and pointed the finger at the "constant barrage from Washington encouraging 'all out production' . . . that productive land must be utilized 'fence to fence' for the purpose of saving the hungry people of the world."[29] The events that led to millions of acres being put back into production—intense, chemically enhanced production at that—took place over the course of just two years, but the recognition of the need to conserve drastically lagged the drivers to exploit. Worse still, the individual exploiter—in this particular case the farmers and the agribusinesses—benefited on the upside, while the public suffered the consequences, such as sediment pollution from soil erosion.

The Watergate scandal and the swift end to the Nixon presidency added yet another element. It drove voters in the midterm elections to send a tidal wave of liberal, reform-minded Democrats to Congress, especially in the House. These liberal Democrats would open 1975 with major reforms to the committee system, including seniority. But most importantly for this discussion, they would bring down the long reign of southern chairmen on the House Ag Committee. Chairman W. R. Poage (D-TX) was voted out of his role and replaced by Representative Tom Foley (D-WA), a northerner and more liberal Democrat better aligned ideologically with the direction of the Democratic caucus. Since 1955 the House Ag Committee gavel had been in southern hands, first Harold Cooley (D-NC) and then W. R. Poage (D-TX). That power was often backed by Jamie Whitten (D-MS), chair of the agricultural appropriations subcommittee. Nearly impenetrable for decades, this factional redoubt in Congress had been breached by the increasingly liberal, northern, and urban Democratic caucus. It began in 1964, when the Food Stamp Act was used to help cotton and wheat policies after a decade of southern resistance to the food assistance program. Further weakened by the 1973 Act's inclusion of food stamps in the farm bill itself, the once imposing, nearly unbeatable southern power in Congress was at a low point two years later.[30]

Dethroning Chairman Poage interestingly coincided with an emphasis on conservation and erosion in the Senate Agriculture Committee, where its chairman, Herman Talmadge (D-GA), acknowledged the risks of erosion resulting from "asking our farmers to go all out for maximum farm production." In 1975 the committee released a report prepared by a team of scientists that painted a bleak and rather damning portrayal of farming at that point in time. Deep in thrall to the technological revolution, farmers were putting more land back into production that was often marginal or less productive and farming it more intensively. Technology provided short-term solutions, with chemicals and synthetic fertilizers helping offset losses in productivity that resulted from erosion, but "modern technology [had] not made agriculture weather-proof" and, in fact, may have contributed to "an agriculture with a high degree of weather sensitivity." Advances in technology enabled farmers to recklessly increase the intensity of their farming while masking the consequences. With "more than one-third of the U.S. cropland . . . suffering annual soil losses in excess of the limit at which soil productivity can be sustained over time," it was the very definition of mining the soils upon which the future of production depended.[31]

The fallout from 1972 and 1973 presented an almost condensed version of that which precipitated the Dust Bowl in the 1930s but with the addition of chemical and synthetic inputs. The young environmental movement, riding the tidal wave of its legislative and political successes, helped feed public awareness of the consequences. As such, 1975 was not 1935 in the fields, nor in the halls of Congress.

Turning Point: Conservation Policy at the End of the Decade

The bicentennial of the Declaration of Independence in 1976 witnessed progressive Iowa Democrats, Senator Clark and Representative Tom Harkin, rising in prominence and driving a focus on conservation and soil erosion. At a field hearing in Iowa on July 6, 1976, Senator Clark explained that "it's been my impression ... that very little has been done on this problem" and that soil conservation has "come to be taken for granted over the years. ... As we continue to demand more of farmers by way of greater productive capacity," the nation "run[s] some great risks in what that does to the soil."[32] Their rise coincided with the exit in disgrace of Secretary of Agriculture Earl Butz, and Gerald Ford's defeat by Jimmy Carter. It was a changing of the guard of sorts, as the environmental movement began to make itself heard in agricultural circles. At a House Ag hearing in August, Tom Barlow of the Natural Resources Defense Council (NRDC) argued that "a heavy use of fertilizers can boost production quite dramatically on a respective acreage," but doing so often meant that "the runoff can pollute streams and receiving lakes to the point that human use is jeopardized."[33] He noted that the problems were concentrated with a few farmers and included the challenges proposed by short-term perspectives of tenants. Subsequent research largely proved the critics correct. Farmers, encouraged by university researchers and extension educators, as well as fertilizer companies, increased fertilizer application rates to chase goals for increasing yields and production.[34]

Not that they were connected, but weeks before he was defeated by Jimmy Carter, President Ford vetoed the Agricultural Resources Conservation Act of 1976, arguing that the "bill would set the stage for the creation of a large and costly bureaucracy . . . in an attempt to insure land use in compliance with the master plan."[35] The Senate had led the effort to enact a law that "would establish a mechanism for making long-range policy to encourage the wise and orderly development of the Nation's soil and water resources." It would have required USDA to "prepare an appraisal of the Nation's land, water, and related resources" and report it to Congress along with a "National Land and Water Conservation Program . . . to set forth the direction for future soil and water conservation efforts on the private and non-Federal lands of the Nation."[36] Ford's veto was notable in that both the House and Senate passed their bills without any debate, let alone objections. It did not change the increasingly troubling realities on the ground but merely handed them off to the next Congress and president. His veto did little to arrest the beginnings of important new developments for agricultural conservation policy; he merely delayed progress.

In the first year of the Carter administration, conservation policy development made an important advance. In early 1977 the U.S. comptroller general sent a report to Congress on soil erosion and conservation policy that provided a needed catalyst. The title of the report pulled no punches: "To Protect Tomorrow's Food Supply, Soil Conservation Needs Priority Attention." GAO found that farmers had plowed up tens of millions of marginal ground, much of which federal conservation dollars had paid them to take out of production. Farmers had also used the local committees to divert millions in funds intended for conservation purposes to help them cover the costs of applying fertilizer and adding tile to drain their fields. It was not a pretty picture, and there was plenty of blame to go around. Assigning blame did little to address the fact that replowed soils loaded with synthetic fertilizers and pesticides continued to wash into waterways. The pressure on Congress to respond grew, but the farm economy was struggling. Once again, economic and erosion problems were consuming American agriculture.[37]

The lure of the promised land proclaimed by former Secretary Butz was the bait for a trap. Farmers borrowed heavily to purchase more land, consolidate operations, and farm more intensively, all to produce for a market that failed to materialize after the artificial events of 1972 and 1973. Pesticides and fertilizers depended on oil, and too many tried to compensate for the damage from mining the soils with chemical treatments at a time when oil had become a serious geopolitical weapon and more expensive. Farmers accumulated debts to the bank and to natural resources in order to plant fencerow to fencerow. The long-term results of this were eroded soils and damaged fields, along with polluted waters by both the soils and the oil-based chemicals. But it also resulted in severely eroded financial situations for many heavily indebted farmers. The mirage of the Butzian promised land was bad for the farmers and for the environment. It was not the first time that lesson had been delivered in farm country.

The late seventies did present some new dimensions to the old problems. Reliance on oil-dependent inputs was one, while unceasing development pressure on farmland from suburban sprawl was another. Soil erosion remained a threat to long-term production, but building houses, strip malls, and roads on farmland was a relatively novel addition. It both consumed existing farmland and drove demand to break open additional and often less productive land. Politically, it also caused a split in conservation policy goals from the traditional one concerning the damage to natural resources from soil erosion. Conservation now took on concerns with damage to production capabilities due to pressures from other, more economic uses of land. As Congress worked to reauthorize the farm bill in 1977, the search for policy solutions was underway. Out of that effort originated a new policy to require that farmers adhere to conservation planning requirements in order to be eligible for federal assistance programs. It would come to be known as conservation compliance and was eventually enacted in the Food Security Act of 1985, but it took nearly a decade of work to achieve.

Much of the credit for this eventual development in conservation policy went to Tom Barlow of NRDC. From the 1977 GAO report, and the fact that erosion was higher in 1977 than it had been in the Dust Bowl, he logically concluded that "new, pragmatic approaches to controlling erosion" were needed. The critical question posed by Barlow was whether USDA efforts to financially help farmers sacrificed the long term for the short term by encouraging production that degraded soils and water. Barlow's argument was that to "protect a landowner's long-term financial security, as well as his short-term security," federal policy needed to help assure "that soil assets remain in place." Federal policies, therefore, should be "offered in return for assurances that [Best Management Practices] were installed and operating" on the fields associated with the assistance. Barlow's argument was clear:

"When we start seeing pictures of the dust building up in roadside ditches and up against barns it's already too late."[38]

The argument and policy concept placed Barlow in the vanguard. He was the first among the environmentalists to call for such action, and he was its most forceful advocate. For agricultural conservation development, his was the first to seek the application of conservation in conjunction with production, not just as a method to remove acres from farming. This was public interest policy; taxpayers helped farmers maintain the food supply, while government incentivized better practices to prevent soil erosion and other forms of natural resource degradation. The concept altered the traditional and preferred philosophy behind farm policy and represented the potential for a real step toward a quasi-regulatory system. It may have been particularly dangerous to the status quo because it held the appeal of simple logic and basic common sense. Why should farmers receive federal taxpayer-funded payments if those payments encouraged soil erosion; why should taxpayers subsidize erosion, suffer the consequences, and pay for cleaning it up? It was a question made more pointed by the 1973 Farm Bill push to expand production. Worse still, much of the acreage put back into production had previously received federal payments to be taken out of production. The logic was lost on, or ignored by, the Agriculture Committees, at least initially.[39]

In contrast to the drastically worsening situation for farmers and natural resources, the Food and Agricultural Act of 1977 was shockingly inadequate. There was little disagreement about the failures and consequences attributable to the 1973 Farm Bill and the push for expanded production. The bill failed to assuage the economic pains farmers were experiencing and accomplished even less for conservation. It was, however, supplemented by the Soil and Water Resources Conservation Act of 1977 (the 1977 RCA). With the second enactment, Congress required USDA to study the problems and report back to Congress with possible policy responses after decades and billions of dollars expended. It was the closest Congress came to direct action for better conservation policy in the seventies. The 1977 RCA was also half measure and more foot dragging. It worked well enough to hold at bay further reforms or stronger policies. It was not much, but it also was not nothing; for agricultural conservation policy, it opened the door to potential developments, new ideas, and a chance at a new beginning.[40]

8

A New Foundation for Conservation Policy

THE FOOD SECURITY ACT OF 1985

Arguably, the most significant words for the development of agricultural conservation policy were in the Food Security Act of 1985. Congress instructed that "any person who in any crop year produces an agricultural commodity on a field on which highly erodible land is predominate" or "on converted wetland shall be ineligible" for federal farm assistance, including payments, loans, and crop insurance.[1] This is conservation compliance. It emerged from the twin crises of the seventies as expanded production recreated massive soil erosion and farm economic collapse. It was the first real development in conservation policy that connected conservation directly to farm production or working lands. The 1985 Farm Bill was written in the depths of the farm economic crisis, a crisis that consumed much of the 1980s and many family farms. To this day the 1985 Farm Bill also remains the most expensive farm bill enacted, but it is considered landmark legislation for achievements in conservation policy.

Conservation compliance policy was seeded in 1977 by the critical GAO report, the advocacy of Tom Barlow of NRDC, and the enactment of the 1977 Soil and Water Resources Conservation Act (RCA). More than anything, however, compliance developed because of the problems on the ground and throughout farming. The concept of conservation compliance took nearly eight years to bear fruit, providing yet another lesson in the difficulties of change in the Madisonian system. Strong resistance from farm interests kept Congress from moving on the issue, but the issue was dormant, not dead. In 1979 Representative Jim Jeffords (R-VT) introduced the first compliance legislation with a bill "that farmers may not participate in the price support program . . . unless they implement plans for conservation and sustained use of soil and water on their farms." Under the RCA process, Secretary Bergland and USDA moved closer to supporting conservation compliance as well. The investigatory work produced more concerns with the consequences of federal policies and the effectiveness of conservation; little actual progress was made during the Carter administration.[2]

Conservation compliance raised concerns—whether actual, perceived, or simply politically convenient—that the policy conflicted with private property rights. Conservation compliance also raised more mundane concerns, such as whether it would be effective, practical, or even possible to implement. All of these were on display before the House Ag Committee's field hearings in 1979.[3] At the hearing, the National Association of Conservation Districts (NACD), among those opposed to mandatory conservation or compliance, proposed a middle ground approach. NACD's opposition was built on the concerns that mandatory compliance could not be effectively implemented because it was likely beyond the capabilities of USDA and the conservation districts. The organization was also concerned that compliance would not be accepted by landowners and thus would be counterproductive for conservation efforts. If mandatory compliance could not be implemented, it would leave a lot of farmers and landowners in the lurch; unable to get assistance to implement the conservation plans required, they would be precluded from federal assistance for reasons based on USDA or NACD staff resources. This, in turn, would turn people off and give them reasons for skepticism or frustration and create real risk of backfire for conservation policy.

As a result, NACD introduced an alternative concept that it called the green ticket approach. This policy would increase or provide better assistance to those farmers who implemented conservation plans, such as a higher target price or loan rate. Fundamentally, NACD's green ticket retained the voluntary and incentive-based policy structure for conservation. It was a novel and notable policy proposal at a critical time. The proposal was not taken up by Congress, however.[4]

The hurdles for conservation policy at the time were significant. For example, Representative Jim Jeffords (R-VT) also introduced a bill in 1980 that would protect farmers and their land from development pressure. The development of strip malls and suburbs was contributing to the loss of millions of acres of farmland, and the issue was gaining much public attention. The policy split farm interests because it involved private property rights and raised concerns about federal land-use planning or regulation.

On the Democratic side, southerners presented the strongest opposition once again. Republican opposition was strongest among westerners such as Representative Ron Marlanee (R-MT) and fiscal conservatives like David Stockman (R-MI).

Opponents of the bill carried the day. The bill's defeat highlighted how the protection of private property rights presented a difficult obstacle. Provided for in the Constitution, private property rights are fundamental and, to some, even sacrosanct. The 1980 bill was an effort to help private property owners, but it could not overcome concerns about the potential creep toward land-use planning and regulation. Opposition to the bill and its defeat also provided a window on the larger property rights movement that became one of the more effective lines of opposition to the environmental movement.[5]

Bigger issues were also unfolding at the time that were difficult and momentous. In 1979 Federal Reserve chairman Paul Volcker sought to break inflation by drastically increasing interest rates. High interest rates damaged cash flows and held down or depressed land values. In January 1980 President Carter imposed an embargo on grain exports to the Soviet Union. In November 1980 heavily indebted farmers helped the conservative Ronald Reagan defeat President Carter; Reagan was sworn in as president in January 1981.

Conservation's Achievement against Politics in the Age of Reagan

There can be little risk of overstating the impact of the 1980s farm economic crisis. Farmers were struggling. Farm incomes, which had spiked in the wake of the 1972 Russian grain deal, plummeted in the 1980s. Worse still for farmers, many had borrowed heavily to expand production, and the income collapse was not accompanied by a corresponding decrease in the expenses of farming. In fact, chemically enhanced agriculture was increasingly more expensive for farmers who competed for better yields to overproduce crops that lacked market demand. Oversupplied markets depressed crop prices and farm incomes, again. Those farmers deep in debt would be collateral damage. Over the next five years, federal farm payments increased drastically. Farm bankruptcies would become such a problem that Congress revised federal bankruptcy laws, enacting a special chapter just for farmers. The Farm Credit System would have to be bailed out by Congress and the American taxpayer as it teetered on collapse.[6]

The false promises of the seventies had done more than drive farmers into bankruptcy and crisis. They had also brought farming to the brink of another reckoning with nature and the political consequences in Congress. That these two matters—financial collapse and natural resource degradation—appear to operate hand in hand was not at all surprising given the lessons from history. Conservation policy would claim its biggest achievements in the middle of it all but also because of it.

If the 1977 Farm Bill and RCA were half measures, they represented some progress. They also exposed real weaknesses for the farm interest faction. It was the 1981 Farm Bill, with its myriad challenges, that turned out to be a turning point for the conservation policy movement itself. The 1981 Act broke little new ground, but it nourished the seeds planted in 1977, which would, in turn, grow into the provisions of the landmark 1985 Farm Bill. Historically, the 1981 Farm Bill provides a useful demarcation. It also presented another missed opportunity. This time the missed opportunity held drastic consequences for both the tons of soil that would be lost from farm fields and the many farmers who would lose their farms to bankruptcy.

The 1981 effort also planted seeds of a more dangerous variety. The Reagan administration and Senate Republicans converted budget disciplines into an offensive weapon to attack the New Deal federal state. This would grow into the go-to strategy for conservative Republicans and, eventually, for southerners when they merged under the same partisan umbrella. That development, too, would take time.

In 1981 Senator William Armstrong (R-CO), a conservative western Republican and budget hawk, introduced legislation to prohibit price supports if the crops were produced on ground that had not been farmed in the ten preceding years. Concerned about irrigation and dust storms, his bill applied only to the areas west of the one hundredth meridian, but it was progress. Eroding land was a problem that crossed property lines, and the abuse of land on one farm impacted the property rights of neighboring farmers and ranchers. They "become the victims of unwise conservation of their neighbors, even their relatively distant neighbors," Armstrong argued. He also added arguments about the wise use of federal funding. Reagan's USDA agreed: "A landowner should not be given taxpayers' funds . . . to do things with the soil that is not beneficial to the general public." For their part, environmentalists pushed for a more robust compliance.[7] With that, conservation compliance crossed ideological lines, where enacting it would protect private property rights and was more budget conscious than paying subsidies to cultivate land that wasn't suited for it.

The Reagan administration had requested that Congress hold off on any real work to develop conservation programs until USDA could complete the RCA effort. The effort began after enactment of the 1977 Act and was a year behind schedule in 1981. The 1977 RCA had required USDA to not only investigate soil and water problems but to also provide Congress with policy alternatives or options for addressing the problems. USDA was under new political leadership but was not matching its efforts to the urgency of the soil erosion problems. The pressure to do something about erosion was relentless, and doing nothing was quickly becoming politically untenable.[8] President Reagan signed the 1981 Farm Bill into law in late December of that year.

In the 1981 Farm Bill, agricultural conservation achieved the status of a separate title within the ever-expanding omnibus legislation. Congress acknowledged the incredible toll that expanded production had taken on the nation's soils and accepted that "some of the most productive agricultural areas of the United States are also those having the most serious and chronic erosion-related problems," such as Iowa. Once again, Congress found itself taking steps to address excesses and heedlessness in farming. It was another turn toward policy to "promote soil and water conservation, improve the quality of the Nation's waters, and preserve and protect natural resources through the use of effective conservation and pollution abatement programs."[9]

First, Congress created a "Special Areas Conservation Program" to conserve soil, water, and other resources in specially designated areas by entering into contracts with owners and operators based on "a conservation plan approved by the Secretary and the soil and water conservation district in which the land on which the plan is to be carried out is situated." The program was in response to "increasing concern about the depletion of the Nation's soil and water resources" and "chronic erosion-related or water management-related problems." Designed in the mode of the Great Plains Conservation Program created in 1956, the 1981 version targeted conservation assistance to where it was needed most. It signaled recognition that farmers had managed to recreate major soil erosion problems under the 1970s push for expanded production and that the problems of pesticides and fertilizers had been added to those of soil erosion.[10]

Second, Congress added the "Farmland Protection Policy Act," a version of which it had rejected the year before. The reversal made the new program more notable than its actual reach because Congress protected farmland only by ensuring that neither federal programs nor other federal efforts contributed to the conversion of farmland to nonagricultural uses. This was rather limited in scope at a time when the U.S. was losing about three million acres of agricultural land each year to development. The "loss of this resource to U.S. agriculture is irreversible," but the committees were also clear about the need to protect private property rights. As such, the protection of farmland was only extended to actions by the federal government.[11] The farmland protection program exemplified the slow, difficult progress for conservation policy development.

In 1982 Senator Armstrong tried again. He offered an amendment to the agricultural appropriations bill that was approved by sixty-nine senators. It was dropped in conference with the House, cut out of the final appropriations bill by Chairman Whitten. The implications for soil erosion and conservation were substantially worsening, however. As the farm economy cratered, heavily indebted farmers under financial duress were left with few good options. Duress and desperation drive bad decisions and leave no room for investing in conservation; captured by short-term thinking, a farmer on the verge of bankruptcy struggled to make it through the year. Those farmers were more likely to push their fields and soils ever closer to the brink, if not over it. If the Agriculture Committees sought to do more to help, they had options. They were bluntly informed that farm programs were "subsidizing improper land use" and "underwriting land damaging activities." These were lessons that had been bought and paid for but largely wasted.[12]

It was upon this political ground that conservation compliance could take root. The farmer who put in the time, effort, and dollars to conserve soils was losing out to the farmers who appeared to care little. Farm policies increasingly looked like they punished the good actors. Worse still, bad practices on a neighbor's farm could cause damage for the farmer who put conservation in place, because soil erosion was not contained to any single field. And more was known about the problem than ever before, with every report, piece of research, or evaluation blaring another alarm. Something more substantive would be needed, and regulatory pressures were building up. The problems were especially acute for clean water as soil erosion spread the consequences of bad decisions in the field to those in surrounding communities. These were not unanimous views, however, and the last holdouts were notable. The Reagan administration's Office of Management and Budget (OMB), where cutting the budget appeared to be the singular priority, lacked interest in investing in conserving soils or combatting erosion. To some in Congress, the administration presented "a very warped view of what really sustains the strength and security of this country." OMB recalcitrance also undercut claims by Secretary of Agriculture John Block as he defended USDA on the Hill.[13]

The 1981 Farm Bill did little to stanch the slide into crisis. Contrary to the budget hawkishness applied to other matters, the Reagan administration recreated a Payment-in-Kind (PIK) emergency program in January 1983. The massively expensive initiative was modeled after the 1960s feed grains programs and paid farmers to remove nearly eighty million acres from production in the 1983 crop year. It was a rather traditional form of conservation policy, paying farmers to divert land into conserving uses for a year in order to decrease production of oversupplied commodities and help prices and incomes. It was not a development in conservation policy but was maybe the strongest statement on the depth of the farm crisis. A conservative Republican administration that sought budget cuts on almost every nondefense item of spending suddenly decided to spend tens of billions of federal taxpayer funds to directly intervene in commodities markets and land-use decisions. The program was also a mess as some farmers and speculators figured out how to game the system. If anything was left of the mirage concocted in 1973, it was washed away in 1983, when Reagan delivered the final death blow to Butz's promised land.[14]

The 1983 PIK program might also have signaled the Reagan administration's retreat on conservation compliance. While the administration persisted in demands for budget cuts to already inadequate conservation programs, that position became harder to maintain when they were spending billions on the PIK program. The administration's position was too incoherent to last, and they lacked allies in Congress. The first retreat from opposition to conservation compliance was peculiar, however. Secretary Block proposed to attach it to the farm loans of the Farmers Home Administration. Doing so would make conservation compliance apply only to the poorest and most distressed farmers for whom the loans were the last resort. Worse still, these programs had long been the source of substantial discrimination against Black farmers and other socially disadvantaged producers. Inexplicably, Reagan's USDA resisted conservation compliance for the largest farmers, who were more likely to be the main source of the problem, but it was willing to further saddle the smallest, poorest farmers with this perceived burden. This was possibly a hint of more troubling problems at the Reagan-era USDA, because nearly two decades later, it was found to have continued and possibly accelerated discrimination through farm loan and other assistance programs that had led to massive land loss by Black farmers.[15]

In March 1983 Senator Armstrong and more than twenty Senate cosponsors introduced an updated version of conservation compliance. It became more commonly known as "sodbuster." It applied to highly erodible land nationwide. Sodbuster precluded any federal assistance to any person who produced a crop on such land unless the farmer had implemented an approved conservation system. By October the Senate Agriculture Committee reported the bill with USDA's support. It was "a complete U-turn, 180 degrees" that appeared to seal the consensus on the issue of conservation compliance. For conservation interests, some of whom had been pushing for some form of compliance since 1976, the sodbuster bill was progress and a recognition that federal assistance policies had an impact on farmers' decisions in managing their lands.[16]

Progress against the status quo in the legislative process can be a tricky matter. Some senators had to be assured that the bill "will not dictate farming practices to farmers" and that compliance "does not mean the Government will dictate what a farmer can or cannot do with this private property." They were also told that the opponents of the bill were "the very farmers who were doing the sodbusting and causing soil erosion." The ease with which sodbuster legislation met political and ideological objectives across the partisan spectrum is notable. It also draws attention to how conservation policy—even something as narrowly tailored as sodbuster—struggled against concerns about property rights. Progress can create its own challenges; as the policy gathered momentum, it was pushed to go further. Senator Armstrong was unhappy with the committee bill because he felt the committee had "strengthen[ed] the bill to death." It was a minor objection, and sodbuster cleared the U.S. Senate. Nothing further would be achieved in 1983, however, as the bill stalled in the House.[17]

The House Agriculture Committee moved the bill to the House floor in May 1984. The only public opposition came from Representative Ron Marlanee (R-MT), an eight-term representative who was also a farmer and rancher from northeastern Montana. Marlanee's opposition was built on land-use regulation and property rights concerns. He was not persuaded by the committee report's assurances that the sodbuster bill would not interfere with property rights or land-use decisions. The committee argued that sodbuster provided protection for taxpayers. The House effort was led by Representative Tom Harkin (D-IA), and the House Ag Committee expanded sodbuster beyond the version passed by the Senate. They attached compliance to all commodities produced by the offending farmer, not just the specific commodity produced on the highly erodible.[18]

When the bill reached the House floor, Representative Marlanee was the lone voice in opposition. He argued that the bill "puts into place a vehicle for forcing conservation measures using farm program benefits to blackmail compliance." Sodbuster, he added, "takes managerial decisions from the greatest conservationist of all, the farmer, and makes those managerial decisions subject to the judgmental decisions of a bureaucrat." He attacked the creation of "regulations when none existed" and claimed that the bill "confiscates property rights." To Marlanee, sodbuster was "a vehicle to land use regulations" that would render the farmer "subservient to local bureaucratic judgmental decisions."[19]

While it proved to be a lonely position on the House floor, Marlanee's arguments effectively summarized the litany of concerns about the policy from property rights hardliners. His argument did fail to address, however, the basic point that compliance only applied to federal subsidies. A farmer did not have to comply with the policy if the farmer elected to not receive federal funds. This was hardly the specter of regulation and property rights violations he claimed.

House Agriculture Committee chairman de la Garza (D-TX) made the point. He informed the House that "nothing in this bill imposes demands on any farmer or rancher concerning what may be or may not be done with his or her land." Instead, he reminded everyone, the federal taxpayer "shared the cost" of conservation practices, which were undone by farmers. Covering his bases, Chairman de la Garza added, "Never will we yield to any temptation or never will we be oblivious or disregard the concerns" about land-use regulation and property rights. Representative Jim Jeffords (R-VT) added that "sodbuster provisions will not deny the farmer the right to farm highly erodible soils" but "will simply deny him the taxpayers' help in doing so unless he agrees to farm them in such a way that will not create unacceptable soil losses." As in the Senate, the House passed its more expansive version of sodbuster without the need for a recorded vote.[20]

House passage of the sodbuster bill turned out to be the high-water mark for conservation policy in 1984. The House had also added a new Conservation Reserve Program to the more expansive version of sodbuster. Combined, the two derailed the bill in conference due to concerns about the cost of the Conservation Reserve Program and its potential return to policies of yore. An unpublished transcript of a single conference committee meeting on June 27, 1984, provides the story of the policy's demise. Near-universal acknowledgment about the seriousness of the soil erosion problems across the country—and the dire need for new policies to address it—was not enough. House and Senate negotiators were unable to overcome the details about how exactly to address erosion through federal policy. From the unpublished transcript, the problem appears to have been with Republican senators who felt the House position on sodbusting went too far. Concerns about costs and the differences over the design of sodbuster proved irresolvable in an election year. Despite its promise and progress, sodbusting legislation died in conference. The last gasp for the policy was an amendment to the agricultural appropriations bill, but even that was disputed and ultimately dropped in conference.[21]

Compared to nearly a decade of work, enacting conservation compliance and the Conservation Reserve Program (CRP) in 1985 would prove to be rather anticlimactic; both were accomplished easily, while also going beyond the scope of what had been killed in conference in 1984. Development began with a near consensus on the issue, which had the look of unstoppable momentum. The last bastion of resistance was Reagan's OMB, which pushed for drastic cuts to the soil conservation budget at USDA.

It was a position so far removed from the consensus political position that it risked becoming farce, subject to open mockery. USDA conceded for the administration and agreed to compliance. USDA also went further and lent its voice to rejecting the land-use regulation and property rights arguments against the policy, which brought House Republicans on board. Compliance was merely Congress "removing the Government incentives for harming land." OMB's strict ideological adherence to budget discipline backfired but provided political lessons. The position strengthened the farm bill coalition and expanded it to include environmental and conservation interests long ignored.[22]

In fact, interest group opposition to adding conservation permanently to the farm bill was difficult to find. Cotton interests arguably represented what was left, giving tentative acquiescence to sodbuster but only on the most limited of terms. Cotton was short of allies, however. Representative Charlie Stenholm (D-TX), for example, exclaimed that "for the first time in my period of time in the Congress, the so-called 'environmental movement,' particularly the Sierra Club and Charlie Stenholm are on the same side."[23] There was more to this story. Real pressure was being brought to bear in the House. For example, Representative Howard Wolpe (D-MI) brought before the House Ag Committee the power of a bipartisan group of members. It was "an informal agricultural conservation coalition" with the "objective to mobilize solid support in the House and Senate for strong legislative action to attack soil and water problems confronting our Nation, and we seek inclusion of the strongest possible conservation measures" in the 1985 Farm Bill.[24]

Further proof of the momentum behind, and political power of, conservation in 1985 can be found in the inclusion of swampbuster, which applied conservation compliance to converted wetlands. The issue did not appear on the policy radar prior to 1985. Swampbuster made its debut in the House Ag Committee hearings on April 4, 1985. Representative Tom Daschle (D-SD) noted "an increasing amount of interest" for including it because of concerns with "the destruction of the wetlands as part of the problem in developing new grounds for production." It was added in the full committee markup in the House in July. Representative Daschle's amendment was adopted on a voice vote and with little reported opposition. It was included along with a sodbuster provision stronger than that rejected in 1984. The bill also included a Conservation Reserve Program to retire highly erodible ground that was bigger and more expensive.[25]

Featuring swampbuster, stronger conservation compliance for erodible lands, and an expanded CRP, the conservation title was roundly praised on the House floor. The chairman of the subcommittee that handled conservation policy, Representative Ed Jones (D-TN), proclaimed it "probably the most comprehensive soil conservation package ever to come before the House." Representative Wolpe noted the strengthened sodbuster provisions and the fact that farmers would have a four-year grace period "to adopt an acceptable conservation plan for their land" if it was highly erodible. The subcommittee's ranking Republican, Representative Thomas Coleman (R-MO) added that "we must take action now to save our land" from soil erosion. He reported that "the committee has adopted very strong sodbuster legislation" that was part of "a consensus package which has been supported by the administration." The momentum behind conservation continued on the floor as the House easily agreed to an amendment to further strengthen sodbuster. If it was a test vote for conservation, especially sodbuster and swampbuster, they all passed with flying colors. It was also a clear signal as to the mood of the House that conservation policy would carry the day.[26]

Opposition was basically left to Representative Marlanee (R-MT) alone. He again trotted out the argument that it all amounted to "land-use planning" and a direct assault on private property rights. He argued, "As bad as the sodbuster legislation is, this is even worse. . . . It confiscates property rights" and would bring the federal government onto private property. It was an argument that no longer held any force because it was out of step with reality. Representative Dan Glickman (D-KS) went straight to the heart of the matter: "The gentleman and his constituents do not have to participate in Federal farm programs." The critical connection was not land-use regulation but the fact that these were "federal tax dollars that are going to subsidize farmers . . . money from the taxpayers of this country, and that money should not be going to subsidize farmers who plant on highly erod[i]ble land."[27] Farmers could ostensibly abuse their land all they wanted; they just couldn't get federal funds to do so. If, however, they wanted the taxpayer-funded payments, they had to abide some basic rules. It was an argument at least as old as the 1942 Supreme Court case that validated federal farm policy under the Constitution.

Reviewing the legislative text that passed the House uncovers the changes that were subtle but important on sodbuster. Strengthening the provisions converted it to a policy of conservation compliance, not just sodbusting. Sanctions would apply to any farmer who farmed highly erodible land without a conservation system in place to control erosion. The development as the policy moved through the House presented a relatively rapid evolution after languishing for years. Facing fallout from failed policy that fueled farmer folly in the fields, conservation's victory was resounding. In 1985 conservation development was a step to begin undoing some of the damage.[28]

The path for conservation in the Senate was surprisingly easier given its opposition in 1984. The Senate had little trouble moving far from the original application of sodbuster to the much stronger, broader application of conservation compliance that included both sodbuster and swampbuster. Conservation ascendant, the Senate included a Conservation Reserve Program with a higher acreage cap on enrollment than the House. The Senate also included authority for conservation easements that could be used to remove highly erodible land and wetlands from production for at least fifty years. Development in the Senate was further demonstration of the degree to which the political environment for conservation policy had changed.[29]

On the Senate floor, the consensus on conservation was bipartisan. There were no opposing or dissenting voices to be found. In fact, the Senate bill struggled due to disagreement over the farm programs and payments for low prices. The debate consumed much of November and nearly derailed Senate passage. The continued fighting over farm payments and the inability to figure out how to help farmers in the middle of an economic crisis contrasted sharply with the consensus on conservation. If it was noticed by senators, it went unmentioned.[30]

Conference did not pose problems for conservation. Conferees agreed to conservation compliance requirements for highly erodible lands and wetlands. They applied ineligibility to any commodity produced by the farmer for farming highly erodible lands without a conservation plan or on converted wetlands unless grandfathered or of minimal impact. The conference report contained strong hints that conferees had grasped the distance that sodbuster had traveled since the year before. Beginning in 1990, a compliance provision would be operational that would apply to all highly erodible lands farmed without a conservation plan and all converted wetlands. Additionally, conferees recreated CRP and provided authority to enroll up to forty-five million acres.

Just before the holidays, on December 23, 1985, President Reagan signed the Food Security Act of 1985 into law. It was landmark legislation codifying far-reaching conservation policy, but President Reagan's signing statement gave little indication he grasped that significance; or, if he did, he didn't choose to acknowledge it. Instead, the president praised the "hard-working men and women who till the fields and tend the herds."[31] Conservation's achievement in the realm of politics and disagreement, however, deserved more. Celebration of the achievement was likely muted given that it was the second consecutive failure of Reagan's ideological warriors to use the budget to bring an end to the farm bill. Instead of a revolutionary change to limit the federal government, Congress went in an opposite direction by adding conservation policies on private farmlands and spending record federal funds on farmers.

Taking the Measure of a Landmark: Concluding Thoughts on the 1985 Act

The Food Security Act of 1985 stands as a landmark in American farm policy history.[32] Congress recreated the Conservation Reserve Program (CRP), which recycled more than just its name from the 1956 version in the Soil Bank. More importantly, Congress conditioned eligibility for federal farm assistance on adherence to conservation requirements. Known as conservation compliance, its inclusion by the Agriculture Committees should be understood in political context. It is that political context—where the economic crisis that was dragging down farmers and the continuing struggle to protect the environment collided with the overwhelming reelection of President Ronald Reagan—that most accurately measures the 1985 Farm Bill's status and delivers critical lessons.

CRP was created as a relatively standard land retirement program, using rental payments to pull land out of production. It was within the lineage of acreage policy in which conservation was deployed to reduce acres in hopes of improving crop prices. It represented some developmental progress because the 1985 version was more focused on environmental concerns and environmentally sensitive lands: enrollment was to include lands that "pose an off-farm environmental threat or, if permitted to remain in production, pose a threat of continued degradation of productivity due to soil salinity." Congress had also learned from the Soil Bank, including a limit that not more than 25 percent of the cropland in any one county could be enrolled unless the secretary determined that it "would not adversely affect the local economy of such county" and adding two years of mandatory funding from the Commodity Credit Corporation (CCC) borrowing authority.[33]

CRP also carried the look of a political deal. Congress authorized a large program, as much as forty-five million acres, but instructed that "the Conservation Reserve be administered to the extent practicable, so as not to reward those who in recent years have converted highly erodible land to cropland uses."[34] This can be credited to the 1970s. Congress invested federal funds to reverse the mistakes of planting fence row to fence row, which had recreated massive erosion problems in a short amount of time. Those mistakes came after having paid to remove millions of acres in previous bills: nearly thirty million acres by the Soil Bank and as many as sixty million acres by the 1960s feed grain programs. Many of the acres previously subject to federal policy and funds were put back under the plow, the funds washing off with the topsoil.

CRP also served as a bridge to conservation compliance, which was designed to remove the "government incentives for harming land" and avoid doing it all over again.[35] Any farmer who farmed highly erodible land was required to control erosion through an approved conservation plan; any farmer who drained a wetland or farmed on a drained wetland would be ineligible unless an exception or exemption applied. Congress applied it to price supports, payments, farm storage facility loans, crop insurance, disaster payments, and any farm loan, as well as any other payment issued by the Commodity Credit Corporation.[36] The degree of political development contained in these provisions measures up despite any loopholes built into the policy or concerns about effectiveness in the hands of local USDA officials.

Much is due to the political circumstances into which Congress enacted the 1985 policies. Critical aspects of the successes by agricultural conservation policies in the Age of Reagan were built on the tough realities of soil erosion and the farm economic crisis, both of which emerged from the mirage of Earl Butz's promised land in the 1970s. Farmers were struggling for the same reasons their soils were eroding: they had ridden exuberance to excess. Assistance was deemed politically necessary for farmers in financial crisis, but so was conservation. A bankrupt farmer cannot invest in conservation or soil erosion control. After the experiences of the 1970s, however, that assistance could no longer simply be a blank check with no strings attached. It made little sense to bail out farmers again, nor was there logic in paying to take land out of production after it was heedlessly brought back in at the first sign of high prices.

By any reasonable, logical political accounting, 1985 should have been the year least likely to produce major changes to farm policy on behalf of conservation and the environment. This holds true most acutely for conservation compliance. Conditioning farm program payments on conservation does not fit neatly with the realities of the moment or the historical storyline of self-interested factional policy outcomes. The farm economy was in the depths of its worst crisis since the Great Depression. Within two years of the 1985 Farm Bill, Congress would provide billions to rescue the entire Farm Credit System as it teetered on the brink of collapse. President Reagan had been overwhelmingly reelected in 1984, and not only was environmental policy not a priority, but it was ideologically antithetical to those accustomed to reading broad mandates in such electoral victories. Both CRP and conservation compliance went against the political views of the administration, but compliance was most out of sync given the relationship between federal actions and private property.

One answer to the question begged by this outcome is that neither the Supreme Court nor the South got in the way. Unlike in 1936, the 1985 authorization debate featured no interference or disruption from outside of the political system. The Supreme Court had not catalyzed action by Congress, and the policy deliberations remained squarely within the political branches. This prevented any factional interests from gaining an advantage from judicial proclamations or directives that were more questionable than clear.

Overwhelmingly reelected, the Reagan administration possessed the political power of its perceived mandate but also the limits on that power in Congress; the president would not face voters again, but senators and representatives would. The Reagan administration continued the push for its ideological agenda, cutting spending and taxes to shrink the federal governmental footprint in the American economy and society. David Stockman remained at the helm of OMB, but the playbook was well known. It also had been unsuccessful in 1981. Extreme ideological positioning was irrelevant to the details of deal making, while making for a convenient common adversary for those in the negotiations. Repeating it in the president's second term offered little chance of success as the farm economy sank further into crisis. As if to prove the point, Stockman resigned as head of OMB in the summer of 1985 and before Congress completed work on the farm bill.[37] His exit punctuated these limits on political power. Attacking both farm payments and food stamps amplified for both the need for each other, while also driving home the value of strengthening the coalition.

If Ronald Reagan's election was a reaction to an unraveling of the New Deal era, it was also unique at the time for its view on addressing public problems with federal authority. Elected federal officials sought to tear down much of the federal apparatus and to take away from, not provide for, the constituents who had elected them. They forged ahead even when the agenda proved unpopular or conflicted with needs driven by realities on the ground. Problems for which people demand some form of response from the officials they entrusted with power do not comfortably fit with an ideological desire to cut spending and government services. While they might get applause on a campaign trail or be lauded by elites and the political pundit class, the Reagan reactionaries had little to offer when inevitable problems arose, or when voters hit upon hard times. Farmers were a case in point: budget cuts would only hurt farmers struggling with economic crisis, many of whom likely voted for Reagan. Their votes would be needed by many Republicans in Congress who would face them after the president had left the political stage.

Without Supreme Court interference and with limited executive branch leverage, policymaking was squarely within the legislative branch. In Congress, Democrats had maintained their majority in the House but not the Senate; the legislative power was divided between the parties. More than partisan splits in Madison's double majority, the 99th Congress may have presented a unique moment because of the diminished power of the southern factional stronghold in Congress. In the House, southerners maintained some degree of political power, but it was not nearly as absolute as in 1936. Southern power was also not as strong as it had been in the late 1950s when Jamie Whitten led the sabotage effort on the Soil Bank. Whitten remained chair of the agricultural appropriations subcommittee in the House, but the House Agriculture Committee was chaired by Representative Eligio "Kika" de la Garza (D-TX), who differed from previous southern chairs, including on environmental issues.[38] Representative Thomas "Tip" O'Neill (D-MA) was Speaker of the House and part of a leadership team that included former House Agriculture Committee chairman Tom Foley (D-WA), who had replaced W. R. Poage (D-TX) as chair in 1975.

The long-dominant southern faction, in fact, appears to have been in a transitional state that was best embodied by the "Boll Weevil Democrats." They were caught between voters who had backed a president seeking to gut federal programs and traditional farm interests demanding federal assistance in an economic crisis. Aligning with ideological warriors like Stockman was not going to help them in the next election. On the other side of the equation, southern Democrats were increasingly out of step with an urban and coastal Democratic caucus. It is not unreasonable to look at the Reagan era, including his vice president and successor in office, and see the political power of the South at its lowest ebb since the mid-1960s, when Congress enacted both civil rights and voting rights legislation. For the farm bill at least, the South simply lacked the power to block, preclude, slow, or complicate policy. And, without that power, they also had less of an ability to dictate its contours, provisions, and direction.

The political realities weighing down on both President Reagan and congressional southerners provide much to explain the 1985 Farm Bill's inclusion of conservation policy, but they are insufficient. The other critical factor—maybe the most critical factor—was the new factional competitor in the process and negotiations. The environmental movement was able to provide a political counterweight that had not existed previously in farm bills. Unlike 1936 and 1956, specifically, environmental and conservation interests had important seats at the negotiating table. They were able to demand policy outcomes through outside interest groups like NRDC and Tom Barlow, as well as the Sierra Club and others. Just as importantly, conservation and environmental interests had strong representation in Congress, such as Representative Wolpe's coalition, and Representatives Jim Jeffords and Tom Harkin. In 1985 the farm faction was writing its version of conservation policy not alone but rather in tandem with environmental and conservation interests.

Filtered through this political lens, the 1985 authorization effort very nearly leaves the impression of a fair and robust political competition among factions that might well have fulfilled the ideals laid out by James Madison. The farm interest faction could fight for expanded subsidies. It was backed by the economic crisis but lacked the overwhelming political power of the South, which was compromised across partisan lines. The conservative faction could compete to achieve its ideological goals for spending reductions but needed important factional interests critical to its power. Liberal or progressive interests with little direct connection to farm subsidies could stand in defense of food stamps and push back against the Reagan demand for cuts. And, finally, environmental and conservation interests were also able to compete for policy outcomes, even as they conflicted to some degree with the demands of the Reagan administration and the interests of the farm faction.

The resulting competition, especially that in reaction to the Reagan reactionary efforts, produced a larger, stronger coalition and the most expensive farm bill in history. Farmers continued to receive their subsidies. Poor families continued to receive food stamps. Agricultural conservation policy—defined and largely confined at the time to CRP and conservation compliance—opened a pathway for including the environmental movement in the coalition. The only interest unable to achieve its goals was the Reagan administration's ideologues, likely due to the extreme positions they continued to take, especially in times of crisis. Ideological purity was of limited power as the electoral dynamics shifted away from a successful reelection to the realities of the next one.

Arguably, this reading of political developments cements the landmark status of the Food Security Act of 1985 at least as much as the substantive policy achievements. The substantive policy enactments were achievements in the "circumstances of politics" rather than the result of a more idealistic vision of policymaking in the public's interest.[39] It is an important lesson of 1985, if not the most important. The complex, difficult, and oftentimes unwieldy process for the development of federal law was designed on the Madisonian theory of political competition among factions. That competition was expected to achieve the best outcome for the greater good, or an approximation of the public interest. In the American system, the legislative power is not exercised by popular referendum. Great ideas and important policy concepts are not achieved by acclamation, nor do they succeed only on their merits. They require interests who will champion them in the process and fight for outcomes in the political arena. Those interests must protect gains and push further developmental progress.

Ultimately, it is that political reality that best explains why the outcome in 1985 differed drastically from the outcomes in 1936 and 1956. It is why that outcome looks different to this day. Title XII of the Food Security Act of 1985 established conservation policy for American agriculture. Each new program or policy, each revision and development, is written as an amendment to Title XII.[40] In both the U.S. Code and the political arena, the 1985 Act remains the foundation for agricultural conservation policies authorized in farm bills. It is upon this foundation that development of those policies is undertaken in Congress. To the extent that the 1985 Act stands as landmark legislation in the history of American farm policy, it is a distinction that is due to conservation policy and, specifically, the conditioning of farm program eligibility on highly erodible land and wetland conservation.

PART 5 Modern Conservation Policy Developments

9

Modern Developments

FARM BILL CONSERVATION POLICY AFTER 1985

The 1985 Farm Bill placed agricultural conservation policy in the farm bill status quo. Proponents no longer had to fight the uphill, asymmetrical battle in the political and legislative processes, and the policies could avail themselves of the protections afforded the status quo, including those offered by the farm bill coalition. Assistance delivered by the policies would also be expected to build political support for them. The total paid to farmers for conservation has generally increased each year since, with very few years counter to that trend; payments exceeded $2 billion in 2003 and $3 billion in 2007; USDA forecasts payments for conservation in excess of $4 billion in 2021. Counted another way, the modern title reauthorizes at least thirteen different conservation programs, subprograms, and pilot programs. The Congressional Budget Office (CBO) reports mandatory funding for seven or eight different conservation programs. These measures demonstrate growth and development.

More than thirty years after Congress created it, CRP continues to be the largest conservation program by spending. The Congressional Budget Office (CBO) projects that over 87 percent of total conservation program mandatory spending will come from three programs. CRP alone accounts for nearly 40 percent of the total. Conservation spending has, however, continued to increase as the acres enrolled in CRP have dropped considerably, indicating shifts in conservation policy and priorities with implications for future developments. This chapter reviews the development of the modern system of agricultural conservation policy after the 1985 Farm Bill.[1] The review covers nearly forty years and six farm bills (1990, 1996, 2002, 2008, 2014, and 2018). It is organized topically by policy rather than chronologically.

Conservation Compliance:
More Continuation Than Development

Compliance is quasi-regulatory in that it works through eligibility for federal farm assistance. As such, it does not show up in the scoring estimates or baseline forecasts of the Congressional Budget Office. Conservation compliance has surely had an impact on combatting soil erosion and possibly preventing further wetlands losses but would require proving a negative. Highly erodible lands compliance has undergone little change or development. The most notable came in the 1996 Farm Bill, which removed crop insurance from compliance, and the 2014 Farm Bill, which reattached compliance to crop insurance, albeit in a limited fashion.[2]

Unlike highly erodible lands compliance, wetlands compliance has experienced a more complex and difficult political situation since its rather surprising inclusion in the 1985 Act. For the 1990 Farm Bill, the conference managers acknowledged that there had been "considerable confusion and controversy generated by attempts to apply" the wetlands compliance provisions. Congress therefore enacted "improvements" to the wetlands compliance provisions, including revisions to the definition for wetlands, while excluding wetlands drained prior to 1990 and revising exemptions.[3] Problems continued after the 1990 Act for wetlands compliance, however, as farmer frustration with its operation led to pressure on Congress and the Agriculture Committees to figure out how to make compliance functional.[4]

Wetlands and highly erodible lands are fundamentally different on the ground. Highly erodible lands are generally always highly erodible and a relatively consistent problem. They consist of visible consequences such as blowing soils or gulley, sheet, or rill erosion. Soil erosion has almost always been a challenge for farming. It has been a persistent consequence of production with highly visible events like the Dust Bowl to provide political salience. Soil lost to erosion can also be framed as a risk to the farmer's economic interests because lost soil can be easily translated in reduced productivity that can hurt a farmer's bottom line.

Wetlands, by contrast, can be inconsistent; ponds or wet areas form when there is an abundance of rain in a season or just after heavy rains, but they may dry out naturally when conditions are drier for longer. More fundamentally, efforts to control, combat, or address erosion have a long history, but the historic treatment of wetlands has been far different, if not the exact opposite. Wetlands have historically been treated as a nuisance or a problem. Federal, state, and local policies have long been designed to eliminate wetlands and encourage tiling or other efforts to transform wetlands into productive farmland—in the case of the Midwest, some of the most productive in the history of humankind.[5]

When Congress included wetlands compliance in the 1985 Act, it was part of a larger shift in the scientific understanding about the value of wetlands that was accompanied by changes in federal policy toward wetlands, from nuisance to protection and restoration. It amounted to substantial, substantive, and relatively abrupt change for farmers who had been encouraged to drain swamps for generations. In fact, federal conservation assistance had long provided cost-share for subsurface drainage tile. Former NRCS chief Paul Johnson put it succinctly, and with no small amount of the wisdom gained from experience, when he told the Senate Ag Committee in 1995, "In the last 10 years we have turned the whole issue of wetlands around and whereas in the past we provided assistance to drain wetlands and we have brought terrific land into production." He added that the "last 10 years have been tough."[6]

The Agriculture Committees revised wetlands compliance in attempts to thread the needle between protecting and restoring this vital natural resource while also not pushing too many farmers over the edge and into a situation where they would be out of compliance or business. After revising it in 1990, Congress again rewrote the provisions in 1996, exempting federal crop insurance from its application just as it did for highly erodible lands but also trying to better balance the delineation of wetlands for violations, exemptions, and mitigation efforts.[7] Despite the problems Congress, to its credit, continued to reauthorize wetlands compliance, which arguably validates the political salience of the issue. The 2014 Farm Bill also reattached wetlands compliance to federal crop insurance on a limited basis and added further limits on its application.[8] Shifting weather and climatic conditions in the northern Great Plains further complicated wetlands compliance issues, with increasing struggles by NRCS to appropriately deal with the issue.[9] Wetlands compliance remains the law of the land, in spite of the numerous challenges it has faced; Congress reauthorized it again in 2018 with barely a change.[10]

Continuation and Development of Land Retirement Conservation Policy

USDA reported that enrollment in the CRP exceeded 32.5 million acres in 1990, and the Food, Agriculture, Conservation, and Trade Act of 1990 (FACT) continued conservation's momentum. Congress reauthorized the program at a high acreage cap—conference settled on at least 40 million acres but not more than 45 million acres—and provided an important measure of the continued momentum behind conservation policies. Congress also explored concerns about the limit on the types of acres the program should cover, as well as the treatment of acres when released from the reserve after the contracts expired. From the vantage point of the 1990 farm bill, the next farm bill (1995) would be when the first contracts would expire and acres could be brought back into farming. Congress made clear that it did not intend for the program "to retire excessive amounts of productive cropland, nor unduly diminish economic activity in rural areas" and that the secretary should consider "the consequences of such enrollments on the economic health and vitality of rural communities." Budget disciplines dominant during the Reagan years were also evident as counterpressures on CRP enrollment. The budget may have been part of the reason the program didn't enroll more than 35 million acres, topping out at 34,997,682 in 1995.[11]

The political environment was very different for the Federal Agriculture Improvement and Reform (FAIR) Act of 1996, and it would magnify the issues for CRP. Republicans returned to power in Congress after forty years of Democratic control of the House of Representatives. The acreage cap was reduced to 36.4 million acres. Congress also allowed early termination of a contract that had been in effect for at least five years. These were the first real brakes on conservation's momentum after 1985. They were the result of budget matters. Actual enrollment would fall below 30 million acres by 1999.[12]

CRP had exceeded its predecessor from 1956 but had yet to face real opposition until 1995. Under new leadership in Congress, Agriculture Committee members wanted to assess not only the program but also difficult questions about the future. Hearing records provide evidence of strong and widespread support for CRP and only a few discordant notes. CRP had been effective in reducing soil erosion from the highly erodible and environmentally sensitive lands enrolled in the program. It was estimated to have saved nearly 700 million tons each year on the acres enrolled. Critical at the ten-year mark were the consequences as contracts began to expire for the land that had been enrolled. USDA reported that most would return to production and soil erosion would increase as a result. More than just soil was at risk because the program had made an investment of nearly $20 billion, which would also be wasted or lost if the program were permitted to expire. It was a "wake-up call" for farm policy. Faced with trade-offs and a tricky balancing act, the committees had to weigh the degree to which situations on the ground had changed since 1985, and whether there were options to control erosion without the large costs of the program and complications from retiring whole fields for ten years.[13]

The few discordant notes were echoes from the Soil Bank about economic damage from retiring acres, orchestrated by the grain and feed industry concerned about their businesses and the ceding of world markets to foreign competitors. Arrayed against these arguments was evidence about the "beneficial impact on local economies from the support industries related to the hunting boom spurred by the CRP habitat" as well as the program's popularity and the benefits for soil erosion and water quality. Opposition was minimal, however, seeking minor changes. CRP faced little actual risk of expiration. The head of the Natural Resources Conservation Service (NRCS) told the Senate Ag Committee that "we have yet to hear a negative about CRP," and research returned favorable, supportive analysis. A decade after the landmark 1985 Act, reauthorization provided an opportunity to reflect on the meaning of those conservation accomplishments for future generations of farmers and Americans. Notably, the partisan political earthquake in 1994 changed little politically for the policies.[14]

If there was a threat to CRP, it was because of the budget, not policy. Since 1981, budget discipline and procedures had increasingly disrupted the legislative processes, impacting each of the three farm bills prior to 1995. The rise of CRP to a $2 billion per year, $20 billion (ten-year), program coincided with this intense focus on federal deficits and debt. The cost became a particular problem as budget discipline became increasingly narrowed to reductions in federal expenditures. Federal policy had entered a new budget-obsessed paradigm.

In 1995 a Congress newly under control of Republicans—and a more aggressive, activist, and reform-driven version of the Republican Party under the leadership of Speaker of the House Newt Gingrich (R-GA) and his allies—undertook a highly partisan, budget-centered fight with the Clinton administration. That fight pulled the 1995 Farm Bill reauthorization effort into its massive budget reconciliation work, which was vetoed by President Clinton and culminated in a government shutdown.

In the middle of the political fights over spending, CRP was put at risk not because it was unpopular or had proven problematic but because of an obscure disagreement between the budget offices of Congress and the executive about how to treat expirations for the 38 million acres of CRP contracts in the budget. Under budget rules, if the committees wanted to reauthorize CRP in 1995 they would need to cut billions from other programs—mostly taking funds from either the farm payment programs or the food assistance programs—to offset the increased cost. The fix said more about the oddities of budget policy than CRP. USDA merely had to announce an extension of expiring contracts, which it did. The whole episode accomplished little but highlighted the kinds of messes Congress was creating for itself under budget disciplines.[15]

From another budget angle, limits on the CRP had already been applied in the four years prior to the 1996 Act. Congressional appropriators had struck again, rolling back the higher acreage cap authorized in 1990.[16] Congress effectively codified this acreage cap (36.4 million acres). More importantly, the Agriculture Committees strengthened CRP in a trade with appropriators. The most important provision for CRP in the 1996 Act was the revision that provided the program with mandatory funding through the Commodity Credit Corporation (CCC). Doing so permitted USDA to run the program by enrolling acres as close to the cap as it could get, but it also effectively limited the ability of appropriators to cap the program. For appropriators, it was a good deal because this relieved them of the burden of having to find funding within their allocations for a massive program like the CRP.

After 1996 the program would operate on an acreage basis with whatever funds were needed for the contracts and paid for out of the CCC. The CRP had passed its first major political test, arguably coming out stronger.[17] For one, it meant that the program was in a good position to benefit from the anomaly that occurred in 2001, when the Ag Committees secured additional baseline funding for the 2002 Farm Bill. One result of the additional funding is that Congress expanded CRP, and the final bill increased the CRP acreage cap from 36.4 million acres to 39.2 million acres.[18]

Having cleared the budgetary hurdles, the next challenge for the CRP came from the Renewable Fuels Standard (RFS) created by Congress in 2005 and increased in 2007. The RFS created a mandate on the domestic transportation fuels industry, requiring them to blend increasing levels of renewable fuels. Most of the renewable fuel in the United States is ethanol produced from corn, and the RFS proved to be a boon for the corn industry, driving up prices, which improved farm revenues. Improved prices from strong demand add pressure on land to be put in the production of increasingly valuable corn, which in turn pulls acres from other crops. It may also serve as an incentive to put land into production after the CRP contract expires or as a disincentive to enroll land in the CRP. There is significant debate about the impacts on the CRP from the RFS.[19]

In 2008 Congress reduced the CRP acreage cap to 32 million acres, a reduction made in conference.[20] The CRP had 36.8 million acres enrolled in 2007, which was the highest enrollment in the history of the program. A year later enrollment had already fallen to 34.6 million acres and would decrease each year thereafter. The Agricultural Act of 2014 reduced the CRP acreage cap even further. Congress stepped down the cap from 27.5 million acres in 2014 to 24 million acres in 2018. The program had only 25.4 million acres enrolled in 2014.[21] The Agricultural Improvement Act of 2018 increased the CRP acreage enrollment cap each year, stepping up from 24 million acres in 2019 to 27 million acres in 2023. The acres enrolled in the program have continued to decline, however. In 2022 USDA reported CRP enrollment had fallen below 22 million acres, the lowest level since the second full year of the program in 1987.[22]

CRP has been and remains the largest land retirement conservation program but is no longer the only one. Congress has expanded reserve or retirement policy to alleviate some pressure on CRP. In 1990 the Senate created the Wetlands Reserve Program (WRP). It was adopted with the goal of retiring one million acres of farmed wetlands and wetlands converted prior to passage of the 1985 Act. The intent of the WRP was to "restore and protect converted and farmed wetlands, achieving as significant an increase in wetland functions and values as are possible and practical" rather than protecting existing wetlands.[23]

In 1996 Congress added the Farmland Protection Program, a modification of the earlier program for farmland protection. Congress authorized the purchase of conservation easements or other interest on prime farmland that was under pressure for development or nonagricultural uses.[24] The 2002 Farm Bill added the Grasslands Reserve Program (GRP) to enroll up to two million acres of restored or improved grasslands (including rangeland and pastureland). GRP also worked through conservation easements, and Congress increased it to three million acres in 2008.[25]

In 2014 Congress combined the three easement programs into a single authorization called the Agricultural Conservation Easement Program (ACEP). ACEP was designed to combine the functions and funding of the existing easement programs into a single authorization.[26] The 2018 Farm Bill reauthorized the programs. While Congress has made some minor revisions and other changes, these programs have remained largely unchanged. In April 2021 USDA announced that it had enrolled 5 million acres in total conservation easements, 2.8 million in wetlands easements.[27]

The Other Side of Conservation Policy: Developing Working Lands Programs

In 1995 then-chairman of the Senate Ag Committee Richard Lugar (R-IN) noted a "conservation revolution on the American landscape" but that "25 percent of the land currently in the Conservation Reserve Program is not highly erodible and probably should not remain in the reserve."[28] He was echoing former secretary John Block from a decade earlier, and his remarks were an indication that conservation policy had been too narrow. As reviewed throughout the previous chapters, agricultural conservation was largely defined only as retiring acres from production. Moreover, the whole field enrollment of CRP was an incredibly blunt policy instrument that took an entire field out of production for ten years. Some argued that conservation could be better tailored to address soil erosion, water quality, and other natural resource concerns without locking whole fields out of farming. To the extent that CRP and compliance addressed the most pressing soil erosion problems on the most problematic acres, it freed Congress to develop other conservation policies.

That development began with the 1990 Farm Bill. Congress created the Agricultural Water Quality Incentives Program to deliver direct assistance to farmers and landowners on land that remained in farming. The program provided incentive payments to farmers for adopting practices designed to protect water quality by reducing the source of farm-level pollution, including "nutrients, pesticides, animal waste, sediment, salts, biological contaminants, and other materials." It also included "minimizing the generation, emission, or discharge of agricultural pollutants or wastes through the modification of agricultural production systems and practices."[29] The main goal was to help farmers meet or avoid environmental regulations regarding water, and it represented a new development in conservation policy. Rather than retire acres from production, it assisted with practices to mitigate or reduce the environmental consequences of production. Now known as working lands conservation, the programs keep land in production but seek environmental and natural resource conservation integrated with farming.

In 1996 Congress created the Environmental Quality Incentives Program (EQIP) as the main working lands conservation policy. Congress combined the functions of the Agricultural Conservation Program, the Great Plains Conservation Program, the Colorado River Basin Salinity Control Program, and the Agricultural Water Quality Incentives Program into a single program. The goal of EQIP was to provide flexible financial and technical assistance to farmers. The assistance was in return for the farmer adopting a practice or practices that addressed natural resource challenges, as well as to help farmers comply with or avoid environmental regulations. Financial assistance is in the form of cost-share payments—in 1996 that was up to 75 percent of the projected cost—for specific practices implemented by the farmer, including structural practices and land management practices. Initially, Congress provided $200 million per fiscal year in mandatory funding for EQIP and required that 50 percent of the funds "be targeted at practices relating to livestock production." EQIP would benefit substantially from the budgetary anomaly that was the 2002 Farm Bill. Congress increased funding for EQIP to $400 million in fiscal year 2002, rising each fiscal year until it reached $1.3 billion in fiscal year 2003.[30]

Congress reauthorized the program in the 2008 Farm Bill with some revisions that mostly expanded it. Congress also increased funding from $1.2 billion in fiscal year 2008 to $1.75 billion in fiscal year 2012.[31] The budget-battled 2014 Farm Bill again reauthorized the program with revisions. Congress provided EQIP funding that began at $1.35 billion in fiscal year 2014 and increased to $1.75 billion in fiscal year 2018.[32] For EQIP the development has been fairly straightforward. A measure of its political popularity, Congress has consistently expanded the scope of policy and increased its funding.

The real story for working lands conservation policy development—and arguably for all of conservation policy since 1985—occurred in the 2002 Farm Bill. Congress created the Conservation Security Program (CSP), a major new conservation program initiated by the Senate Agriculture Committee. CSP was a complex program that provided for three tiers of multiyear (between five and ten years) contracts with producers in return for adopting conservation practices and addressing natural resource concerns on their farms. In this way it was a complicated conservation or green alternative to the annual fixed payments created by Congress for farmers in the 1996 Farm Bill. CSP in 2002 was funded through the Commodity Credit Corporation's mandatory funds rather than annual appropriations.[33] But CSP was unique and provides unique insights into the conservation question. No conservation program before or since has provided direct payments to farmers in a manner similar to farm payment programs; the difference being that CSP payments were in return for implementing conservation on the farm.

CSP was formed as the recognition and understanding about the value of conservation improved. Experience had built through fifteen years and three farm bills of the modern conservation era. On the other side of the new millennium, policy and policymakers alike emerged from the pressing emergencies that had consumed earlier debates and looked ahead to a longer-term vision for conservation policy. To some, it was time to "go beyond damage control to actually encourage widespread enhancement of the environment across the country" through a bigger, bolder investment in agricultural conservation. CSP was a pivot from traditional conservation policies of land retirement or annual acreage set asides. CSP represented a major new policy to invest in the adoption of conservation on working or producing farmland. It held a vision of a more environmentally focused farming that reduced the externalities or consequences from production. In the Senate Agriculture Committee there was an almost palpable sense of the possibilities for policies that blended environmental benefits with economic benefits to the producer. It was a new millennium dawning of policy that for once rewarded good farming behavior. It was drawn from the deeper well of a land ethic in agriculture.[34]

Senate Agriculture Committee chairman Tom Harkin (D-IA) was CSP's undisputed champion, its author, and the force behind getting it enacted; there is no CSP without Senator Harkin. He had first introduced the "Conservation Security Act" in 1999, proposing "incentive payments to producers for the application of appropriate conservation measures on land that is currently and likely to remain in production . . . a way to reward both those who have undertaken the establishment of conservation practices in the past and those who implement future activities." Harkin considered CSP "the beginning of a new approach for conservation in the next farm bill." His cosponsor, Senator Gordon Smith (R-OR), added that the bill represented "new ways to address the needs of American farmers in the area of conservation" in advance of the farm bill debate.[35]

In historical context, conservation policy had been born of catastrophe, called on to remedy the damage of human error and hubris. Emergency responses offer little room to develop policy. The years 2001 and 2002 were anomalous. The farm economy was again struggling against economic headwinds due to an Asian financial crisis. With CRP and conservation compliance, the economic crisis was not accompanied by an environmental one. The time presented a real opportunity for expansive thinking that pushed beyond the narrow land retirement policy and into supporting practices in conjunction with farming. Adopting conservation practices on the farm came with real costs. These costs were financial and managerial, including the time and effort required to implement, maintain, and manage the practices. Farmers also had to manage the risks added to farming by the practices. The realities of conservation in farming could easily place a farmer at a competitive disadvantage. Too often, these fundamental challenges were made worse by federal farm policy that was designed to align with production and prices, punishing a farmer who had taken on conservation voluntarily. Without an emergency, a more sober assessment of the state of conservation policy could be had; neither land retirement nor the sharing of part of the costs of a practice was sufficient for the challenges that a farmer actually faced, especially with insufficient funding for the need and the demand.[36]

The moment was unique in other ways. The 1996 Act had produced major reforms to farm policy but had stumbled when the reality of international markets and geopolitics put the reforms to the test. More importantly, a brief moment of budget reprieve when revenues exceeded spending had led key players in Congress to negotiate additional baseline for the farm bill. It was the first and only occurrence of a funding increase in the budget discipline era, and the additional funds made it possible to think big. Such rarified political air was fuel for CSP. The concept received support from the traditional farm interests who saw real value in the working lands conservation concepts. The most traditional of the farm interests cautioned that they were not willing to trade any of the usual farm program benefits and other subsidies.

The fluid partisan politics of the time likely pushed CSP over the finish line. Senator Jim Jeffords (R-VT), who had been a leader on conservation issues in the House in the seventies and eighties, left the Republican Party to caucus with Democrats, and control of the Senate changed partisan hands. Senator Harkin became chair of the Senate Agriculture Committee. He would explain later that conservation funding had been concentrated on land retirement through the CRP. While EQIP provided cost-share for specific practices, it was too limited and didn't cover all that conservation truly cost the farmer. Harkin concluded that policy needed to reward farmers who were actually living out the standard talking points of politicians that lionized farmers as stewards of the land and natural resources. There was also, however, evidence of a split among conservation and environmental interests at the program's creation.[37]

It did not take long, however, for the promise of CSP to be dashed upon the rocks of political realities. Senate Ag Committee chairman Tom Harkin (D-IA) reportedly had to fight with House Ag Committee chairman Larry Combest (R-TX) and others throughout the process to get CSP included in the final bill. It was a difficult fight, but from the legislative record it appears to have taken place in behind-the-scenes negotiations, not floor debates. The record does indicate the extent to which Chairman Harkin was focused on conservation. The Senate bill provided an additional $21.5 billion to conservation and included the CSP.[38]

Getting CSP included in the farm bill was only the beginning of the fight, however. USDA did not announce the first sign-up until 2004, as the promise of CSP went unfulfilled. Critically, CSP would be attacked over budget and spending concerns. CBO's cost estimates for the final 2002 Farm Bill were that the bill would increase spending by just over $80 billion for the ten-year window (fiscal years 2002–11). More than 70 percent of the new funding was for farm program payments, mostly direct payments and countercyclical payments. CBO estimated that CSP would increase spending by $2 billion. By comparison, the direct payments would increase spending by $9.5 billion, and the new countercyclical program would increase spending by $35.3 billion. When it came time to make cuts and reductions, however, it was CSP that was placed in the crosshairs of budget hawks. Senator Harkin had to fight with appropriators to protect the program.[39]

The 2002 Farm Bill authorized CSP as an uncapped, mandatory program that was unique among conservation programs. CSP was designed without limits on the total amount of funding available (like EQIP) or the total acres that could be enrolled (like CRP). According to a 2006 Government Accountability Office (GAO) report on the program, CBO officials lacked sufficient information for the initial scoring of CSP. Beginning in January 2003, CBO dramatically increased the program's costs to $7.8 billion and then $8.9 billion in March 2004. According to GAO's investigation, the massive increases in estimated spending were technical in the budget scoring process; with each passing year the ten-year budget window shifted out and pulled in later years in which CBO logically estimated more participation and costs. This was especially true given that the program didn't begin enrolling farmers until two years after enactment. But there was more to it; the revised estimates bore all the markings of political work by officials in the Office of Management and Budget (OMB) in the White House and at USDA. Before it had rolled out the program or conducted a single sign-up, USDA officials suddenly informed CBO that they anticipated much greater program participation. USDA's new expectations led to CBO's drastic increase in estimated program costs.[40]

With program cost estimates suddenly more than quadruple, Congress enacted a variety of legislative provisions that instituted caps on CSP funding. Beginning in February 2003, most controversially, Congress placed the first caps on the program to offset the costs of emergency legislation to assist farmers suffering from drought. In 2004 congressional appropriators—led by Senator Thad Cochran (R-MS), who, during his long tenure, chaired both the Agriculture and Appropriations Committees—revised the cap to once again offset the costs of emergency disaster appropriations, this time for hurricanes. The implementation of funding caps altered the vision for CSP. A series of complicated, uncertain funding issues also complicated implementation. For example, when USDA finally opened the program, it did so on targeted watersheds rather than as a nationwide program as intended in 2002. The decision caused an uproar from farmers and damaged the program in the eyes of those it was supposed to help.[41]

The entire episode cast a cloud over CSP and helped generate frustration. Frustration, in turn, fueled speculation about partisan politics and sinister motives by the Bush administration and congressional Republicans. The fighting created very real risk that CSP, and possibly even conservation policy, could become another partisan, political matter. Partisan politics or not, Senate Democrats did have reasons for their arguments. Republicans used an unprecedented move to cut mandatory farm bill program spending to provide an offset for emergency disaster assistance. Doing so would set a precedent that was concerning for farm bill programs but a precedent that was unique to conservation policy and to CSP.[42]

Adding intrigue, CSP was designed more like direct payments in the commodities title, and thus it carried a strong resemblance to the policies long supported and defended by the farm interest faction. The only difference between CSP and direct payments was that CSP required the farmer to implement conservation on the farm to get the annual contract payments. Direct payments were issued merely for having historical planting records for the crop. One could search the historical record in vain for an example where appropriators cut farm program payments to offset the costs of emergency assistance to farmers. The treatment of CSP provided an unfortunate recycling of the sabotage efforts aimed at the Soil Bank. It was as if the dead hand of Jamie Whitten took up the pen once again to strike at the heart of innovations in agricultural conservation policy.

The CSP fight explains key changes to the program in the 2008 Farm Bill. Senator Harkin returned to the chairmanship of the Senate Agriculture Committee in 2007, when Democrats recaptured majorities in both the House and Senate during the final years of the George W. Bush administration. His committee undertook the "biggest single change in the conservation title" by "restructuring the Conservation Security Program into the Conservation Stewardship Program." The task would not be easy, especially after the 2002 Farm Bill had provided more than $70 billion in additional baseline. Reauthorization in 2007 would be undertaken within very different budgetary parameters. Projections for better crop prices in the wake of stronger demand driven in part by the Renewable Fuels Standard meant much less spending on the countercyclical payments program created in the 2002 Farm Bill. As just one example, the CBO baseline for 2007 provided that spending on farm programs was estimated to drop off drastically. Outlays went from above $18 billion in fiscal year 2006 to under $10 billion in fiscal year 2007, with continuing decreases thereafter.[43]

The challenge for CSP involved congressional budget rules that required any increase in spending to be offset, creating a zero-sum game under the CBO baseline. If Chairman Harkin or others wanted to increase CSP, other farm bill programs would have to be reduced to offset the increased cost. This pitted CSP against other conservation programs or, most problematically, against the farm payment programs in Title I. The challenge was magnified by the extraordinarily high estimates for the program put forward by USDA. Budget and farm bill politics—the limited perspectives of the farm interest faction—placed incredibly difficult hurdles in the way of CSP. The farm interests were clear that while they supported more funding for CSP, they did not want that funding offset out of the farm programs in Title I.[44]

Republicans in Congress, turned out of power in the 2006 midterm elections, made sure to reinforce this point. Southern Republicans in particular knew they had a powerful tool to limit the options for conservation and CSP, as well as an effective method for tying Chairman Harkin's ambitious legislative hands. Traditional (and relatively conservative) Democrats in the House, like Chairman Collin Peterson (D-MN), were willing to comply. The common claim was that the budget hurdle was too much to overcome for conservation and for CSP. The budget challenges appeared to be a convenient excuse that applied only to conservation. The House Agriculture Committee claimed to revise the Conservation Security Program, reporting a bill that read more like sleight-of-hand on this topic. The committee reported a replacement CSP "with a new, simplified program," but the new program did not start until 2012, which would have been after the farm bill was scheduled to expire.[45]

Chairman Harkin would again have to bear the brunt of the burden. It added challenges to the complicated farm bill reauthorization process in the Senate. The debate would drag through 2007 and into 2008 before the House and Senate committees could agree to a final legislative product. What Congress agreed to was too much for President Bush, however, and he vetoed the final bill; Congress easily overrode his veto. It took substantial effort and time, but Chairman Harkin was able to rewrite the CSP in 2008. Most notably, Congress created another unique feature for the program, requiring the secretary to enroll 12.769 million *additional* acres in the program each fiscal year. In a relatively status quo farm bill challenged by budget rules and limitations, Harkin's success with CSP stands out. In fact, CBO's scoring estimate of the final legislation reported an increase in conservation spending and a decrease in farm program outlays.[46]

While it remained quite a distance removed from the high aspirations of its authors and the potential for the policy, CSP had survived its first reauthorization test. More importantly, the program was growing. In 2011 CBO estimated CSP outlays to top $1 billion by fiscal 2013 on the way to more than $2.5 billion by fiscal year 2020. The projections for CSP in the baseline moved it to the largest conservation program in the farm bill, the first to surpass CRP spending. A growing conservation program would also be an expanding target, which was a concern for CSP in the troubled and budget-scarred Agricultural Act of 2014. Congress focused on improving the program's operation and effectiveness, as well as farmer acceptance, but also scaled back its acreage enrollment cap to ten million acres per year at a national average of eighteen dollars per acre. With congressional partisan budget hawks circling the farm bill, CBO projected that the outlays for CSP would be reduced by an estimated $512 million for the five-year reauthorization. More important were efforts to trim back some of the program's complexity and focus the payments on farmers who were undertaking "additional conservation activities" and "improving, maintaining, and managing existing conservation activities."[47]

Coming out of 2014 relatively unscathed would appear to be the high-water mark for CSP. The troubled 2018 Farm Bill reauthorization debate reinforced that conclusion. The program's congressional champion, Senator Tom Harkin (D-IA), retired at the end of his term in 2015. As Congress prepared for reauthorization, the Agriculture Committees were informed that CSP had grown to become the "nation's largest conservation program with over eighty million acres." USDA under President Trump and Secretary Sonny Perdue signaled an intent to align it with EQIP, however. Taking the cue, the House Agriculture Committee, under the leadership of Chairman K. Michael Conaway (R-TX), eliminated CSP as a stand-alone, acreage-based program and combined it with EQIP.[48] Most remarkable for a program of its size and scope, let alone the high expectations for it in 2002, the House committee's move to eliminate the program received a relatively quiet response. The attack on CSP was mostly drowned out by controversies raised by the cuts to food assistance programs and the initial defeat of the bill on the House floor.[49]

By comparison, the Senate farm bill took a bipartisan and much less controversial route. Senators had little interest in the partisan cuts to food assistance that had consumed the House. Senators also noted that they were protecting the conservation title, although not CSP specifically. The future of CSP rested in conference, which was locked in stalemate for months. Ultimately, the more reasonable Senate position on CSP would largely win out. The basic five-year contract for annual payments continued, but the program lost considerable funding. It also lost the acreage basis from its original design and as modified in 2008. Arguably, the acreage basis was a critical policy feature that made it unique among conservation programs. Finally, CSP was also placed in the EQIP program in the statute, meaning it was no longer a stand-alone program. This might well jeopardize its reauthorization in the future. If this was survival for CSP, it wasn't much of one.[50]

Whether CSP survived the attack in 2018 remains an open question and one subject to debate. It emerged a different program in very important ways, even further removed from the original vision and intent of the policy. CSP in 2002 arguably presented the potential to replace or supplement the annual direct payments that some farmers received. The 2018 Farm Bill didn't just cut its funding, it eliminated fundamental policy concepts that had originally underwritten the program. From Whitten to Cochran, Poage to Conaway, a pattern emerges for conservation policy, seen most clearly in the treatment of the Soil Bank and CSP. Midwestern policy proposals have to overcome hurdles put up by the farm interest faction, led by southern interests. Achieving enactment, the programs are then subject to sabotage by southerners using budget and appropriations processes as weapons.

The rise and fall of CSP may leave the wrong impression about the current state of agricultural conservation policy. While CSP remains in real jeopardy, the Agriculture Act of 2014 included a new version of conservation policy that may hold important potential for the future. Created by Chairwoman Debbie Stabenow (D-MI) in the Senate Ag Committee in 2012, the Regional Conservation Partnership Program (RCPP) was designed to work across existing conservation authorities but on a regional basis and with the addition of nonfederal partnership funding. The final program authorized $100 million in annual mandatory funds, as well as 7 percent from the existing conservation programs from which it drew its basic authorities and functions. Conservation policy had officially entered the realm of the public-private partnership concept, as USDA was instructed to match the federal funds for farmers with private funds.[51]

The House Ag Committee in 2018 reauthorized RCPP and expanded the program, increasing funding and acknowledging that it was "innovative and popular." The final bill largely adopted revisions that sought to streamline and improve the program. Congress continued the basic functions but intended for RCPP to become "a stand alone program, with its own rules and regulations . . . increasing producer access, improving conservation outcomes, and simplifying procedures."[52] In the Senate Agriculture Committee, Senator Debbie Stabenow (D-MI), ranking member, touted the RCPP authorization for "matching public conservation dollars with private dollars." It was one of the 2014 Farm Bill success stories, and RCPP, "a new and innovative approach to voluntary conservation" had by 2017 "leveraged more than $1.2 billion in private funding and brought together over 2,000 diverse partners to address local conservation goals."

There were warning signs as well, and they were familiar for conservation programs. Insufficient funding to meet demand and continuing implementation challenges dogged the program, causing "considerable frustration among partners regarding the use of technical assistance by NRCS in implementing the program." In RCPP, Congress kept pushing the need for more flexibility and simplified administration. Among these were issues with contracting and applications, as well as designing innovative features for partnerships and funding arrangements. A combined program, RCPP could also serve as a reminder of the challenges and struggles for much of the suite of conservation policies.[53]

As of this writing, the single biggest development for agricultural conservation policy since the 1985 Farm Bill occurred in 2022. Significant in its own right, the importance of this development was magnified because it took place outside of the farm bill reauthorization process. After the troubled 2020 elections, Democrats in Congress reversed the long trend for using budget reconciliation procedures to force cuts in program spending. With narrow majorities in the 117th Congress (2021–22), Democrats turned to those special budget rules and procedures to pass the Inflation Reduction Act of 2022. This allowed them to enact the legislation on party-line votes, without any Republican support.

Critical for farm policy, the Inflation Reduction Act contained an unprecedented investment in the conservation programs of the farm bill. Congress added over $18 billion through a multiyear appropriation to EQIP, CSP, RCPP, and ACEP. Potentially more consequential, Congress extended the statutory authorizations for conservation programs through 2031. This places conservation programs on a different expiration schedule than farm programs and alters the political and negotiating dynamics in uncertain ways.[54] Congress also directed that the additional funding be available only for a limited set of conservation practices that help address climate change, such as reducing, capturing, avoiding, or sequestering greenhouse gas emissions. Specifically, Congress required the Inflation Reduction Act conservation funds to be available only for "agricultural conservation practices . . . that the Secretary determines directly improve soil carbon, reduce nitrogen losses, or reduce, capture, avoid, or sequester carbon dioxide, methane, or nitrous oxide emissions, associated with agricultural production."[55]

Arguably, the Inflation Reduction Act represents the most significant policy effort by Congress to address climate change through agriculture since 1990.[56] Since the first faltering steps in the Soil Erosion Act of 1935, the Inflation Reduction Act of 2022 represents the first and only time that Congress has gone outside of the farm bill process to increase the investments in conservation policies or programs. Putting the 2022 effort in more historical perspective, it took Congress fifty years to add conservation policy to the farm bill status quo. Twenty years after reaching the status quo, conservation policy innovations were still victimized by partisans and southerners in the appropriations process. It wasn't until the budget battles in 2011 and 2012 that farm programs were targeted more than conservation programs for spending reductions. Another decade passed before Congress took the positive step of investing in conservation programs without adding more to farm payment programs. The priorities represented in the Inflation Reduction Act have been a long time coming, but their ability to weather the increasingly polarized and partisan Congress remains a major concern and a vast uncertainty.

10

Of Congress and the Conservation Question

A WORKING THEORY AND CLOSING ARGUMENT

Between soil and society, the U.S. Department of Agriculture reports that society spent a grand total of $1.2 trillion (adjusted for inflation) on all farm payments from 1933 to 2020, of which only 19 percent went for conservation.[1] During that time the relationship between soil and society has changed in significant ways. By one measure, the U.S. farm population has fallen from almost one quarter of the U.S. population in 1930 to barely 1 percent today.[2] As it becomes an increasingly smaller portion of the general population, farm interests are at greater risk of being disconnected from the general public and its interests. The general public is also at increasing risk of being disconnected from farming, food production, and related interests. As the distance between soil and society grows, the divergence becomes extraordinarily counterproductive for both.

In the span of history covered here, American society has suffered a Dust Bowl and countless other dust storms during droughts; immense tonnages of soil have been lost and vital water resources depleted. Important habitat has been consumed or destroyed, some of which has been lost for any foreseeable future. Increasingly in recent years, dead zones, poisonous algal blooms, and other water quality degradation problems have plagued major water bodies like the Gulf of Mexico, the Great Lakes, the Chesapeake Bay, and others. Two farm economic crises have combined with some of these natural resource crises, demonstrating a very direct, important connection between the economic and the environmental. It is a connection that arguably should be better understood and be given a greater consideration in national policy. At no point during this time, however, has American society invested an equal amount of its funds on efforts to conserve natural resources or preserve its soil. In this basic fact, the conservation question is rooted.

To be clear, nothing covered in this book is meant as an attack on conservation policy, nor is it intended to diminish the myriad accomplishments on the ground and for natural resources. Much outstanding, and in some cases even heroic, work has gone into improving conservation in American agriculture. But that reality also goes toward proving the point: rather than a collective priority, too much of the conservation effort in this country has depended on individual efforts or has been limited to brief, heroic interventions. The conservation question takes as its most basic presumption that it is in society's interest to invest in the conservation, preservation, and protection of natural resources and the environment. It is also in the interest of the individual farmer and of agriculture generally. It is reasonable to ask why we have not done better considering all that we know and have learned, given our scientific and technological advancements during these years.

The conservation question focuses on Congress and the dynamics between narrow, private, or special interests and the broader public interest. For policymaking, it compares policies that exist on the same jurisdictional turf but with different benefits for the public and private interests. Both farm program payments and conservation assistance go to farmers and, in many cases, to the very same farmers. Conservation payments to farmers also benefit the public interest but have always been a secondary priority to payments that merely benefit the farmer. The public interest has been consistently sacrificed or subordinated to the private interest as a priority for the expenditure of public funds.

The conservation question comes full circle to face squarely the theories embedded in the constitutional design and the paradoxes at the heart of our experiment in self-government. But it is also worth remembering the extent to which the framers of the U.S. Constitution designed a system of self-government that was unique in its time. There was little precedent and few models to guide their work. They were conflicted and fallible, which unavoidably impacted their design. No critical evaluation should lose sight of the enormity of their achievement, however, or of the fundamental political theories they attempted to apply to the profound task of designing a government. If nothing else, their design created a first-of-its-kind system of self-government that could change, evolve, and grow with the nation—no matter how difficult, slow, frustrating, or imperfect it may be.

Toward a Working Theory on Congress, Part 1: The Design Revisited

Reviewing the advocacy essays by Alexander Hamilton and James Madison after exploring the legislative history and political development for conservation policy helps develop a working theory on Congress that can answer the conservation question. The framers' concerns began with the risks of concentrated political power and its tendency to be misused and abused, or worse, to descend into tyranny. They identified the primary avenue for these threats as special interest groups in society known as factions: the "disorders that disfigure the annals" of previous societies, according to Hamilton, or the "mortal diseases under which popular governments have everywhere perished," in Madison's writing.[3]

Factions can possess real political power, and when coupled with the motivation to manipulate the system of self-government, factions can achieve narrow interests. In general, this involves extracting or capturing benefits only for themselves. Worse still, such factions could violate the rights of individual citizens, groups of citizens, or society in general. In reality, however, no system of government can prevent factions without destroying self-government and liberty, nor can factions simply be controlled. The framers' solution to this conundrum was structural: design a system that channeled factional political power toward better ends and outcomes, defined as those that better represented the public interest, by requiring them to compete against each other. The legislative branch was of primary concern. Congress is the only branch capable of initiating the making of laws or amending existing laws, and only Congress can transform ideas, demands, or policies into national laws where none existed before. The framers designed a complex, complicated legislating process and divided the branch to create a double majority hurdle for enactment because of the potential for factions to accumulate and actuate political power.

According to Madison, the constitutional design was built upon a theory of political competition among factions channeled through the legislative process. That competition was expected to achieve the best outcome for the greater good, or an approximation of the public interest. Hamilton and Madison envisioned a large, diverse, and highly fractionated society from which a myriad of interests would compete to achieve preferred outcomes in the government. To succeed, factional interests would need to form governing coalitions that could achieve the double majority. This would force negotiation and compromise to reach at least some version of the public interest or the general welfare. The competition was expected to produce an amalgamation of factions and interests that would best represent a larger interest or greater good for the nation.

Two basic, fundamental challenges with this theory become immediately obvious. First, there must be competitors in the process representing the various factions with interest in the legislation. The design included nothing to ensure that all interests were represented in the competition or had representatives in the fight. Second, the competition must be fair and a functional contest among relative equals, or equally matched competitors. In such a competition, the various factions could counter and check each other in order to prevent or minimize the negative and harmful effects of the exercise of factional power. A third challenge emerges from these two: there is no actual provision for the public interest because the theory assumes competing factional interests will sufficiently approximate or protect the public interest. To say the least, this is a massive assumption and one with substantial risks.

Madison defined faction based on having common interests or possessing a common "impulse of passion," but not whether those were held by a majority or a minority of the citizens.[4] He did not distinguish between private interests and the public interest or public good but rather conflated them both as factions that were cause for concern. Arguably more concerning, his arguments present an aversion to majorities. By including majority interests in his definition of a faction, Madison's theory risks being burdened by overly pessimistic or cynical views about government. Views that are too skeptical or concerned with majorities and how they could operate as a faction risk skewing the design against majorities and jeopardize competition in the political arena. The result would be a design that places too much emphasis on preventing bad legislation while discounting the need for accomplishing lawmaking in the public interest.

A problem with this design can be understood through Mancur Olson's theory that smaller factions would be more likely to form and actuate their interests, while larger interests would not be able to. This would create asymmetrical political power benefiting the smaller interests.[5] More importantly, Herbert Hovenkamp explained how this design only protects against legislation demanded by a faction that is considered harmful to the general public interest. It is counterproductive when the general public interest demands legislation but a faction is opposed.[6] Much of this is because it is easier to prevent a bill from being enacted than it is to succeed in the process; there are far more opportunities to stop, block, or kill legislation—and they are easier to execute—than there are ways to get a bill passed. The design limits factions by achieving nothing, which is overly protective of the status quo, and the status quo likely benefits some factions more than others, as Hamilton pointed out in the first essay in *The Federalist*. Too much of that outcome, however, is not realistic for a system of government.

Asymmetries in a system built on the control of factional power through competition, but stacked against achievement, end up favoring factions. First, it is very likely that the status quo favors some factions. Second, the design empowers factions because, ultimately, the public (or at least significant portions of it) demands legislation, and the system's legitimacy depends on being able to achieve something in response to those demands.[7] The legislative power, after all, is distinguished by initiating the making of law. This design provides factional interests substantial negotiating leverage in those times when the public demands action from its government because they can hold legislation hostage and extract benefits. Moreover, to the extent that the system has previously delivered for a factional interest, favoring the status quo further empowers those factional interests to protect their existing benefits. Hamilton, in particular, warned about the substantial obstacles to self-government from entrenched factions who would exert power to resist changes that they perceived might diminish their power, status, or benefits.

Factions also demand legislative outcomes, but those outcomes may be opposed by the general public or congressional majorities. In that situation, the majority can simply vote down the factional demand, something that Madison may have overvalued in his theory on the design. And yet, Madison recognized that the primary method by which a faction would operate is through a coalition. The problem was that his theory to address this was to make it more difficult for a majority coalition to achieve its goals. In reality, an effective faction can co-opt or manipulate a coalition to leverage its political power, holding hostage the demands of the larger coalition until the faction gets what it wants.

The experience of agricultural conservation policy adds another important piece to the working theory by highlighting the consequences for public policy when there is no competition in a system built upon it. This observation extends Olson's in that interests representing the public good are likely larger, more diffuse, and much less likely to voluntarily organize and act. The existence of a competitor on behalf of the environment and natural resources was one of the most important factors behind the different outcome for conservation policy in the Food Security Act of 1985. The modern environmental movement demonstrated incredible political power with its major legislative achievements throughout the 1970s. It turned its attention to agriculture belatedly, and the Reagan era presented a unique political opportunity. The twin farm crises of the 1980s sealed the deal.

By comparison, the Dust Bowl did not produce an organized interest on behalf of the environment or even the natural resources (soil) upon which farmers themselves depended. When Secretary Wallace went looking for suggestions, he called in the various members of the farm coalition, and they were the only ones at the table. A narrow version of conservation policy was written largely as a means to work around a poorly reasoned Supreme Court decision. The policy was effectively abandoned at the first sign of problems. Neither the 1950s drought nor a self-destructing farm program was sufficient to catalyze a competing interest on behalf of the environment and natural resources. Instead, Midwest farm interests combined with the Eisenhower administration to craft and push for the Soil Bank. Little surprise, then, that it was another version of the basic farm commodity policy at the time; the 1936 program design was modified slightly in 1956. No conservation or environmental faction was represented in the deliberations at the committee level, nor on the floors. When the Soil Bank came under attack, neither the midwestern farm interests nor the administration were able (or willing) to effectively defend it against southern sabotage. The general public interest is not an effective competitor. Ambition cannot check ambition if there is no ambition on the other side.

By launching a second assault on both farm payments and food stamps, the Reaganites inadvertently encouraged the farm bill coalition to strengthen its political power, and it expanded to include environmental interests through conservation policy. As discussed in chapter 8, the achievements for conservation policy and the environment in the Food

Security Act of 1985 were maybe the closest in this history to Madison's theoretical ideal of an open competition among relatively equal factional interests, with the result an adequate approximation of the public interest. The 1985 and 1990 Farm Bills stand as testaments to the importance of having competing interests—especially representation on behalf of interests more public or general—at the negotiating table and in the deliberations.

The 1985 Act provided cautionary tales as well. It served reminders of the extraordinary difficulties in seeking change to the status quo, and the amount of time, energy, and resources that must be devoted to the effort. It was also costly across the board, including the cost of more than a decade of natural resource degradation and the loss of many farmers. Conservation policy in 1985 was also relatively limited in scope to CRP and conservation compliance. Its most important achievement was taking its place in the farm bill status quo, which was critical for the opportunities and possibilities it opened for future development. It put in place a new foundation for developing agricultural conservation policy in a farm bill.

From sharecroppers to soil erosion and the Soil Bank to CSP, factional interests do not voluntarily relinquish power. They neither sacrifice their narrowest self-interests for altruistic purposes nor share benefits of their own volition. Conservation policy teaches that these statements hold true even when the direct benefits go to the factional interests and indirect benefits go to the general public. The method matters as well, and factions are predisposed to having things their own way. The "clamors of an impatient avidity for immediate and immoderate gain," Madison warned, too frequently drown out "the mild voice of reason, pleading the cause of an enlarged and permanent interest."[8] The problem is that the design of Congress tends to encourage this and even favors it.

Power, interests, benefits, and other outcomes are won through political competition in the process, shared only when forced to by competing interests. The rival interests either defeat problematic factions or force compromise—generally by defeating them or threatening to—and this replaces "the defect of better motives."[9] Madison's dose of realpolitik is important as far as it goes, but it conveniently overlooked the asymmetries that exist between smaller, better organized factions and the larger public interest, and the congressional design failed to correct for them. The public interest is too diffuse and often too indirectly impacted to serve as an effective competitor. Conservation policy demonstrates that to achieve even a modicum of success, the public interest or general welfare needs dedicated, interested, and effective representation in the legislative competition. Working on the public interest's behalf must be sustained and relentless because factional interests certainly will be.

Toward a Working Theory on Congress, Part 2: The Southern Citadel

The history and political development of farm bill conservation policy exposes another matter of realpolitik that is not fully accounted for in the above discussion. Throughout much of American history, a singular faction has possessed outsized influence and power. Madison, for his part, acknowledged that the "most common and durable source of factions has been the various and unequal distribution of property."[10] His observation applies unequivocally to American farm policy. The most successful and powerful among the property owners who benefit from its programs, moreover, has been the southern planter elite.[11] The experiences for conservation policy serve ample reminder that farm policy was among this faction's most protected and prioritized political ground. Southerners were so entrenched in power over agricultural policies that they arguably exceeded the standard definition or understanding of the term "faction." Many of their names were synonymous with both power and farm policy through the decades—Jones, Cooley, Poage, Cotton Ed Smith, Ellender, Talmadge, Helms, Whitten, Russell, Bankhead (both Senator John H. Bankhead II and Speaker William Bankhead)—and this faction would be more appropriately understood as the southern citadel in Congress. Madison's theories do not properly account for this reality.

By providing strong protections for the status quo, many features of the congressional design more than helped southern interests consistently capture outsized benefits and protect or advance their narrowest goals. Between those features and the path-dependent nature of policy, the South was able to hold onto its outsized share of federal benefits long after its power had waned. In addition to the lack of a competitive factional interest, much of the fifty-year delay in the development of conservation policy can be blamed on the southern citadel. Suffice it to say, if the conservation of soil and other natural resources had been a priority for that faction, it would have been enacted in law. As dictated by the South, either conservation was disfavored as an unacceptable interference with private property and power, or it was discounted as welfare or unnecessary expenditures of federal funds.

Much about political power can be difficult to discern or can even appear shadowy, hidden. When viewed through the lens of conservation policy history, it is possible to trace critical developments in the political power of the South. The difficult progress for conservation intersected with critical moments in the trajectory of southern political power and was impacted by them. When that power was at its apex, conservation policy was nowhere to be found until it was needed to work around a misguided Supreme Court decision (1936). The first indications that southern power was weakening coincided with conservation policy being used by the South's primary competitors (midwestern corn interests) as an alternative to the South's preferred policy design. And conservation attained membership in the farm bill status quo after southern power had declined precipitously and was transitioning from the Democratic caucus to the Republican caucus (1985). Look again at these three episodes in conservation history.

First, when the southern faction was at its strongest, the structural devices in the constitutional design of Congress offered little impediment even where the legislation they demanded was socially harmful. Their reputation and the threat they posed to his agenda were sufficient reasons for the extremely popular and powerful Roosevelt to acquiesce.[12] The Supreme Court was the only real center of power in opposition, but the court's intrusion was exogenous. It arguably overstepped its bounds and authority in a misguided attempt to exert power where it had little. With gavels and pens in southern hands, all the Supreme Court accomplished was to provide the catalyst for emergency legislation in an election year. The decision was folly. It exposed the court's weakness and precipitated a rather rapid retreat. Worse than that, the court's overreach likely overrode whatever the public interest might have demanded in response to the Dust Bowl. The 1936 Act was mere legal and political cover even if reformers at USDA envisioned much more. The evidence was most clear in the long-term program designed to operate through the states, which would have devolved the policy and funding to local control in those areas of the South most threatened by federal interference.

Second, the first cracks in the South's stronghold coincided with the Soil Bank saga. The 1956 enactment process demonstrated the extent to which southern factional power was beginning to weaken under pressures from a changing American society. President Eisenhower became the primary competitor in the process and a power center who represented some version of the public interest. He demanded action to address the failed parity system and the increasingly costly surpluses piling up in USDA storage. Eisenhower was also aligned with the South's factional adversaries in the Midwest and among Republicans.

The need to deploy weapons in defense indicated a loss of political power and was costly in draining that power further.[13] In Congress, southerners attempted a standard powerplay and dared the president to veto it; he called their bluff. The veto loss on farm policy in an election year was a serious setback and a costly expenditure of power. It made clear that the threat and reputation of the South was insufficient to force a popular president's hand. They retreated and resorted to sabotage, rather than fight a losing battle further. Eisenhower's veto also demonstrated the power of the structural elements in the design of the legislative process. The veto prevented a faction from achieving policies viewed to be harmful to the public's interest. But deploying it was also a sign of political weakness—the president's threat was not enough to deter the southern sabotage—and doing so was costly.

Appropriations provided an effective weapon but not without cost. A popularly reelected president backed down rather quickly, and any blame for the problems they caused did not fall on Whitten and southerners. They expended significant effort and energy across multiple years, however. Each time provided a reminder to their competitors and adversaries, which may well have fueled intensifying reactions. Factional feuding within the farm coalition escalated and worsened, further draining southern political power. This is not to claim that the Soil Bank or farm bills in general were the cause of the decline in southern political power in the sixties; they were symptoms of it that also contributed. From school desegregation to voting rights, reapportionment, and redistricting, the Supreme Court inflicted the most damage. While President Eisenhower signed the first Civil Rights Act since reconstruction in 1957, it was the southern Democratic president Lyndon Johnson who effectively razed the citadel with the Civil Rights Act of 1964 and the Voting Rights Act of 1965.[14] After these tectonic shifts in American politics, the South began its slow transition out of the Democratic Party and into the Republican Party.

It is notable that the South was also forced to compete on a second front during the 1950s and into the early 1960s. After Eisenhower's time in office ended, southerners in Congress redirected the acreage reserve policy against midwestern feed grains farmers. At the same time, southerners were repelling efforts by northern and urban Democrats to enact policy providing food assistance to low-income families. The need for successive deployment of the tools and weapons of political power was draining and costly, while the competing interests gained strength to the point that they retaliated against southern commodity policies. When Congress finally enacted the Food Stamp Act of 1964, food stamps quite literally saved farm policy from self-destruction and offered a very clear statement on the decline in southern power. It also cracked open what had previously been an enclosed policy space. The opening from 1964 was opened much further by the 1973 Farm Bill, which included food stamps, and the 1977 Farm Bill, which substantially expanded the program by removing the purchase requirement. In turn, these developments were critical to assisting conservation policy. The farm bill was no longer an exclusive farm interest redoubt when conservation and environment interests finally joined the competition.

Arguably, a citadel is a useful metaphor. A citadel is generally a commanding fortress or a defensive stronghold for an area or a city. It is an enclosure within the larger area that commands or defends. It is of value to think about what this concept means in terms of self-government and public interests, as well as in Madison's theories for the American system of government. The challenge goes straight to the design of Congress, built not only on competition but on the need to form functional, governing coalitions to clear the double majority in the legislative process. A factional enclosure—a political citadel—could struggle with an open competition among relative equals. Conceptually, coalition-building is expansive and additive, but an enclosure is the opposite. A coalition can change and evolve as it competes. It can adapt over time. An enclosure is more insulated and resistant to change or evolution, and it is much less likely to adapt over time. With path dependency, these matters tend to trend; a coalition will grow as it evolves and adapts, but an enclosure likely shrinks or restricts.

The southern citadel was built upon a system of racial oppression and voter suppression to defend one-party rule in the region and on behalf of the planter elite. It was, in nearly every sense of the word, an anticompetitive political system. Southern Democrats used anticompetitive power in the region to vastly increase competitiveness in Congress. Its most straightforward example was the seniority system for chairmanships of committees and often caucus leadership. The problem for the southern citadel was that enclosed, concentrated factional power in an open, competitive system was nearly impossible to sustain and maintain over the long term. A citadel's defenses work by retreating within them when under attack or unable to advance offensively. In a system of competition and coalition-building, this tends to drive coalitional partners away. From the history of conservation policy, there is evidence in the fight to prevent the dairy amendment in 1936 and the fights with midwestern corn interests over the parity acreage system and the Soil Bank. The South drove away its coalitional allies, and especially with corn in the 1950s, it did so when it likely needed them the most.

Political enclosures and concentrated factional power centers are likely to unleash extremely self-destructive tendencies in their defense, enclosing further and becoming more self-destructive. Consider that even as farm policy failed in open and obvious ways, the South blocked change or sabotaged it. The destruction spread to larger interests and the general public, which, in a system of competition, increases opposition and requires more defensive, destructive actions. These are among the self-destructive tendencies of faction as they operate an enclosure movement that trends more and more narrowly until its only options are defeat or tyranny. If the enclosing faction is powerful enough, or willing to fight to the bitterest of ends, it can inflict harm far and wide with repercussions that impact large swaths of the public. By preventing change, evolution, and adaptation, factions magnify problems. As problems build without being addressed, they can reach breaking points that are incredibly damaging and often require drastic, radical responses that themselves become status quo, difficult to correct. If the key to the system is competition, then concentrated factional power is a problem of the first order.

Toward a Working Theory on Congress, Part 3: The New Budget Regime

Scanning history from a contemporary vantage point, the year 1990 materializes as a high-water mark: the point where a wave of potential broke and receded. The farm bill reauthorization process that year built upon conservation policy's gains from 1985, and budget negotiations reduced farm program payments. The Food, Agriculture, Conservation and Trade (FACT) Act of 1990 also included the first, and thus far only, title in a farm bill dedicated to global climate change.[15] Expanding the field of vision, President George H. W. Bush pursued environmental policy goals that included declaring an end to wetlands loss nationwide, calling for federal policies to address climate change, and signing legislation that established the National Climate Assessment. Most notably, he signed the Clean Air Act Amendments of 1990 into law that, among other things, successfully regulated the pollution that was causing acid rain.[16] That 1990 turned out to be a high-water mark rather than a new trend for conservation, and environmental policy reemphasizes the conservation question and tests the working theory.

The third part of an answer to the conservation question and the final piece of this working theory on Congress begins with a brief return to the 1970s. During the final weeks of the Nixon administration, when the president was less than a month away from resigning from office, he signed into law the Budget Control and Impoundment Act of 1974. It was partially the result of battles between Congress and President Nixon in his second term but was also a reform that had been in the works for some time. Congress centralized its intrabranch budgetary management and created tools to discipline itself on fiscal matters, such as standing Budget Committees in each chamber and the nonpartisan expertise of the Congressional Budget Office (CBO).[17]

A new budget regime began, however, with the inauguration of Ronald Reagan in 1981, which brought to power a cohort of reactionary conservatives who upended the political competition. David Stockman, the director of the Office of Management and Budget (OMB) in the Reagan administration, worked closely with Senate Republicans to convert budget procedures from the 1974 Act into an offensive weapon that required committees to reduce spending in their jurisdiction, including for mandatory or entitlement spending, compile it in a single, omnibus bill, and fast track bill passage through Congress. By his own admission, Stockman's goals were nothing less than revolutionary and a direct assault on the constitutional system. He proclaimed it a "stunningly radical theory of governance" by which the "constitutional prerogatives of the legislative branch would have to be, in effect, suspended" so that Reagan's policy goals could be achieved by "rubber stamp approval, nothing less." He bragged that the "world's so-called greatest deliberative body would have to be reduced to the status of a ministerial arm of the White House."[18] Stockman embodied the ideological nature of the new budget regime, where the "purpose was to gain power."[19] If budget discipline had appeared to be innocuous good government reform in 1974, the new offensive use of it eliminated that appearance quickly.

In 1985 Republicans in Congress, allied with conservative Democrats, went even further by enacting the Gramm-Rudman-Hollings balanced budget legislation. The new law focused almost solely on deficits via reductions in spending. The core policy intended to block Congress from changes that would increase the deficit, requiring spending cuts (or revenue increases) to offset any new spending.[20] It was followed by the Budget Enforcement Act of 1990, which went further still and strengthened the requirement that any new spending be offset by cuts in existing spending or other offsets (e.g., tax increases).[21] In 1997 Congress went further in the Balanced Budget Act. Among a myriad of specific program spending cuts, it sharpened the tools and weapons used to enforce budget discipline into the future through items like the Congressional Budget Office baseline provisions, pay-as-you-go requirements, and other controls on spending.[22]

Without pursuing a full review of budget law development here, the key takeaway for this discussion is that application of budget discipline became entirely one-sided. It now applies almost exclusively to federal spending and not revenues. This point was most recently reinforced in the Budget Control Act of 2011, which implemented spending caps across the federal government for ten years, enforced by a process known as sequestration that makes indiscriminate cuts in all nonexempt programs if those caps are exceeded.[23] The one thing that the budget regime has not accomplished is its central reason for existence: budget discipline has completely failed to control the budget, decrease deficits, or reduce the national debt.[24] The budget regime's failure to control federal budgets stands in stark comparison to what it has wrought in the legislative process and for the theory of congressional design. The consequences have been substantial.

Budget reform was a process revolution in government, a strategic siege on the legislative power and the methods for achieving success. It was undertaken by ideological warriors armed with questionable economic theories. The vast consequences can easily be lost to obscure procedural matters and the esoteric details of federal budgeting, but rule changes "often work to shift power from one group of legislators to others."[25] Budget discipline, however, was not a simple rule change because the power of the purse is one of the most consequential powers of the government. The budget "goes to the roots of governmental will" and impacts "all the other commitments government makes and would like to keep." The budget "brings to a head questions about what kind of government we will have and, therefore, what kind of a people we will be," and control over the federal budget is a "synonym for power."[26]

Critical to the working theory on Congress, the new budget regime altered the competition by adding a procedural superstructure that magnified existing asymmetries and further limited competition in the process.[27] It makes building and maintaining coalitions much more difficult, with higher political costs. For one, it created "a strong prejudice on behalf of existing allocations" in existing programs, so long as Congress did not revise them.[28] In that way, the budget rules and enforcement mechanisms provided further protection for the status quo by creating a zero-sum dynamic in policymaking. Under mechanisms requiring offsets and adherence to spending baselines, any new program or any increases in existing programs require that other policies or programs must decrease. In effect, a new policy or program is blocked unless and until existing policies or programs are cut or eliminated to make room for it.

The annual Congressional Budget Office (CBO) baseline, which estimates spending over ten years into the unknown future, becomes the measure by which factional interests evaluate their success or failure in the competition to capture benefits. Any policy proposal that changes the allocation is perceived as a factional loss or another faction's gain. Any policy proposal that CBO estimates would alter the division of benefits is perceived as taking from one or more factional interests. Changes under these circumstances face substantial resistance in the process, beginning at the committee level.[29]

The budget regime can leave policymakers with limited options and little room to maneuver, negotiate, and compromise. It has curtailed entrepreneurialism in the policy and political arena, stifling creativity and adaptability to the dynamics of an ever-changing nation and reality. It causes the self-government system to lose the dynamism it needs to survive. By locking existing coalitions in place, the new budget regime blocks expansion or addition; it precludes the formation of new coalitions and severely restricts any factional interests that are not part of the status quo. It also limits innovation, adaptation, adjustment, and growth; it permits only narrow paths for change.

In 2000 Elizabeth Garrett warned that the "structure of congressional budgeting leads almost inevitably to stalemate" because budget decisions "implicate fundamental ideological divisions between the political parties" and influence negotiations, such as whether the "government [should] be larger or smaller" or whether it should "raise taxes" and on whom.[30] The governed—the people who vote and expect or demand things from those they elect—are the victims but have little understanding of the reasons why; they have fewer and fewer places to turn. This fuels extremism that can escalate into a recipe for some of the worst impulses of the people and the biggest fears of the framers.[31] The result is policy stasis and stagnation, a system that becomes sclerotic, unable to adapt. Or, worse, it degenerates into self-destruction.

In 1988, three years after Congress and the Reagan administration pushed through the Gramm-Rudman-Hollings budget enforcement law, Aaron Wildavsky predicted that the new budget regime would lead to an "increase in partisanship" and minimize "Congress' capacity to cope with conflict." He added that it "transforms inaction into a form of action," causing a game of "budgetary chicken" that "does not bode well for the ability to govern."[32] Polarization increases as the scope of policymaking narrows in a zero-sum game in which "few programs are considered solely on their substantive or political merits" but instead on the "degree [to which] programs contribute to the deficit"—or simply what they cost rather than what they achieve or for whom they achieve it.[33] The concept of policy in the public interest gets sacrificed to battles over the budget. The purpose of the policy and the people on the other end of policy decisions become lost to the process.

As matters degenerate further, they become "heavily minus-sum—each side ends up feeling like losers—there is no way to do much without angering enough interests to block a proposal," and factional competition can spiral into mutual destruction. In a minus-sum world all outcomes are negative; factions compete less on gain and more on hurting other factions. From there, the divisions widen and deepen; extremism is rewarded and increases. The system of self-government turns against itself when "political factions are more eager to hurt each other than to help themselves."[34] Under those circumstances, there is little opportunity to achieve an amalgamation of factions, or a working political coalition, that approximates a governing majority intended to represent the public interest. Significant aspects of these outcomes were not an accident; they were a feature, not a bug.

Budget reform was an inherently conservative ideological position that gained political ground with increased entitlement spending. Balancing the budget became "a conservative position used by lower spending Republicans as a weapon against the higher-spending liberal Democrats," and that "conflicted with the widespread desire that the government ensure a healthy economy."[35] The government's responsibility for the economy was, of course, the core of the Keynesian view that held sway for most of the era that began with the New Deal. More important than any clash of theories, the new budget regime altered the competition so that "opponents of domestic spending would have powerful institutional advantages over supporters" of domestic spending, and the burdens would "fall disproportionately on Democratic programs and constituencies" given the reality of "the greater preponderance of social welfare" spending in the budget.[36]

Budget discipline provided an ideological trap for economic policies that provide for the distribution of resources for the benefit of the larger public interest. The entire effort was aimed at public policy as it had built from Roosevelt to Johnson. It aligned with academic and economic theories being adapted to explain politics based on very negative views about the distortive effect of collective action. Coalition-building was not viewed as deliberation in the public interest but rather a problem of factional interests being bought off in the legislative process known by the pejorative term "logrolling." The origins of these theories notably coincided with reactionary opposition to school desegregation and voting rights.[37] The new budget regime was a triumph for the conservative faction, empowering it with procedural weapons that produced real consequences for public policy. It created a process-based factional trump card on all policy decisions and completely distorted the competition among various interests in the system.[38]

The new budget regime did not remedy any of the legacy problems with the designs and instead provided factional interests with more tools and weapons. The obscure and esoteric features of the budget regime, coupled with its privileged position in the legislative process, have helped obscure the deployment of these methods and decreased the cost to, or expenditure of, political power. In turn, this empowers and entrenches existing factional interests, which precludes new competitors or competing factional interests.

Under the new budget regime, the structural designs that empowered factions before, empower them further. The smallest and most organized factions gain additional advantages and increased power. Entrenched factional interests are gifted defenses that are sturdier and stronger; the barriers to new policies, reform, or change are steeper and more treacherous. The baseline of achieving nothing has become more prevalent and far likelier. With the budget, the entire system is more than just counterproductive when the public interest requires legislation; it is dysfunctional and destructive.

A primary method by which small factional interests can achieve narrow goals in legislation is through outsized negotiating leverage. A focus on the budget emphasizes the bottom line or the baseline under which large benefits to small numbers of beneficiaries cost less than small benefits to large numbers of beneficiaries, without regard to (or contrary to) the public interest. Budget discipline further stacks the process against the public interest. The focus on spending empowers the smallest, best organized, and most entrenched factional interests to capture the largest benefits and protect their narrow interests.

Large benefits for small interests are relatively cheap at the level of topline budget estimates, but the ability of small interests to block progress or kill legislation can also cost the public interest. The costs of rejecting even socially harmful legislation become too high, and the public interest is too easily held hostage. Thus, the narrowest factional interest must be appeased rather than rejected in the competition. Rather than depleting political power, success by the faction tends only to strengthen its power. Furthermore, the budget regime limits entrants to the competition by restricting options for coalition-building. Examples abound in the farm bill, beginning with farm program payments. The smallest factions within the coalition capture the largest benefits per member.

Farm bill history since 1981 has been defined by recurring battles on ground defined by the budget and those partisans eager to use it. The 1981 farm bill reauthorization was among Director Stockman's targets. His strategy was to use the budget to destroy the coalition in order to cut or even eliminate all programs and spending, which he deemed wasteful intrusions on the private sector.[39] His strategy backfired due in part to his eagerness to brag about it to the press. It backfired in 1985, when the farm bill coalition was strengthened by the fight and added environmental interests through conservation policy. The fallout from Stockman's farm bill strategy carried through to 1990 but did little to slow the rapid advancement of the new budget regime. Since the high-water mark in 1990, Republicans took control of the House of Representatives for the first time in forty years and turned to budget procedures to force massive policy changes and spending reductions that President Bill Clinton vetoed. Farm bill reauthorization was ensnared initially but subsequently authorized the biggest changes in farm policy since 1933. After Republicans recaptured the majority in the House in the 2010 midterm elections, they turned to budget procedures to fight against President Barack Obama and Democrats in the Senate. The farm bill was again consumed by partisan fights, which eventually produced cuts and changes to farm policy in the Agricultural Act of 2014.

The most prominent and damaging example in the farm bill comes from partisans using the budget and focusing on the largest spending item, the Supplemental Nutrition Assistance Program (SNAP, formerly Food Stamps). The sheer size of its outlays makes SNAP the biggest target, without regard to the fact that such a program provides relatively smaller benefits to a far larger (and needier) constituency. Each of the 2014 and 2018 Farm Bill reauthorization debates featured partisan attacks on SNAP, and each suffered an initial defeat on the House floor. Partisan fighting over the budget also obscured hypocritical policy positions, such as seeking to kick families and schoolchildren out of SNAP while expanding loopholes in farm payment limitations that would allow some farms to add family members to increase payments.[40]

The consequences for conservation policy are more difficult to discern, requiring counterfactuals and what-if questions. For most of the post-1985 era, conservation has been popular and generally supported on a bipartisan basis, across the states and regions. Conservation policy has developed and advanced new programs, authorities, and policies. While conservation policies and programs have expanded, conservation interests have splintered and formed factions. Conservation and the environment are massive in scope with a broad-based interest over natural and public goods. Smaller factional interests form within the movement, focused on specific and narrower policies. Such factions exert power mostly on protecting existing turf and funding rather than on expanding the reach of conservation or the coalition in support of it. Rather than developing new policies or programs, these interests also begin to enclose around maintaining the status quo; at most, they seek expansion only in terms of additional funds for the faction's preferred policy. Any new competitor or new conservation policy requires offsets from existing programs, and factional interests supporting any of the existing programs would see this as a threat. Moreover, any efforts to increase spending on conservation policies or programs generally would require a politically difficult reduction in spending from other programs or priorities of other interests. This is the zero-sum game created by the new budget regime, by which members of a coalition no longer work together or negotiate mutually beneficial outcomes but must fight each other for limited resources or avoid any changes that would risk such a fight.

Since 1985, conservation has attained more money and more programs but remains a distant second to farm support. Funding increases have taken place slowly and incrementally. Conservation policy has also been purposefully limited by Congress. The modern CRP has been subject to an acreage cap that Congress has reduced multiple times to achieve savings or appease opponents of the program. The main working lands program, EQIP, has operated under a fixed amount of funding that has steadily increased since 1996 but remains limited by Congress, with consequences. In fiscal year 2020 USDA was able to fund only 27 percent of eligible program applications that it received, a pattern in which tens of thousands of farmers seeking assistance with conservation are turned away from the programs.[41]

The Conservation Security Program (CSP) was cut by appropriators in a method very reminiscent of Whitten's sabotage of the Soil Bank. Importantly, these features of conservation programs are unique to conservation policy. By comparison, neither farm payment programs nor crop insurance have acreage or funding caps applied. They are open-ended; while good years may reduce the amount spent, these programs respond with unlimited funding in bad years. Conservation programs do not have, nor have they ever had, this kind of open-ended and unlimited policy design.

The budget regime has substantially modified the basic operation of the congressional system, and in doing so, it provides the final piece of the working theory on Congress and completes an answer to the conservation question. The Madisonian theory created an approximation of the public interest through a competition to form coalitions of factional interests. The concept of factional competition produces that approximation through mutually beneficial outcomes; those that joined had won the competition, helping the coalition win in the process. The new budget regime has changed the rules of the competition and drastically complicated building and maintaining a coalition. It has magnified and exacerbated existing asymmetries in the system. It has further empowered the smallest factions at the expense of the general welfare and the public interest. The budget further silences the mild voice of reason but does little to calm the clamors of impatient avidity.

Climate Change and Applying the Lessons of the Past: A Closing Argument

Conservation policy in farm bills offers tempting connections to climate change or other environmental challenges, with the Dust Bowl a tantalizing historical analogue. Contemplating those connections also risks despair; the history and development of farm bill conservation policy offers rather tough lessons. The massive dust storms of the 1930s may have grabbed national attention, but they faltered against factional powers in Congress and the Supreme Court. Environmental catastrophe was not able to surpass economic crisis as the legislative priority. The conservation question and the working theory's answers to it present extraordinary challenges to policies that seek to address climate change.

Climate change presents massive, widespread challenges that operate on temporal and global scales that do not align with political boundaries or timetables. No matter how compelling the public interests are in addressing it, the consequences are likely to continue unfolding too slowly and relatively imperceptibly (at least until it is too late). The consequences are also likely to be too diffuse, global, and general, while entrenched interests that would have to sacrifice are wealthy, concentrated, and powerful. Adding to the problems, the issue has increasingly become partisan and consumed by the metastasizing polarization in the United States. Staring into this abyss risks being swallowed by it, however. History counsels caution and the realistic setting of expectations, not surrender. The smallest factions have always possessed advantages from the asymmetries in the system, but they haven't always won; no citadel is indestructible or impregnable. Advocates for addressing climate change have mountains of challenges in front of them, but much work, creativity, and innovation is rising in response.

What is most clear from the lessons of conservation policy and the working theory on Congress is that the key to addressing climate change in the legislative process is healthy, effective competition. Climate change is a collective action problem, and experience teaches that interests on behalf of such problems rarely form or will struggle to operate and compete effectively. These challenges are never more obvious than when they are competing against concentrated, smaller, and better-organized factional interests. From Madison and Hamilton to Olson and Hovenkamp, the design of Congress and the legislative process are stacked in favor of the status quo, with substantive asymmetries that empower factional interests.

Conservation proves the need for active, powerful factions that can compete in the system and represent interests at the negotiating table. For climate change, strong factional competitors are needed for which climate is the top priority and not a bargaining chip for other priorities. It also teaches that even with such a faction, the effort will be difficult and take a long time, requiring substantial effort. Competing interests are also likely to need ideal political opportunities that open existing coalitions and complicate the power dynamics for entrenched factions. Achieving success in Congress amid the circumstances of politics constitutes only one level. Policies and programs also need to be implemented and operated effectively. Looming on the horizon is an increasingly conservative ideological and activist Supreme Court willing to impose its views by judicial fiat.[42]

In Congress, a problem-solving approach would prioritize addressing the systemic challenges the working theory identifies. Climate change interests might find allies and coalitions with other reform interests seeking to improve the competition in Congress, adjusting the rules of the legislative game to open the process for innovation, adjustment, reform, and change. Setting aside revolutionary amendments to the Constitution as unrealistic, conventional ideas, such as reforming or eliminating the filibuster and cloture requirements in the Senate, are important starting points.[43] Other reforms that seek to improve competition throughout the political system and government will also benefit climate change policy efforts, as well as more generally improve legislation in the public interest. These include expanded voting rights and more competitive congressional districts at the state and federal levels. More specifically to climate change policy, mechanisms that can be built into or added to existing programs that work to shift competition toward ecosystem benefits or services are also important.

Confronting factional citadels in the legislative process is also needed to open competition. To the extent that southerners have rebuilt a citadel in the Republican caucuses, they have much less hegemony over policy. Southern farm interests occupy political ground that is not as favorable or accepting, given the modern Republican Party's antipathy to federal policies that operate in the market or economy and that issue payments from the federal treasury. The southern farm faction alone is unlikely to present too much of an obstacle to climate change policies, but they have a coalition from which to seek refuge and political strength. They benefit mightily from being able to hide behind other factional interests, such as the fossil fuels industry fighting climate and environmental policy or fiscal conservatives' obsession with low-income food assistance. These are reminders that coalition-building can work in multiple ways.

As the working theory demonstrates, a priority area must be the budget regime and how it has altered legislation and policymaking. This will also help improve competition in the legislative arena. Eliminating it outright is unlikely in the extreme, but it cannot be left in place as currently constituted. It should be subject to an intensive reform effort built for the greater good and general public interest, rather than continue as a partisan vehicle for narrow attacks on spending. Where the budget regime can be altered or adapted through creative revisions, it should be revised. These could include changes to how policies are scored and costs calculated, potentially implementing scores that represent public interests, noneconomic priorities, and other more dynamic mechanisms.[44] Where it cannot be altered or revised, any climate change policy interest would be advised to seek workarounds and other methods with a focus on those that can be deployed to help rather than hinder policy innovation.

The 117th Congress provided a powerful recent example, the implications of which have yet to fully reverberate through the political system. On August 16, 2022, President Joe Biden signed into law the Inflation Reduction Act of 2022. Democrats in Congress passed the bill through the complicated budget procedures that have long been used to force spending reductions. In a near-reverse of the Reagan-Stockman weaponization of budget procedures that began in 1981, the Inflation Reduction Act featured Democrats using narrow majorities in both chambers to increase spending by hundreds of billions for policies to combat climate change.[45] It was a demonstration that creative use of the new budget regime can work around budget discipline and unified Republican opposition.[46]

Critical for conservation policy and the farm bill, the Inflation Reduction Act contained an unprecedented investment in conservation programs. Congress added over $18 billion through a multiyear appropriation to the Environmental Quality Incentives Program (EQIP), the Conservation Stewardship Program (CSP), the Regional Conservation Partnership Program (RCPP) created in the 2014 Farm Bill, and the Agricultural Conservation Easement Program (ACEP). Congress also directed that the additional funding provided to conservation programs in the Inflation Reduction Act be available only for "agricultural conservation practices . . . that the Secretary determines directly improve soil carbon, reduce nitrogen losses, or reduce, capture, avoid, or sequester carbon dioxide, methane, or nitrous oxide emissions, associated with agricultural production."[47] Potentially more consequential, Congress extended the statutory authorizations for conservation programs through 2031. This places conservation programs on a different expiration schedule than farm programs and alters the political and negotiating dynamics.

Throughout the history explored in this book, Congress had never invested specifically in farm bill conservation programs outside of a farm bill authorization process. Congress had also not made such investments without also providing funding—and often much more funding—to farm programs and payments. The additional funds could alter baseline budget constraints and help initiate a real step toward major changes in farm policy going forward. The potential for driving a change in direction is significant and substantial, but much depends on the next Congress and those after it. Certainly, the reforms to the competition in the legislative process are still necessary to protect the gains and push them further, as are the needs for more competitive interests and coalition-building around these items in the public interest. Today's achievements can be easily reversed in tomorrow's Congresses, but building on them will take significant work over an extended period of time. The Inflation Reduction Act presents the potential for an inflection point in agricultural conservation policy, but history has taught that potential can dissipate all too quickly. The competition ahead will be fierce.

The Inflation Reduction Act might well provide the strongest closing argument. It also presents an exemplar of the single greatest strength and most important aspect of a system of self-government: the capacity for self-correction. Effective political competition can drive innovation, adjustment, and adaptation. The ambitions of factional interests mix with the need for functional governing coalitions to produce an alchemy of sorts, by which narrow self-interests can be transformed into outcomes for the public interest and society's general welfare.[48] This theoretical ideal has proven extraordinarily difficult and rare in practice. While there is great risk that the system as designed works to reduce competition or restrict it, and usually by powerful factions, the ability to self-correct remains. So long as it does, that capacity is key, and every effort expended to preserve it is valuable.

It is possible that early disasters and catastrophes galvanize the public interest and drive changes before society crosses the point of no return. It is possible that technological advances and great leaps in innovation will provide solutions that are not within our imaginations today. Just prior to the dawning of the modern environmental movement, for example, a historian wrote critically that "surely no creature other than man has ever managed to foul its nest in such short order."[49] Silent springs, oil spills, and burning rivers galvanized the nation and spurred the farthest-reaching action to clean up the environment and protect natural resources. So potentially powerful and politically lucrative was this movement that Richard Nixon sought to harvest its votes. The political competition that unfolded resulted in landmark environmental legislation for clean air, clean water, and environmental protection; the most far reaching of such policies ever achieved in Congress, and they cleared Madison's double majority by wide margins.

Between soil and society there are ample opportunities for a self-governing society to achieve change in its relationship to the natural resources upon which it depends for its success and survival. In the final analysis, the words of Aldo Leopold continue to echo loudly. "A land ethic," he wrote, "reflects the existence of an ecological conscience, and this in turn reflects a conviction of individual responsibility for the health of the land." We have an individual as well as a collective responsibility for the health of the land but also of society. Most importantly, he defined health as the capacity for self-renewal. For society, self-renewal is ultimately the responsibility of the governed. For both soil and society, conservation "is our effort to understand and preserve this capacity."[50] We as society have accomplished much; failure to achieve against the challenges presented by climate change guarantees that much more than conservation policy will be written in dust.[51]

NOTES

A note on congressional source materials: for committee hearing records, numbering protocols or reference numbers were added relatively recently. Beginning with the 99th Congress, House and Senate hearings have reference numbers (e.g., House serial numbers or S. Hrg.). For ease of reference and better organization of the notes, I have applied the modern numbering system to the hearing records prior to the 99th Congress, using the chamber (H. Hrg. or S. Hrg.), Congress and year (H. Hrg. 74-1935 or S. Hrg. 74-1935), and additional summary information after a comma (e.g., H. Hrg. 83-1953, "Long Range, part 8"). The notes also contain short citations for all legislation, statutes, reports, and other congressional documents. Full citations for all congressional source materials are contained in a separate section of the bibliography, arranged numerically within each category. The only exceptions are citations to the Congressional Record, which are included in the notes and not the bibliography. For all congressional primary sources I have used the *ProQuest Congressional* database, accessed online via the University of Illinois Library. See *ProQuest Congressional* (Bethesda MD: Congressional Information Service, 1997), https://congressional.proquest.com.

1. Of Farming

1. Hopkins, "Bread from Stones." My thanks to Dr. Andrew Margenot, assistant professor in crop sciences, University of Illinois at Urbana-Champaign, for introducing to me the phrase and the philosophy behind it and for sharing the article.
2. Güldner, Larsen, and Cunfer, "Soil Fertility Transitions"; Montgomery, *Dirt*, 236; Coulborn, *Origin*, 11–12; Sampson, *Farmland or Wasteland*, 1; Worster, "Transformations," 1087–106, esp. 1095–96; Janzen, "Soil Remembers," 1431; Swidler, "Social Production of Soil," 7; Cunfer, "Introduction to Soils and Sustainability," 617–18; Helms, "Soil and Southern History," 723. See generally Diamond, *Collapse*; Cunfer, "Soil Fertility," 733–34; Dotterweich, "History of Human-Induced Soil Erosion"; Weiss and Bradley, "What Drives Societal

Collapse?"; Scholes and Scholes, "Dust Unto Dust"; Bardgett, *Earth Matters*; McNeill and Winiwarter, "Breaking the Sod"; Minami, "Soil and Humanity"; Bergh et al., "Time in a Bottle"; Holleman, *Dust Bowls of Empire*; Frossard, Blum, and Warkentin, *Function of Soils*; Yaalon and Arnold, "Attitudes Toward Soils"; Amundson et al., "Soil and Human Security in the 21st Century."

3. See U.S. Department of Agriculture, Natural Resources Conservation Service (hereafter USDA, NRCS), "Soil Erosion—About the Data."

4. According to the most recent report, nearly 82 percent of cultivated cropland in the United States was experiencing water erosion at T or less, but the remaining 18 percent constituted over fifty-seven million acres. See, e.g., USDA, NRCS, *Summary Report: 2017 National Resources Inventory*. See also Bennett, *Soil Conservation*, 94–95; Morgan, *Soil Erosion and Conservation*, 12, 27–29, 35–38, 40–41, 81; Sampson, *Farmland or Wasteland*, 110–32 (chapter 6); Thaler et al., "Rates"; Brevik, Homburg, and Sandor, "Soils, Climate, and Ancient Civilizations."

5. See Cunfer and Krausmann, "Sustaining Soil Fertility," 42–43; Urban, "An Uninhabited Waste"; McCorvie and Lant, "Drainage District Formation"; Pollans, "Drinking Water"; Coppess, "Return to the Crossroads"; Stuart et al., "Need for a Coupled Human"; Vos, "Agricultural Drainage"; Crawford, "Nutrient Pollution"; Berardo, Turner, and Rice, "Systemic Coordination."

6. See, e.g., McGuire et al., "Farmer Identities"; Bennett, *Soil Conservation*, 5.

7. Leopold, "Land Ethic," 208.

8. See, e.g., Bennett, *Soil Conservation*. See also Sampson, *Farmland or Wasteland*; Dotterweich, "History of Human-Induced Soil Erosion"; Nelson, "To Hold the Land"; Milde, "Roman Contributions"; Held and Clawson, *Soil Conservation*, 3–5, 12–13, 15–16, 25–27, 29; Morgan, *Soil Erosion*, 162.

9. See, e.g., Hansen and Libecap, "Small Farms"; Deaton and Hoehn, "Social Construction"; Heller and Starrett, "Externalities," 9–27; Ruhl, "Farms"; Angelo, "Corn"; Grossman, "Agriculture"; Schutz, "Agricultural Discharges"; Chen, "After Agrarian Virtue"; Finney, "Agricultural Law"; Ruhl, "Three Questions"; Chen, "Get Green"; Coppess, "Perspective"; Carolan, "Do You See"; Gao and Arbuckle, "Examining"; Eubanks, "Rotten System"; Eubanks, "Future"; Eubanks, "Sustainable Farm Bill"; Angelo and Morris, "Maintaining"; Morath, "Farm Bill"; Grossman, "Good Agricultural Practice"; Smith, "Corn, Cows, and Climate Change"; Frump, "Up to Our Ears"; Breggin and Myers, "Subsidies"; Williams, "Soil Conservation."

10. It is not intended to be a complete history of environmental policy or even conservation. See, e.g., Taylor, *Conservation Movement*; Bates, "Fulfilling"; McConnell, "Conservation Movement"; Pinchot, "How Conservation Began."

11. Public Laws (hereafter P.L.) 115-334.
12. All spending data is compiled by the author from reports by the Congressional Budget Office (hereafter CBO) or USDA (where indicated). See CBO, "Baseline."
13. Data from USDA. See U.S. Department of Agriculture, Economic Research Service (hereafter USDA, ERS), "Major Land Uses"; USDA, Risk Management Agency, "Summary of Business Reports."
14. See Subedi, Giri, and McDonald, "Commercial Farms"; McFadden and Hoppe, "Evolving Distribution"; Banker, MacDonald, and Hoppe, "Growing Farm." See, e.g., Anne Schechinger, "Under Trump, Farm Subsidies Soared and the Rich Got Richer; Biden and Congress Must Reform a Wasteful and Unfair System," Environmental Working Group, February 24, 2021, https://www.ewg.org/interactive-maps/2021-farm-subsidies-ballooned-under-trump/; Environmental Working Group, Farm Subsidy Database, https://farm.ewg.org/. See also Smith, Glauber, and Goodwin, *Agricultural Policy in Disarray*; Smith, *Economic Welfare*.
15. USDA, NRCS, "Environmental Quality Incentives Program"; USDA, NRCS, "Conservation Stewardship Program"; USDA, NRCS, "Easements." CRP is operated by the Farm Service Agency, which generally administers the farm subsidy programs. See U.S. Department of Agriculture, Farm Service Agency (hereafter USDA, FSA), "Conservation Reserve Program."
16. Coppess, *Fault Lines*.
17. See Katznelson and Lapinski, "At the Crossroads"; Katznelson, Geiger, and Kryder, "Limiting Liberalism." See also Orren and Skowronek, *The Search*; Lapinski, *Substance of Representation*; Katznelson, *Fear Itself*; Katznelson, *Affirmative Action*; Skocpol, *Protecting Soldiers and Mothers*; Bensel, *Sectionalism*.
18. See Congressional Research Service Report (hereafter CRS Report) IF12024, "Farm Bill Primer"; Scott Faber and Jared Hayes, "Growing Farm Conservation Backlog Shows Need for Congress to Spend Smarter," Environmental Working Group, News & Insights, August 18, 2021, https://www.ewg.org/news-insights/news/2021/08/growing-farm-conservation-backlog-shows-need-congress-spend-smarter.
19. See Katznelson and Lapinski, "At the Crossroads"; Katznelson, Geiger, and Kryder, "Limiting Liberalism." See also Orren and Skowronek, *The Search*; Lapinski, *Substance of Representation*; Katznelson, *Fear Itself*; Katznelson, *Affirmative Action*; Skocpol, *Protecting Soldiers and Mothers*; Bensel, *Sectionalism*.
20. See, e.g., Bradshaw et al., "Underestimating"; Ben Ehrenreich, "We're Hurtling Toward Global Suicide," *New Republic*, March 18, 2021, https://newrepublic.com/article/161575/climate-change-effects-hurtling-toward-global-suicide. See also Intergovernmental Panel on Climate Change, "Climate Change 2022."

21. See, e.g., Hudson, "World View," 338; Wilkinson, "Soil Conservationists," 310. See also Rose, "Joseph Sax"; Sax, "Public Trust Doctrine"; Sax, "Liberating."
22. H. Ex. Doc. 37-78, 1863, at 5–6. I am indebted to Secretary of Agriculture Tom Vilsack for this bit of wisdom from history. He brought it to my attention during a briefing as part of the transition in 2021, and he mentioned it in his confirmation hearing. It has stuck with me as I wrote this book.

2. Of Congress: Preliminary Discussion

1. See, e.g., Tomer et al., "Decade"; Tomer and Locke, "Challenge"; Wu et al., "Microlevel Decisions"; Feng et al., "Environmental Conservation"; Prokopy et al., "Adoption"; Hellerstein, Higgins, and Roberts, "Options"; Ribaudo et al., "Conservation Programs"; Ribaudo, "Limits"; Anderson, "Conservation"; Smith et al., "Assessing"; Gleason et al., "USDA Conservation"; U.S. Government Accountability Office/General Accounting Office (hereafter GAO), "Agricultural Conservation: USDA Needs"; GAO, "Agricultural Conservation: USDA's Environmental"; GAO, "Farm Programs: USDA Should Take"; Environmental Working Group, "Fooling Ourselves"; Environmental Working Group, "Here Today, Gone Tomorrow"; Leah Douglas, "The USDA Wants to Make Farms Climate-Friendly. Will It Work?" *Mother Jones*, July 19, 2021, https://www.motherjones.com/environment/2021/07/farmers-climate-change-conservation-reserve-program/; Virginia Gewin, "Why Aren't USDA Conservation Programs Paying Farmers More to Improve Their Soil?" *Civil Eats*, January 12, 2021, https://civileats.com/2021/01/12/why-arent-usda-conservation-programs-paying-farmers-more-to-improve-their-soil/.
2. See, e.g., Eskridge, Frickey, and Garrett, *Legislation and Statutory Interpretation*, 86 (quoting Earl Latham, *The Group Basis of Politics: A Study in Basing-Point Legislation* [New York: Octagon Books, 1952], 36) and 87 (quoting E. E. Schattschneider, *The Semisovereign People: A Realist's View of Democracy in America* [New York: Holt, Rinehart and Winston, 1960], 34–25); Hovenkamp, "Appraising," 1097; Eskridge and Ferejohn, "Article I," 530–33.
3. Madison, *Federalist*, no. 62 (raising "the propensity of all single and numerous assemblies to yield to the impulse of sudden and violent passions, and to be seduced by factious leaders into intemperate and pernicious resolutions").
4. U.S. Constitution, art. I, sec. 1, https://www.archives.gov/founding-docs/constitution-transcript.
5. See, e.g., Marbury v. Madison, 5 U.S. 137, 177 (1803); McCulloch v. Maryland, 17 U.S. 316, 426 (1819); Gibbons v. Ogden, 22 U.S. 1, 187 (1824).
6. See, e.g., Feldman, *Three Lives*; Chernow, *Alexander Hamilton*; Dahl, *How Democratic*, 4–5.

7. See, e.g., Nourse, "Toward," 499–505.
8. See Hamilton, *Federalist*, no. 1.
9. Madison, *Federalist*, no. 10.
10. Hamilton, *Federalist*, no. 9.
11. See Shane, *Madison's Nightmare*.
12. Nourse, "Toward," 499–505.
13. Madison, *Federalist*, no. 10.
14. Madison, *Federalist*, no. 47.
15. See, e.g., Madison, *Federalist*, no. 37, 42, 48, 51, 58, 62, and 63; Hamilton, *Federalist*, no. 33. See also Vermeule, "Constitutional Law"; Eskridge, Frickey, and Garrett, *Legislation and Statutory Interpretation*, 69–116 (chapter 3, "Theories of the Legislative Process"); Waldron, *Dignity*; Waldron, *Law and Disagreement*; Bessette, *Mild Voice*; Waldron, "Legislation"; Roberts, "Are Congressional Committees Constitutional"; Levinson and Pildes, "Separation"; Whittington, "Place"; Marbury v. Madison, 5 U.S. 137 (1803); McCulloch v. Maryland, 17 U.S. 316 (1819); Gibbons v. Ogden, 22 U.S. 1 (1824).
16. Immigration and Naturalization Service v. Chadha, 462 U.S. 919, 951 (1983); Clinton v. City of New York, 524 U.S. 417 (1998).
17. U.S. Constitution, art. I, sec. 7.
18. Madison, *Federalist*, no. 51.
19. Bessette, *Mild Voice*, 57–58. See also Joanne B. Freeman, "The Violence at the Heart of Our Politics," *New York Times*, September 7, 2018, https://www.nytimes.com/2018/09/07/opinion/sunday/violence-politics-congress.html.
20. Waldron, *Dignity*, 156; Eskridge and Ferejohn, "Article I." See also Waldron, *Law and Disagreement*, 16; Bessette, *Mild Voice*, 45–46; Eskridge, Frickey, and Garrett, *Legislation and Statutory Interpretation*, 86.
21. See, e.g., Eskridge, Frickey, and Garrett, *Legislation and Statutory Interpretation*, 86 (quoting Latham, *Group Basis of Politics*) and 87 (quoting Schattschneider, *Semisovereign People*, 34–25).
22. Madison, *Federalist*, no. 58.
23. Madison, *Federalist*, no. 10.
24. Madison, *Federalist*, no. 62.
25. Madison, *Federalist*, no. 58.
26. See, e.g., Madison, *Federalist*, no. 51.
27. Madison, *Federalist*, no. 62.
28. See U.S. Constitution, art. I, sec. 5; P.L. 79-601; P.L. 91-510; Senate Rules XXV and XXVI, *Rules of the Senate*, https://www.rules.senate.gov/rules-of-the-senate; House Rules X, XI, and XIII, *Rules of the House of Representatives*, https://rules.house.gov/rules-house-representatives. See also Kravitz, "Advent"; Ford,

"Legislative"; Galloway, "Operation"; Davidson, "Advent,"; Roberts, "Are Congressional Committees Constitutional," 522; Bruhl, "Using Statutes," 397.

29. See, e.g., Gluck and Bressman, "Statutory"; CRS Report 98-242, "Committee Jurisdiction"; CRS Report 98-175, "House Committee Jurisdiction." Technically, the markup is a meeting of the committee for the transaction of business. See, e.g., Senate Rule XXVI; House Rule XI. For more on markups, see, e.g., CRS Report 98-244, "Markup in Senate Committee"; CRS Report R41083, "House Committee Markups."

30. See, e.g., Talbert and Potoski, "Setting," 864; Berry and Fowler, "Congressional Committees," 2; Shepsle and Weingast, "Committee Power," 86; Shepsle and Weingast, "When Do Rules," 217. For more on amendment process, see, e.g., CRS Report 98-853, "Amending Process," at 2; CRS Report R43424, "Considering Legislation"; CRS Report 98-995, "Amending Process."

31. See, e.g., Goodwin, "Seniority."

32. See House Rule XVIII; CRS Report 98-612, "Special Rules"; CRS Report 98-995, "Amending Process," at 15. See also Zelizer, *On Capitol Hill*; Bruhl, "Using Statutes," 400.

33. See, e.g., Senate Rule XXII and XIX; CRS Report RL30360, "Filibuster"; CRS Report RL33939, "Rise of Senate Unanimous Consent"; CRS Report RS20594, "How Unanimous Consent"; CRS Report 98-310, "Senate Unanimous." See also Jentleson, *Kill Switch*.

34. See, e.g., Shepsle and Weingast, "Committee Power," 86, 89–90.

35. While conference committees were long a practice in Congress, the 1946 Act provided statutory authority for convening conference committees to resolve differences, as well as codified some basic rules; the 1970 Act went even further. See P.L. 79-601; P.L. 91-510. See also Nourse, "Decision Theory"; Tiefer, "Reconceptualization," 232–33; Roberts, "Committees," 546–48; Shepsle and Weingast, "Committee Power."

36. See, e.g., Roberts, "Are Congressional Committees Constitutional," 547–48; Shepsle and Weingast, "Committee Power," 91; Nourse, "Decision Theory," 96.

37. See, e.g., Eskridge, Frickey, and Garrett, *Legislation and Statutory Interpretation*, 70–72.

38. I am indebted to Dr. Adam Sheingate for this analogy and a very helpful, insightful discussion about this entire chapter as I struggled to pull it together.

39. Bessette, *Mild Voice*, 27, 45–46, and 57–58. Eskridge, Frickey, and Garrett, *Legislation and Statutory Interpretation*, 86 and 99.

40. See, e.g., Grofman, "Public Choice"; Rubin, "Beyond Public Choice"; Hovenkamp, "Legislation"; Kelman, "Public Choice"; Froomkin and Shapiro, "New

Authoritarianism." See also MacLean, *Democracy in Chains*; Coppess, "High Cotton."
41. Madison, *Federalist*, no. 51.
42. Eskridge, Frickey, and Garrett, *Legislation and Statutory Interpretation*, 88–89 (citing Olson, *Logic of Collective Action*).
43. See, e.g., Bickel, *Least Dangerous*; Friedman, "Countermajoritarian, Part One"; Friedman, "Countermajoritarian, Part Two"; Friedman, "Countermajoritarian, Part Three"; Friedman, "Countermajoritarian, Part Four"; Friedman, "Countermajoritarian, Part Five."
44. See, e.g., Katznelson, Geiger, and Kryder, "Limiting Liberalism"; Farhang and Katznelson, "Southern Imposition." See also Katznelson, *Fear Itself*.
45. Congressional Record (hereafter Cong. Rec.), October 4, 2001, at 18825–26.
46. The Congressional Budget Office (CBO) estimated that $49.8 billion (68 percent) would be spent on farm commodity program payments and $12.5 billion (17 percent) would be spent on conservation programs. See CBO, "H.R. 2646" (2001); H. Rept. 107-191, part 2, at 171. CBO reported a total increase in outlays of $69.526 billion over the ten-year window (FY2002 to FY2011); of that, $49.828 was for Title I and $12.502 for Title II. If looked at just over the five-year window of the actual authorizations, the imbalance is even worse: of the $33.404 billion increase, $25.809 billion (77 percent) was for Title I and just $4.714 (14 percent) was for Title II.
47. See, e.g., CRS Report RL31195, "2002 Farm Bill." See also Damon Franz, "Agriculture: Opposition Mounts as Farm Bill Heads to Floor," *Greenwire*, May 2, 2002. See also USDA, ERS, *Data Files*, "Federal Government Direct Farm Payments." The Commodity Credit Corporation budget data report approximately $19.4 billion from 1998 to 2001. See U.S. Department of Agriculture, Commodity Credit Corporation, "CCC Budget Essentials." See also Mercier and Halbrook, *Agricultural Policy*, 262–63.
48. Cong. Rec., October 4, 2001, at 18818–20 (Kind and text of the Boehlert amendment); Cong. Rec., October 4, 2001, at 18821 (Representative Holden [D-PA]), 18824 (Representative Olver [D-MA]), and 18835 (Representative Peter DeFazio [D-OR]).
49. See Cong. Rec., October 4, 2001, at 18819 (Representative Boehlert).
50. Specifically, the amendment instructed USDA to reduce the amount to be paid to farmers through the farm payment programs by $1.9 billion each year ($19 billion over the ten-year budget window) and to shift that money into increasing the spending on conservation programs. See Cong. Rec., October 4, 2001, at 18813–18, 18820–21; Cong. Rec., October 4, 2001, at 18828–29 (Representative

Petri [R-WI]). See also *Congressional Quarterly Almanac* (hereafter CQ *Almanac*), "Farm Bill Delayed a Year."
51. See Cong. Rec., October 4, 2001, at 18820–21; Cong. Rec., October 4, 2001, at 18828–29 (Representative Petri [R-WI]).
52. See, e.g., Cong. Rec., October 4, 2001, at 18838 (Representative Mark Kennedy [R-MN]), 18841 (Representative Stenholm), and 18842 (Representative Lucas [R-OK]).
53. Cong. Rec., October 4, 2001, at 18830. See also Cong. Rec., October 4, 2001, at 18819 (Representative Boehlert's comments about Combest's threats).
54. See Cong. Rec., October 4, 2001, at 18844-45 (amendment defeated 200–226 [5 not voting]); CQ *Almanac*, "Farm Bill Delayed a Year."
55. See CBO, "H.R. 2646" (2001); 107 H.R. 2646 (reported in the House), at Sec. 100(4).
56. Cong. Rec., October 3, 2001, at 18646–67 (adding it would "totally devastate the bill" and jeopardize the "deal with American agriculture"); Cong. Rec., October 4, 2001, at 18819 (Combest) and 18831 (Representative Saxby Chambliss [R-GA]).
57. Cong. Rec., October 4, 2001, at 18829–30 (Representative John Thune [R-SD]); Cong. Rec., October 4, 2001, at 18821 (Representative Greg Ganske [R-IA]); Cong. Rec., October 4, 2001, at 18822 (Representative Collin Peterson [D-MN]); Cong. Rec., October 4, 2001, at 18831 (Representative Earl Pomeroy [D-ND]).
58. Cong. Rec., October 4, 2001, at 18841 (Representative Thune); Cong. Rec., October 4, 2001, at 18835 (Representative Berry [D-AR]); Cong. Rec., October 4, 2001, at 18835 (Representative Jerry Moran [R-KS]).
59. Cong. Rec., October 4, 2001, at 18828.
60. Cong. Rec., October 4, 2001, at 18842 (Representative Frank Lucas [R-OK]).
61. Cong. Rec., October 4, 2001, at 18835 (Moran) and 18830 (Ranking Member Stenholm). See also Cong. Rec., October 4, 2001, at 18823 (Representative Peterson).
62. Cong. Rec., October 4, 2001, at 18839 (Representative Richard Pombo [R-CA]).
63. Cong. Rec., October 4, 2001, at 18821 (Representative Ganske) and 18829–30 (Representative Thune); Cong. Rec., October 4, 2001, at 18835 (Representative Moran [R-KS]).
64. See, e.g., Cong. Rec., October 4, 2001, at 18839 (Representative Sam Graves [R-MO]); Cong. Rec., October 4, 2001, at 18826–27 (Representative Tom Osborne [R-NE]); Cong. Rec., October 4, 2001, at 18824 (Representative Gutknecht [R-MN]); Cong. Rec., October 4, 2001, at 18830 (Ranking Member Stenholm [D-TX]); Cong. Rec., October 4, 2001, at 18835 (Representative Marion Berry [D-AR]).

65. Cong. Rec., October 4, 2001, at 18823; Cong. Rec., October 4, 2001, at 18826–27 (Representative Tom Osborne [R-NE]); Cong. Rec., October 4, 2001, at 18839 (Representative Graves).
66. Cong. Rec., October 4, 2001, at 18835; Cong. Rec., October 4, 2001, at 18835 (Representative Moran [R-KS]); Cong. Rec., October 4, 2001, at 18836 (Representative Rehberg [R-MT]).
67. Cong. Rec., October 4, 2001, at 18832.
68. Cong. Rec., October 4, 2001, at 18822.
69. P.L. 101-624, at Sec. 2501. Members of the category were those "subjected to racial or ethnic prejudice because of their identity as members of a group without regard to their individual qualities." CRS Report R46727, "Defining a Socially Disadvantaged Farmer."
70. See Key, *Southern Politics*, 5.
71. See GAO, "To Protect," at 6.
72. Madison, *Federalist*, no. 62.
73. Madison, *Federalist*, no. 10.
74. *Black's Law Dictionary*, 6th ed. (St. Paul: West Group, 1990), 1176. See, e.g., Page, "Path Dependence"; Hathaway, "Path Dependence"; Bell, "Path Dependence"; Prado and Trebilcock, "Path Dependence." See also King, *Budgeting Entitlements*, 208.
75. Hamilton, *Federalist*, no. 1.
76. Olson, *The Logic of Collective Action*.
77. Hovenkamp, "Appraising."
78. See Bosso, *Pesticides and Politics*, 166.

3. Out of Dust

1. H. Hrg. 74-1935, H.R. 7054, at 73.
2. H. Hrg. 74-1935, H.R. 7054, at 17.
3. H. Hrg. 74-1935, H.R. 7054, at 73 (statement of Dr. Hugh H. Bennett, Director Soil Erosion Service, Department of the Interior).
4. Cong. Rec., April 15, 1935, at 5644–45; Cook, "Hugh Hammond Bennett"; Helms, "Hugh Hammond Bennett." See also S. Hrg. 74-1935, "Protection."
5. P.L. 74-46.
6. See, e.g., Sampson, *For Love of the Land*, 1.
7. Katznelson, *Fear Itself*, 20.
8. Gilbert, *Planning Democracy*, 12.
9. See, e.g., Egan, *Worst Hard Time*, 225–28; Morgan, *Governing Soil Conservation*, chapter 1.

10. See, e.g., Taylor, *Rise of the American Conservation Movement*; Bates, "Fulfilling"; McConnell, "Conservation Movement"; Pinchot, "How Conservation Began."
11. See, e.g., Holleman, *Dust Bowls of Empire*; Montgomery, *Dirt*; Diamond, *Collapse*; Worster, "Transformations"; Weiss and Bradley, "What Drives Societal Collapse?"; Scholes and Scholes. "Dust Unto Dust."
12. See, e.g., Bennett, *Soil Conservation*; Holleman, *Dust Bowls of Empire*, 77–93; Cunfer, "Scaling," 110–12, 117. See also Cronon, "Place for Stories"; Lee and Gill, "Multiple Causes"; Cordova and Porter, "1930s Dust Bowl"; McLeman et al., "What We Learned"; Cook, Miller, and Seager, "Amplification."
13. See, e.g., S. Hrg. 87-1961, S.J. Res. 98; Edwards, "Changing Perceptions"; Laycock, "History"; Gates, "Homesteading"; Stephens, "Why the Dust Bowl"; Shannon, "Homestead"; Alston, Harris, and Mueller, "Development," 741–42. See also Libecap and Hansen, "U.S. Land Policy," 4, 11.
14. See, generally, Libecap and Hansen, "Rain Follows"; Hargreaves, "Land-use"; Hargreaves, "Dry Farming"; Smith, "Rain Follows."
15. H. Rept. 49-848, at 2.
16. See H. Ex. Doc. 1, 53-1895, at 199–200; H. Ex. Doc. 355, 53-1895, at 25–26. See also H. Ex. Doc. 399, 56-1900; H. Ex. Doc. 655, 57-1902.
17. S. Hrg. 70-1928, S.3484, at 6, 9, and 17.
18. S. Rept. 70-1211.
19. See Cong. Rec., May 23, 1928, at 9583.
20. See Cong. Rec., June 7, 1929, at 2493 (S.1444); S. Rept. 71-1784. See also Morgan, *Governing Soil Conservation*, 5; H. Hrg. 71-1929, "Farm Relief."
21. Morgan, *Governing Soil Conservation*, 2.
22. See Cunfer, "Scaling," 117. See also Cronon, "Place for Stories."
23. Egan, *Worst Hard Time*; Hurt, *Dust Bowl*; and Worster, *Dust Bowl*; Phillips, *This Land*; Holleman, *Dust Bowls of Empire*; Cunfer, "Scaling," 108–9. See also Burns, "Dust Bowl."
24. See U.S. House of Representatives, History, Congress Profiles, 72nd Congress (1931–33), https://history.house.gov/Congressional-Overview/Profiles/72nd/; U.S. Senate, History, Party Division, https://www.senate.gov/history/partydiv.htm.
25. See Hargreaves, "Land-use"; Hurt, "National Grasslands"; Lewis, "National Grasslands"; Gardner, "Trees"; Karle and Karle, "200 Million Trees."
26. See U.S. House of Representatives, Congress Profiles, https://history.house.gov/Congressional-Overview/Profiles/74th/; U.S. Senate, Majority and Minority Leaders, https://www.senate.gov/artandhistory/history/common/briefing/Majority_Minority_Leaders.htm.

27. See Katznelson, *Affirmative Action*; Katznelson, *Fear Itself*, 194; Katznelson, Geiger, and Kryder, "Limiting Liberalism"; Farhang and Katznelson, "Southern Imposition"; Katznelson and Mulroy, "Was the South Pivotal."
28. Breimyer, "Agricultural Philosophies," 334; Breimyer, "Conceptualization." See also Gilbert and Howe, "Beyond"; Hooks, "New Deal"; Rasmussen, "'Never a Landlord'"; Biles, "Politics."
29. Morgan, *Governing Soil Conservation*, 14.
30. See Phillips, *This Land*, 107–32. See also USDA, ERS, "Land Utilization," 1, 4–6 (quote), 8–11, 40–41.
31. Gilbert, *Planning Democracy*, 81–83 (first quote) and 87 (second quote). See also Baldwin, *Poverty and Politics*; Held and Clawson, *Soil Conservation*, 64; Morgan, *Governing Soil Conservation*, 1; Leopold, "Conservation Ethic," 635; Hines, "Land Ethic." For examples of the New Deal thinking on private property, see, e.g., S. Hrg. 74-1935, "Protection," at 54, 56–57.
32. See, e.g., S. Hrg. 74-1935, S.1800, at 19 (testimony of L. C. Gray, Chief, Land Policy Section, Agricultural Adjustment Administration).
33. See Nelson, *King Cotton's Advocate*; Irons, *Lawyers*, 156–57; Conrad, *Forgotten Farmers*, 45, 116. See also Nelson, "Oscar Johnston."
34. See Irons, *Lawyers*, 156–57; Conrad, *Forgotten Farmers*, 71, 116, 185; Volanto, "AAA"; Nelson, "Art of the Possible," 421–23; Gilbert, "Eastern Urban," 179–80; Lowitt, "Henry A. Wallace"; Hoffsommer, "AAA," 497–98.
35. See Beckert, *Empire of Cotton*, xvii–xviii; Daniel, *Breaking the Land*, 3–22; Fite, *Cotton Fields*, 150; LeRay, Wilber, and Crowe, "Plantation Organization," 7, 11; Burford, "Federal Cotton Programs," 225–26; Whatley, "Labor," 908; Vance, "Human Factors," 262, 266; Daniel, "Crossroads of Change." See also Grove, "Economics," 148, 159–60.
36. Daniel, *Shadow of Slavery*, 67. See also Woodman, "Post Civil-War." See also Johnson, Embree, and Alexander, *Collapse of Cotton Tenancy*, 1.
37. See, e.g., Alston and Ferrie. "Social Control," 146; Woodman, "Post Civil-War," 322–24; Vance, "Human Factors," 266. See, generally, Billings, "Class Origins"; Bowman, "Antebellum Planters"; Hahn, "Class and State."
38. See Daniel, *Shadow of Slavery*; Conrad, *Forgotten Farmers*, 4–7; Woodman, "Post Civil-War," 324–25; Vance, "Human Factors," 266.
39. See Clyatt v. U.S., 197 U.S. 207, 215 (1905); U.S. v. Reynolds, 235 U.S. 133 (1914); Taylor v. Georgia, 315 U.S. 25 (1942); Pollock v. Williams, 322 U.S. 4 (1944). See also Daniel, *Shadow of Slavery*; Conrad, *Forgotten Farmers*, 4–7; Daniel, *Breaking the Land*, 160–61; Hoffsommer, "AAA," 494; Whatley, "Labor," 905; Daniel, "Crossroads of Change," 431; Woodman, "Post Civil-War," 322, 324–25;

Vance, "Human Factors," 266–68; Alston and Ferrie, "Paternalism," 865–66; Kirby, "Transformation," 268; Mann, "Sharecropping," 418. See also Cobb, *Most Southern Place*; Woods, *Development Arrested*; Mitchell, *Mean Things*; Grubbs, *Cry from the Cotton*; Kester, *Revolt*.

40. See, e.g., Hoffsommer, "AAA," 494; Williams, "Perpetual," 144; Alston and Ferrie, "Social Control," 134; Alston and Ferrie, "Labor Costs"; Alston and Ferrie, "Paternalism."

41. See Fite, *Cotton Fields*, 141–42, 184, 188–89; Daniel, *Breaking the Land*, 100; Whatley, "Labor," 918, 920–22, 924, 927–28; Daniel, "Crossroads of Change," 429, 440, 447; Burford, "Federal Cotton Programs," 226, 236; Kirby, "Transformation," 265–68; Day, "Economics," 441–42; Vance, "Human Factors," 265, 273; Frey and Smith, "Influence," 505; LeRay, Wilber, and Crowe, "Plantation," 330.

42. See, e.g., Holmes, "Whitecapping."

43. See Fite, *Cotton Fields*, 186; Daniel, *Breaking the Land*, 66–70; Irons, *Lawyers*, 161, 166; Vance, "Human Factors," 261; Whatley, "Labor," 907; Hoffsommer, "AAA," 496.

44. See Irons, *Lawyers*, 164–65; Nelson, "Art of the Possible," 420; Kester, *Revolt*, 55, 58–59, 63–64.

45. See Daniel, *Breaking the Land*, 94–95, 98; Conrad, *Forgotten Farmers*, 120–21, 124–33, and chapter 7; Irons, *Lawyers*, 160–62, 166–73, 177; Nelson, "Purge," 416–19, 421, 425–26, 431–33; Lowitt, "Henry A. Wallace," 611; Gilbert, "Eastern Urban"; Ernst, "Mr. Try-it," 1812. See also Saloutos, *American Farmer*, 110–14, 117–18.

46. See Irons, *Lawyers*, 178; Conrad, *Forgotten Farmers*, 188; West v. Norcross, 80 S.W.2d 67 (1935). See also 73 H.R. 9967; 74 H.J. Res. 270; Cong. Rec., February 19, 1935, at 2210 (article by Dr. William Amberson, "The New Deal for Share Croppers" entered into the record). See also Daniel, *Breaking the Land*, 66–70, 99, 103, 242; Coppess, *Fault Lines*, 44–45; Whatley, "Labor," 914; Hoffsommer, "AAA," 496; Kirby, "Transformation," 269–70.

47. See Baldwin, *Poverty and Politics*; Phillips, *This Land*, 107–32. See also USDA, ERS, "Land Utilization," 1, 4–6 (quote), 8–11, and 40–41.

48. H. Ex. Doc. 75-149, 1937, at 7. In July 1937 Congress finally enacted the Bankhead-Jones Farm Tenant Act to provide for lending and other technical assistance to farm tenants for the purchase and operation of farms. See P.L. 75-210; H. Hrg. 75-1937, H.R. 8, at 215–16 (Secretary Wallace testimony delivering the special committee's report and recommendations).

49. See, e.g., Cong. Rec., February 11, 1935, at 1782; S. Hrg. 74-1935, S.1800.

50. S. Hrg. 74-1935, S.1800, at 6–7.

51. S. Hrg. 74-1935, S.1800, at 17.
52. See H. Hrg. 75-1937, H.R. 8, at 60–62 (L. C. Gray); H. Ex. Doc. 37-78, 1863.
53. See H. Hrg. 74-1935, H.R. 7054; Cong. Rec., April 1, 1935, at 4803–8; see Cong. Rec., April 15, 1935, at 5644–45 (Senate Agriculture Committee reported the bill to the full Senate on April 11, 1935, and the Senate made only minor or technical amendments to the House bill); Cong. Rec., April 23, 1935, at 6195, 6223. See also S. Hrg. 74-1935, "Protection"; P.L. 74-46.
54. P.L. 74-46.
55. Morgan, *Governing Soil Conservation*, 23.
56. See, generally, Phillips, *This Land*, 107–32. See also USDA, ERS, "Land Utilization"; Patch, "Soil Conservation"; Morgan, *Governing Soil Conservation*, 14, 34–37; Held and Clawson, *Soil Conservation*, 29–38, 42–43; Egan, *Worst Hard Time*, 225–28. Compare, H. Rept. 74-528; S. Rept. 74-466; Cong. Rec., April 15, 1935, at 5644–45; H. Hrg. 74-1935, H.R. 7054, at 3 (Bennett testimony) and 81–82 (Rufus G. Poole, Assistant Solicitor in the Department of the Interior); S. Hrg. 74-1935, "Protection," at 7 (Bennett testimony) and 52–57 (testimony of C. W. Collier, Special Assistant to the Director of the Soil Erosion Services).
57. H. Rept. 74-528, at 2; S. Rept. 74-466, at 3.
58. P.L. 74-46, at Sec. 1(4); P.L. 74-461, at Sec. 8(b); H. Rept. 74-528, at 3; S. Rept. 74-466, at 2.
59. H. Rept. 74-528, at 2; S. Rept. 74-466, at 3.
60. See P.L. 74-46, at Sec. 1 and Sec. 5; Morgan, *Governing Soil Conservation*.
61. See P.L. 74-46, at Sec. 1, Sec. 4, and Sec. 5; Morgan, *Governing Soil Conservation*, chapter 1.
62. See Morgan, *Governing Soil Conservation*, 34–37; Held and Clawson, *Soil Conservation*, 29–38, 42–43.
63. See Rodell, *Nine Men*, 217; Leonard, *Search for a Judicial Philosophy*, 49, 62; Metzger, "1930s Redux"; Rankin, "Supreme Court," 46–47. See also Panama Refining Co. v. Ryan, 293 U.S. 388 (1935); Railroad Retirement Board v. Alton Railroad Co., 295 U.S. 330 (1935); Louisville Joint Stock Land Bank v. Radford, 295 U.S. 555 (1935); A.L.A. Schechter Poultry Corp. v. U.S., 295 U.S. 495 (May 27, 1935). The appellate decision against the 1933 AAA was handed down on July 13, 1935. See Butler v. U.S., 78 F.2d 1 (1st Cir. 1935).
64. U.S. v. Butler, 297 U.S. 1 (1936); P.L. 73-10, at Sec. 8(1).
65. See Irons, *Lawyers*, 183; Soifer, "Truisms."
66. Receiver William A. Butler was a leading Republican politician, lawyer, and businessman; he had been appointed U.S. senator for Massachusetts by President Calvin Coolidge. Receiver James A. McDonough was on the board of directors of one of the country's largest meat packers. See Irons, *Lawyers*, 182–83;

Soifer, "Truisms," 821. See also Franklin Process Co. v. Hoosac Mills Corp., 8 F. Supp. 552, 553–54 (D. Mass. 1934).

67. See Rudolph, "American Liberty League," 27; Goldstein, "American Liberty League," 300; Richman, "Matter of Degree," 148; Kyvig, "Road Not Taken"; Wolfskill, *Revolt of the Conservatives*, 138. See also Patton, "Standing," 148.

68. For example, he had convinced President Coolidge to appoint Roberts as a special prosecutor in the Teapot Dome scandal in 1924. See, e.g., *Butler*, 297 U.S. 1; Soifer, "Truisms," 821. See also Pepper, *Philadelphia Lawyer*, 197–98, 231, 244; Leonard, *Search for a Judicial Philosophy*, 7; Bates, "Teapot Dome," 312. Pepper had been appointed to the U.S. Senate on January 9, 1922 (presidency of Warren G. Harding) but was defeated in the 1926 election. See Biographical Directory of the United States Congress, "Pepper, George Wharton," https://bioguideretro.congress.gov/Home/MemberDetails?memIndex=P000219. Butler was appointed by Coolidge in November 1924 after having served as national campaign manager for Coolidge in the presidential election of 1924. Soifer, "Truisms," 821; Biographical Directory of the United States Congress, "Butler, William Morgan," https://bioguideretro.congress.gov/Home/MemberDetails?memIndex=B001196.

69. See Shughart, "Bending," 76; Leonard, *Search for a Judicial Philosophy*, 50–51; Rodell, *Nine Men*, 222–23, 236.

70. See Hoosac Mills, 8 F. Supp., at 556–57; Butler v. U.S., 78 F.2d 1, 11–12 (1st Cir. 1935; referring to *Panama Refining* and *Schechter Poultry*). See also Irons, *Lawyers*, 182; Currie, "Constitution," 531. See also U.S. Constitution, art. I, sec. 8; Hamilton, *Federalist*, no. 30–36; Veazie Bank v. Fenno, 75 U.S. 533, 540–42; Massachusetts v. Mellon, 262 U.S. 447, 486 (1923).

71. U.S. v. Butler, 297 U.S., at 53–57, 59–61, and 74; P.L. 73-10, at Sec. 8 and Sec. 9; Pusey, *Hughes*, 743–44; Soifer, "Truisms," 824; Currie, "Constitution," 533. See also Butler v. U.S., 78 F.2d 1, 7 (1st Cir. 1935).

72. See *Butler*, 297 U.S., at 74 and 76 (respectively); *Butler*, 297 U.S., at 79, 87–88 (Stone, J., dissenting). See also Glick, "Soil and the Law," 310–11; Benedict, *Farm Policies*, 348; Howard, "Supreme Court," 301; WJW, "Constitutional Law."

73. *Butler*, 297 U.S., at 75.

74. See Metzger, "1930s Redux."

75. *Butler*, 297 U.S., at 79, 80 (first quote), and 87–88 (second quote) (Stone, J., dissenting).

76. See, e.g., West Coast Hotel v. Parrish, 300 U.S. 379 (1937); Currin v. Wallace, 306 U.S. 1 (1939); Mulford v. Smith, 307 U.S. 38 (1939); Wickard v. Filburn, 310 U.S. 111 (1942).

77. See, e.g., Page, "Path Dependence"; Hathaway, "Path Dependence"; Bell, "Path Dependence"; Prado and Trebilcock. "Path Dependence." See also King, *Budgeting Entitlements*, 208.

4. Reform amid Ruin

1. See U.S. House of Representatives, "Party Divisions of the House of Representatives, 1789 to Present," https://history.house.gov/Institution/Party-Divisions/Party-Divisions/; U.S. Senate, "Party Division," https://www.senate.gov/history/partydiv.htm. For a great interactive map of Congress, the one by the Digital Scholarship Lab at the University of Richmond and the Department of History at Virginia Tech has few equals. See Winling and Nelson et al., "Electing the House of Representatives."
2. *Black's Law Dictionary*, 6th ed. (St. Paul: West Group, 1990), 1176. See, e.g., Page, "Path Dependence"; Hathaway, "Path Dependence"; Bell, "Path Dependence"; Prado and Trebilcock, "Path Dependence." See also King, *Budgeting Entitlements*, 208.
3. P.L. 74-461.
4. Cong. Rec., January 9, 1936, at 191 (Senator McNary); Cong. Rec., January 9, 1936, at 203-4 (Senator John H. Bankhead).
5. Kirkendall, *Social Scientists*, chapter 8; Kirkendall, "Howard Tolley," 29. See also Schapsmeier and Schapsmeier, "Henry A. Wallace," 134; Ezekiel, "Henry A. Wallace," 796.
6. See Associated Press, "Another Farm Meeting Called," *New York Times*, January 7, 1936, https://timesmachine.nytimes.com/timesmachine/1936/01/08/87892562.html?pageNumber=14; Turner Catledge, "Allotment Idea to Fore," *New York Times*, January 7, 1936, https://timesmachine.nytimes.com/timesmachine/1936/01/08/87892407.html?pageNumber=1; Turner Catledge, "AAA Chiefs Draft Substitute Plans," *New York Times*, January 9, 1936, https://timesmachine.nytimes.com/timesmachine/1936/01/10/88624092.html?pageNumber=1. See also Eugene Schult, "New Farm Aid Plan to Replace the AAA," *New York Times*, January 10, 1936, https://timesmachine.nytimes.com/timesmachine/1936/01/12/87895221.html?pageNumber=68. Volanto, "Ordered Liberty," 161-64.
7. "Farm Plan Hinges on New Court Test," *New York Times*, January 10, 1936, https://timesmachine.nytimes.com/timesmachine/1936/01/11/87893032.html?pageNumber=7; "Farm Leaders Unanimous for Roosevelt Soil Plan," *New York Times*, January 11, 1936, https://timesmachine.nytimes.com/timesmachine/1936/01/12/87894010.html?pageNumber=1. "Crop Insurance Included," *New York Times*, January 10, 1936, https://timesmachine.nytimes.com/timesmachine

/1936/01/11/87892954.html?pageNumber=1; Felix Belair Jr., "New AAA Bill Sets Scientific Tilling as Payment Basis," *New York Times,* January 12, 1936, https://timesmachine.nytimes.com/timesmachine/1936/01/13/87897328.html?pageNumber=1.

8. See, e.g., "Farm Leaders Unanimous," January 11, 1936; Schult, "New Farm Aid Plan."

9. See S. Hrg. 74-1936, "Substitute."

10. S. Hrg. 74-1936, "Substitute," at 8–9; "President Insists on New Farm Bill," *New York Times,* January 14, 1936, https://timesmachine.nytimes.com/timesmachine/1936/01/15/88625898.html?pageNumber=6 (reporting it as a "bombshell"); "Strife Threatens over Farm Relief," *New York Times,* January 13, 1936, https://timesmachine.nytimes.com/timesmachine/1936/01/14/88624280.html?pageNumber=11; Felix Belair Jr., "Plan to Continue Farm Aid Under Conservation Law; Soil Law of 1935 Is Key," *New York Times,* January 15, 1936, https://timesmachine.nytimes.com/timesmachine/1936/01/16/85200821.html?pageNumber=1; "A New Farm Policy," *New York Times,* January 17, 1936, https://timesmachine.nytimes.com/timesmachine/1936/01/17/93517133.html?pageNumber=18; "New AAA Plans Indorsed," *New York Times,* January 17, 1936, https://timesmachine.nytimes.com/timesmachine/1936/01/18/88626047.html?pageNumber=5.

11. See Kirkendall, "Howard Tolley"; "President Holds Soil Act Is Basis of Long Range AAA," *New York Times,* January 17, 1936, https://timesmachine.nytimes.com/timesmachine/1936/01/18/88625899.html?pageNumber=1; Frank Lynn, "Substitute for AAA Offered to Farmers," *New York Times,* January 17, 1936, https://timesmachine.nytimes.com/timesmachine/1936/01/19/85208397.html?pageNumber=77; "The New Farm Plan," *New York Times,* January 23, 1936, https://timesmachine.nytimes.com/timesmachine/1936/01/23/88627525.html?pageNumber=20.

12. See Kile, *Farm Bureau,* 228–29; Kirkendall, "Howard Tolley," 29; Schapsmeier and Schapsmeier, "Henry A. Wallace," 134; Hunter, "AAA," 554; Ezekiel, "Henry A. Wallace," 796. See also Catledge, "AAA Chiefs" (January 9, 1936). See also May, "Marvin Jones," 432.

13. 74 S.3780, *Introduced in the Senate.* See also 74 H.R. 10500, *Introduced in the House.*

14. See, e.g., 74 H.R. 10500, at Sec. 8; 74 S.3780, at Sec. 8.

15. See "Plan for State Farm Aid with Federal Payments Backed by AAA Officials," *New York Times,* January 23, 1936, https://timesmachine.nytimes.com/timesmachine/1936/01/24/87902033.html?pageNumber=1; "Agreement is Near on AAA Substitute," *New York Times,* January 24, 1936, https://times

machine.nytimes.com/timesmachine/1936/01/25/87902709.html?pageNumber=4; "Senators Dubious on New Farm Bill," *New York Times*, January 25, 1936, https://timesmachine.nytimes.com/timesmachine/1936/01/26/85215373.html?pageNumber=35; Frank Lynn, "Permanent Farm Aid Program is Drafted," *New York Times*, January 25, 1936, https://timesmachine.nytimes.com/timesmachine/1936/01/26/85215702.html?pageNumber=77; "Farm Plan to Rely on States' Rights," *New York Times*, January 27, 1936, https://timesmachine.nytimes.com/timesmachine/1936/01/28/87903230.html?pageNumber=1; "Plan for State Farm Aid with Federal Payments Backed by AAA Officials," *New York Times*, January 23, 1936, https://timesmachine.nytimes.com/timesmachine/1936/01/24/87902033.html?pageNumber=1; "States Free to Join Plan Meanwhile House Votes Adopt Amended Repealer of Cotton, Tobacco, Potato Acts," *New York Times*, February 6, 1936, https://www.proquest.com/docview/101971708.

16. See "Senate Body Nears Vote on Farm Bill," *New York Times*, January 28, 1936, https://timesmachine.nytimes.com/timesmachine/1936/01/29/88627064.html?pageNumber=4; "Senate Committee Acts on Farm Bill," *New York Times*, January 29, 1936, https://timesmachine.nytimes.com/timesmachine/1936/01/30/85221909.html?pageNumber=1; "Senators Rewrite New Farm Aid Bill," *New York Times*, February 5, 1936, https://timesmachine.nytimes.com/timesmachine/1936/02/06/88630604.html?pageNumber=4.
17. S. Rept. 74-1481; H. Rept. 74-1973.
18. See Cong. Rec., February 6, 1936, at 1564–68 (Senators McNary and Richard Russell [D-GA]).
19. See 74 S.3780, *Reported in the Senate with amendments*.
20. Cong. Rec., February 6, 1936, at 1578.
21. See Cong. Rec., February 15, 1936, at 2162–63 (McNary's motion to recommit); Cong. Rec., February 15, 1936, at 2164 (motion defeated 21–54 and bill agreed to by 56–20). See also "Senate, 56 to 20, Passes Stop-Gap Farm Aid Bill," *New York Times*, February 15, 1936, https://timesmachine.nytimes.com/timesmachine/1936/02/16/294274652.html?pageNumber=1; "Farm Bill Changes Blocked in Senate," *New York Times*, February 13, 1936, https://timesmachine.nytimes.com/timesmachine/1936/02/14/87905486.html?pageNumber=6.
22. Cong. Rec., February 6, 1936, at 1570–73, 1576.
23. See Cong. Rec., February 15, 1936, at 2149–50, 2152, 2153–54, 2156–57 (Senator George Norris [R-NE] and Senator Walter George [D-GA]), 2157–58 (Senator William Borah [R-ID]), and 2160 (Senator Joseph O'Mahoney [D-WY]).
24. 74 S.3780, *Ordered to lie on the table in the House*.
25. 74 S.3780, *Ordered to lie on the table in the House*.

26. See, e.g., Cong. Rec., February 19, 1936, at 2361–62 (debate over the rule) and 2374; Cong. Rec., February 19, 1936, at 2370–71 (Representative Francis Culkin [R-NY]); Cong. Rec., February 20, 1936, at 2465–66 (Representative Charles Tobey [R-NH]) and 2560 (Representative Culkin).
27. Cong. Rec., February 20, 1936, at 2486.
28. See, e.g., Cong. Rec., February 21, 1936, at 2563 (Representative Phillip Ferguson [D-OK]); Cong. Rec., February 21, 1936, at 2563–64 (Representative Arthur Mitchell [D-IL]); Cong. Rec., February 21, 1936, at 2564 (Representative Kleberg); Cong. Rec., February 21, 1936, at 2565 (on tellers, 111–144); Cong. Rec., February 21, 1936, at 2566–68; Cong. Rec., February 21, 1936, at 2570 (Chairman Jones amendment); Cong. Rec., February 21, 1936, at 2577–78 (Boileau motion to recommit defeated 146–225 [1 present; 58 not voting] followed by passage by vote of 267–97 [1 present; 65 not voting]).
29. See S.3780, *Ordered printed in the Senate* (showing differences between the bill and the House amendment).
30. See H. Rept. 74-2079; Cong. Rec., February 26, 1936, at 2804 (conference report submitted to the Senate) and 2855 (conference report submitted to the House); Cong. Rec., February 27, 1936, at 2931 (House debate on conference report began). See also Cong. Rec., February 24, 1936, at 2611–12 (Senator Robinson [D-AR]).
31. President Roosevelt, "Statement on Signing," 1936, 1–2; Cong. Rec., February 27, 1936, at 2937 (House agrees to the conference report); Cong. Rec., February 27, 1936, at 2925 (Senate agrees to the conference report); P.L. 74-461; Held and Clawson, *Soil Conservation*, 3–5, 12–13, 15–16, 25–29; Morgan, *Soil Erosion*, 162. See also Cong. Rec., February 20, 1936, at 2470 (Representative Clifford Hope [R-KS]). See also "Bill for New AAA Passes Congress," *New York Times*, February 27, 1936, https://timesmachine.nytimes.com/timesmachine/1936/02/28/87916763.html?pageNumber=1; "Roosevelt Signs New Farm Aid Bill," *New York Times*, March 1, 1936, https://timesmachine.nytimes.com/timesmachine/1936/03/02/88638520.html?pageNumber=1; Felix Belair Jr., "Machinery Is Set Up for New Farm Plan," *New York Times*, February 29, 1936, https://timesmachine.nytimes.com/timesmachine/1936/03/01/85253312.html?pageNumber=80.
32. P.L. 74-461, at Sec. 8(b).
33. P.L. 74-461, at Sec. 8(b).
34. 74 S.3780, *Introduced in the Senate*; 74 H.R. 10500.
35. S. Rept. 74-1481, at 5–6; H. Rept. 74-1973, at 4–5; H. Rept. 74-2079, at Sec. 8(b).
36. Cong. Rec., January 9, 1936, at 203–4.
37. See S. Hrg. 74-1936, "Substitute," at 9 (discussion with Senator George Norris [R-NE]).

38. See, e.g., Cong. Rec., February 19, 1936, at 2377 (Representative Gerald Boileau [R-WI]); Cong. Rec., February 20, 1936, at 2464 (Representative Harold Cooley [D-NC]) and 2497 (Representative Scott Lucas [D-IL]); Cong. Rec., February 21, 1936, at 2545 (Representative Whittington) and 2571 (Representative Lyndon Johnson [D-TX]). See also "Farm Plan Hinges," January 10, 1936.
39. See Cong. Rec., February 6, 1936, at 1567.
40. Cong. Rec., February 19, 1936, at 2374.
41. President Roosevelt, "Statement on Signing," 1936, 1.
42. Cong. Rec., February 7, 1936, at 1651.
43. See, e.g., *Butler*, 297 U.S., at 73.
44. Glick, "Soil, II." But see Conrad, *The Forgotten Farmers*, 201–4 (citing memo from Tolley to McFarlane, April 6, 1936, Landlord-Tenant File). See also Felix Belair Jr., "10,000,000-Acre Cut in Cotton Planned," *New York Times*, March 5, 1936, https://timesmachine.nytimes.com/timesmachine/1936/03/05/87921551.html?pageNumber=1.
45. P.L. 74-461, at Sec. 8(b). Compare 74 H.R. 10500, at Sec. 8(b); 74 S.3780, at Sec. 8(b).
46. Cong. Rec., February 20, 1936, at 2524.
47. Cong. Rec., February 20, 1936, at 2556–59 (Boileau and Representative William Bankhead [D-AL]).
48. Cong. Rec., February 20, 1936, at 2486–87.
49. Cong. Rec., February 21, 1936, at 2562.
50. Cong. Rec., February 21, 1936, at 2557 (Representative William Bankhead), 2558 (Chairman Jones), and 2562 (Representative Kleberg). See also Cong. Rec., February 20, 1936, at 2480 (Representative Kleberg) and 2467-87 (Chairman Jones); February 21, 1936, at 2545 (Representative Whittington).
51. Cong. Rec., February 21, 1936, at 2570.
52. Cong. Rec., February 19, 1936, at 2370–71, and February 21, 1936, at 2560 (respectively).
53. Cong. Rec., February 20, 1936, at 2483–85.
54. Cong. Rec., February 19, 1936, at 2376–78, 2384–86 (Representative August Andresen [R-MN]). But see Cong. Rec., February 19, 1936, at 2388–89 (Representative Frank Hook [D-MI]).
55. See Madison, *Federalist*, no. 10 and 58.
56. P.L. 74-461, at Sec. 8.
57. Cong. Rec., February 19, 1936, at 2393–94 (Representatives Tarver [D-GA], Maverick [D-TX], and Whelchel [D-GA]). See also Baldwin, *Poverty and Politics*.
58. Cong. Rec., February 21, 1936, at 2552 (Representative Maverick) and 2554–55 (Representative Whelchel).

59. See Cong. Rec., February 21, 1936, at 2552 (Jones and Representative Maury Maverick [D-TX]). See also Cong. Rec., February 20, 1936, at 2495–98 (Representative Scott Lucas [D-IL]); Cong. Rec., February 20, 1936, at 2482–85 (Representative Gilchrist [R-IA]). Cong. Rec., February 20, 1936, at 2465 (Representative Harold Cooley [D-NC]); Cong. Rec., February 19, 1936, at 2362 (Representative Martin [R-MA]); Cong. Rec., February 19, 1936, at 2394 (Representative Malcolm Tarver [D-GA]).

60. Cong. Rec., February 19, 1936, at 2372–74; Cong. Rec., February 19, 1936, at 2390 (Representative Aaron L. Ford [D-MS]); Cong. Rec., February 20, 1936, at 2464 (Representative Cooley).

61. See Cong. Rec., February 21, 1936, at 2552. See also Cong. Rec., February 20, 1936, at 2480 (Representative Colmer [D-MS]); Cong. Rec., February 21, 1936, at 2569 (debate between Representative Hope and Representative Wittington); Cong. Rec., February 20, 1936, at 2524.

62. Cong. Rec., February 21, 1936, at 2551–52; Cong. Rec., February 21, 1936, at 2555 (Tarver amendment as modified adopted without a vote) and 2556 (Representative Martin Dies [D-TX]).

63. See H. Rept. 74-2079, at 7–13 (statement of the Managers on the Part of the House: the provision was revised to protect their interests only to the degree that it was "practicable" [at 10]); Cong. Rec., February 22, 1936, at 2599; Cong. Rec., February 26, 1936, at 2804 (conference report submitted to the Senate) and 2855 (conference report submitted to the House); Cong. Rec., February 27, 1936, at 2931 (House debate on conference report began). See also Cong. Rec., February 24, 1936, at 2611–12 (Senator Robinson [D-AR] submits USDA technical review of the bills).

64. See Cong. Rec., February 27, 1936, at 2923–24 (Chairman Smith and Senator McNary).

65. See Cong. Rec., February 24, 1936, at 2699–700 (Representative Tarver attacked Chairman Smith); "AAA Fight Centers on Big Income Goal," *New York Times*, February 22, 1936, https://timesmachine.nytimes.com/timesmachine/1936/02/23/87914269.html?pageNumber=23 (quoting Smith); "Conferees Choose House Farm Bill," *New York Times*, February 25, 1936, https://timesmachine.nytimes.com/timesmachine/1936/02/26/88637093.html?pageNumber=6. See also Irons, *Lawyers*, 160 (Alger Hiss, dated July 12, 1978, reporting his memory of Smith's comments that came up in an earlier fight over the AAA cotton contract; Hiss, one of the liberal attorneys at USDA, reported that during the issue, Senator Smith "burst unannounced into his office" and demanded, "You're going to send money to my n——s, instead of me?").

66. Cong. Rec., February 27, 1936, at 2936.
67. In addition to the discussion in chapter 3 and the sources cited therein, sharecropper problems were certainly in the news. See, e.g., F. Raymond Daniell, "AAA Piles Misery on Share Croppers," *New York Times*, April 14, 1935, https://timesmachine.nytimes.com/timesmachine/1935/04/15/95066306.html?pageNumber=6; F. Raymond Daniell, "Arkansas Violence Laid to Landlords," *New York Times*, April 15, 1935, https://timesmachine.nytimes.com/timesmachine/1935/04/16/95066896.html?pageNumber=18; "Tenant Law Class Roils Cotton Belt," *New York Times*, April 17, 1935, https://timesmachine.nytimes.com/timesmachine/1935/04/18/94599640.html?pageNumber=24; F. Raymond Daniell, "Farm Tenant Union Hurt by Outsiders," *New York Times*, April 18, 1935, https://timesmachine.nytimes.com/timesmachine/1935/04/19/94600049.html?pageNumber=18; F. Raymond Daniell, "The Share-Cropper: His Plight Revealed," *New York Times*, May 5, 1935, https://timesmachine.nytimes.com/timesmachine/1935/05/05/95073185.html?pageNumber=147.
68. See, e.g., "Cotton Pickers May Quit," *New York Times*, September 3, 1935, https://timesmachine.nytimes.com/timesmachine/1935/09/03/93702596.html?pageNumber=33; Associated Press, "Share-Croppers End Strike in Arkansas," *New York Times*, July 9, 1936, https://timesmachine.nytimes.com/timesmachine/1936/07/10/87961121.html?pageNumber=1. See also Conrad, *Forgotten Farmers*, 173.
69. See Cong. Rec., February 19, 1936, at 2394 (Representative Tarver); H. Hrg. 74-1935, H.R. 557, at 75 (testimony of Cully Cobb, AAA Cotton Section).
70. See Conrad, *Forgotten Farmers*, 204. See also "Charge Terrorism to AAA Field Aides," *New York Times*, March 27, 1936, https://timesmachine.nytimes.com/timesmachine/1936/03/28/88647260.html?pageNumber=2; "Relief Gap Scored by Sharecroppers," *New York Times*, March 28, 1936, https://timesmachine.nytimes.com/timesmachine/1936/03/29/85270055.html?pageNumber=25; "Farm Strike Stories Denied in Arkansas," *New York Times*, May 21, 1936, https://timesmachine.nytimes.com/timesmachine/1936/05/22/85398117.html?pageNumber=4; "Ask Farm Strike Inquiry," *New York Times*, May 23, 1936, https://timesmachine.nytimes.com/timesmachine/1936/05/23/88664500.html?pageNumber=29; "Tenant Farm Strike in Federal Inquiry," *New York Times*, June 3, 1936, https://timesmachine.nytimes.com/timesmachine/1936/06/04/87943502.html?pageNumber=4; "Woman Flogged in Cotton Strike," *New York Times*, June 16, 1936, https://timesmachine.nytimes.com/timesmachine/1936/06/17/85403661.html?pageNumber=3.
71. See, e.g., Gilbert, *Planning Democracy*.

72. Madison, *Federalist*, no. 42.
73. Phelps, "Vision," 354 (quote contained in D. Harper Simms, *The Soil Conservation Service* [New York: Praeger, 1970], 75-76); House Doc. 75-149, at 7; P.L. 75-210; H. Hrg. 75-1937, H.R. 8, at 215-16 (Secretary Wallace testimony).
74. Bennett, *Soil Conservation*, v (preface).
75. Benedict, *Farm Policies*, 351. See also Sampson, *For Love of the Land*, 14; Held and Clawson, *Soil Conservation*, 72; Hardin, *Politics of Agriculture*, 108; Talbot and Hadwiger, *Policy Process*, 215.
76. Rasmussen, "'Never a Landlord,'" 74.
77. P.L. 75-430; H. Rept. 75-1767.
78. Compared to 1935, cotton acres in 1937 were up more than 120 percent and wheat acres more than 115 percent. Planted acres data from USDA's NASS Quickstats database, https://quickstats.nass.usda.gov/.
79. U.S. Department of Agriculture, Agricultural Adjustment Administration (hereafter USDA, AAA), "Agricultural Conservation 1936," 1 (announced by FDR on October 25, 1935); USDA, AAA, "Agricultural Adjustment, 1937–1938," 17.
80. USDA, AAA, "Agricultural Adjustment, 1937–1938," 17–18, 41–43; USDA, AAA, "Agricultural Adjustment, 1938–39," 8, 16.
81. See, e.g., Cong. Rec., November 18, 23, 24, 29, and 30, 1937; Cong. Rec., December 7 and 8, 1937; Cong. Rec., February 10, 1938; P.L. 75-430, at Sec. 303. See also Kile, *Farm Bureau*, 239–41.
82. P.L. 75-430.
83. See Katznelson, *Fear Itself*; Bateman, Katznelson, and Lapinski, *Southern Nation*; Freeman, *Field of Blood*; Mickey, *Paths Out of Dixie*; Bensel, *Sectionalism*; Key, *Southern Politics*. See also Katznelson, Geiger, and Kryder, "Limiting Liberalism"; Farhang and Katznelson, "Southern Imposition"; Katznelson and Mulroy, "Was the South"; Bateman, Katznelson, and Lapinski, "*Southern Politics* Revisited."
84. See, e.g., Gilbert and Howe, "Beyond"; Hooks, "New Deal"; Rasmussen, "'Never a Landlord,'" 82–83. See also Biles, "Politics."
85. See, e.g., Breimyer, "Agricultural Philosophies"; Breimyer, "Conceptualization."
86. Kirkendall, *Social Scientists*, 136.
87. S. Hrg. 74-1935, S.1800, at 6–7.
88. See H. Ex. Doc. 75-149, 1937.
89. See, e.g., Phillips, *This Land*; Cong. Rec., February 20, 1936, at 2469 (Representative Hope).
90. See, e.g., Cong. Rec., February 20, 1936, at 2492–94.
91. See Cong. Rec., February 21, 1936, at 2569–70 (Hope amendment defeated 81–130 [on a division] and 75–127 [on tellers]). See also Seegers, "Cooley."

92. P.L. 74-461, at Sec. 8 ("In carrying out the provisions of this section, the Secretary shall, in every practicable manner, protect the interests of small producers").
93. P.L. 74-461, at Sec. 7(a).
94. Cong. Rec., February 20, 1936, at 2462–64. See also Cong. Rec., February 19, 1936, at 2390–93 (Representative Maury Maverick [D-TX]); Cong. Rec., February 21, 1936, at 2575–76 (Representative Martin [D?-CO]).
95. See Morgan, *Governing Soil Conservation*, 39–40. See also McDean, "M. L. Wilson," 444–49.
96. See "Senate Body," January 28, 1936; "Senate Committee," January 29, 1936 (reporting that Wallace considered it a form of economic democracy); "Senators Rewrite," February 5, 1936.
97. See, e.g., Bensel, *Sectionalism*, 153. See also Katznelson, *Fear Itself*, 15–16, 25, 163.
98. See, generally, Morgan, *Governing Soil Conservation*. See also Rasmussen, "'Never a Landlord,'" 80; Worster, *Dust Bowl*, 219; Ferguson, "Nation-Wide"; Frarey, Jones, and Pratt, "Conservation"; Massey, "Land Use"; Comito, Wolseth, and Morton, "State's Role." See also National Association of Conservation Districts, "About NACD," NACD History, https://www.nacdnet.org/about-nacd/nacd-history/; D. Helms, "Getting to the Roots," NRCS History Articles, Conservation Districts, p. 25, https://nrcs.prod.usda.gov/sites/default/files/2022-08/Readings-In-The-History-Of-The-Soil-Conservation-Service.pdf.
99. See, e.g., Linder, "Farm Workers"; Perea, "Echoes." See, e.g., "Tenant Law," April 17, 1935; Daniell, "Share-Cropper," May 5, 1935; Daniell, "AAA," April 14, 1935; Daniell, "Arkansas Violence," April 15, 1935; Daniell, "Farm Tenant Union," April 18, 1935; "Cotton Pickers May Quit," *New York Times*, September 3, 1935; "Charge Terrorism," *New York Times*, March 27, 1936; "Relief Gap," *New York Times*, March 28, 1936; "Farm Strike," *New York Times*, May 21, 1936; "Ask Farm Strike Inquiry," *New York Times*, May 23, 1936; "Tenant Farm Strike," *New York Times*, June 3, 1936; "Woman Flogged," June 16, 1936; Associated Press, "Share-Croppers End Strike," *New York Times*, July 9, 1936.
100. Katznelson, *Fear Itself*, 25. See also Katznelson, *Fear Itself*, 15–16, 163; Bensel, *Sectionalism*, 153.
101. See, e.g., Jones, "Trojan."
102. See Baldwin, *Poverty and Politics*.

5. Out of Surplus

1. Eisenhower, "Special Message to Congress on Agriculture," January 9, 1956, 38, 42.
2. Eisenhower, "Statement by the President Upon Signing," May 28, 1956; P.L. 84-540. According to the U.S. Bureau of Labor Statistics CPI Inflation Calculator,

$1 billion in 1956 would be the equivalent of $9.76 billion in 2021 and $10.49 in 2022. See https://www.bls.gov/data/inflation_calculator.htm.

3. H. Ex. Doc. 85-266, 1959 Budget, at M49. See also H. Hrg. 86-1959, 1960 Appropriations, part 3, at 2135–36 (statement of Assistant Secretary of Agriculture Marvin L. McLain); H. Rept. 86-2219, at 69.

4. Note that this era was covered extensively in chapter 4 of *The Fault Lines of Farm Policy*, including the difficult enactment of the Agricultural Act of 1956. This chapter and the next seek to fill in some of the gaps regarding the Soil Bank and especially how it was sabotaged and how it ended. There is much overlap, and I have attempted to not repeat the details from the earlier book, while supplementing it significantly and answering questions left unanswered previously. See Coppess, *Fault Lines*, 95–134.

5. See Cochrane and Ryan, *American Farm Policy*; Blakley, "Effects," 1164. See also J. Coppess, "The Conservation Question, Part 5: Seeds of the Soil Bank," *Farmdoc Daily* 10, no. 3, Department of Agricultural and Consumer Economics, University of Illinois at Urbana-Champaign, January 9, 2020, https://farmdocdaily.illinois.edu/2020/01/the-conservation-question-part-5-seeds-of-the-soil-bank.html; J. Coppess, "The Conservation Question, Part 6: Development of the Soil Bank," *Farmdoc Daily* 10, no. 13, Department of Agricultural and Consumer Economics, University of Illinois at Urbana-Champaign, January 24, 2020, https://farmdocdaily.illinois.edu/2020/01/the-conservation-question-part-6-development-of-the-soil-bank.html; J. Coppess, "The Conservation Question, Part 8: Interregnum and Acres," *Farmdoc Daily* 10, no. 36, Department of Agricultural and Consumer Economics, University of Illinois at Urbana-Champaign, February 27, 2020, https://farmdocdaily.illinois.edu/2020/02/the-conservation-question-part-8-interregnum-and-acres.html.

6. See, e.g., Erisman et al., "How a Century"; Braun et al., "From Inert"; Martín, Shinagawa, and Pérez-Ramírez, "Electrocatalytic." See also Cochrane, *Development of American Agriculture*; Orden, Paarlberg, and Roe, *Policy Reform*, 24–45; Cochrane and Ryan, *American Farm Policy*.

7. See Hurt, *Dust Bowl*, 144–55. See also National Oceanic and Atmospheric Administration, National Centers for Environmental Information, "Historical Palmer Drought Indices," https://www.ncdc.noaa.gov/temp-and-precip/drought/historical-palmers/maps.

8. See H. Hrg. 83-1953, "Long Range, part 8," at 1114–17 (testimony of Melvin Gehlbach); S. Hrg. 84-1956, "Price-Support," part 8, at 3346. See also Schapsmeier and Schapsmeier, "Eisenhower," 375 (referencing Melvin P. Gehlbach).

9. See S. Hrg. 83-1953, "Commodity Inventories," at 3; H. Hrg. 83-1954, H.R. 7339, at 3 (statement of Secretary Ezra Taft Benson); S. Hrg. 84-1955, S.2604, at 26; H. Hrg. 84-1955, 1956 Appropriations, at 12 (reporting stocks in July).
10. H. Hrg. 83-1954, "Long Range," part 21, at 3931 and 3925, respectively.
11. Planted acres are available from USDA's National Agricultural Statistics Service (NASS) Quickstats database: https://quickstats.nass.usda.gov/.
12. See, e.g., Ross, "Man"; Hines, "Land Ethic"; Beatty, "Conservation"; McConnell, "Conservation Movement"; Morgan, "Pressure"; Buttel, "Environmental"; Bosso, *Pesticides and Politics*; Switzer, *Green Backlash*. See also Leopold, "Land Ethic"; Carson, *Silent Spring*.
13. See Speigner, "Analysis," 409 (quoting Eisenhower from "Watershed Protection," *Congressional Quarterly Almanac*, vol. 10, 1954, p. 135). See also S. Hrg. 83-1954, S.3052, at 835, 848–49 (Gehlbach's testimony); 83 S.379; 83 H.R. 1809; 83 H. Con. Res. 189; 83 H. Res. 454.
14. P.L. 83-566. See also Woodward, "Look Back"; Speigner, "Analysis."
15. See 83 S.3036.
16. See, e.g., H. Hrg. 83-1954, "Long Range," part 21, at 3819, 3924, and 3931 (Representative Cooley); P.L. 83-690. See also Coppess, *Fault Lines*.
17. See, e.g., 84 H.R. 2646; 84 H.R. 3912; 84 H.R. 5942; 84 H.R. 6052; 84 H.R. 8137; 84 H.R. 8151; 84 H.R. 8156; 84 H.R. 8369. See also 84 S.1396; Cong. Rec., March 10, 1955, at 2572 (Senator Hubert H. Humphrey introduced S.1396).
18. Cong. Rec., April 25, 1955, at 4979–80 (Senator Humphrey also referenced efforts within the administration to address soil conservation concerns). See also Case, "Proposed," 1.
19. See Lacey, *Farm Bureau in Illinois*, 207–9.
20. Hansen, *Gaining Access*, 133–34. See also H. Hrg. 83-1954, "Long Range," part 21, at 3819, 3924, and 3931 (Representative Cooley).
21. See Brown v. Board of Education of Topeka, 347 U.S. 483, 494–95 (1954) (rejecting Plessy v. Ferguson, 163 U.S. 537 [1896]).
22. See Daniel, *Lost Revolutions*, 181, 188–91; Cobb, *Most Southern Place*, 218–19. See, generally, Bensel, *Sectionalism*; MacLean, *Democracy in Chains*; Woods, *Development Arrested*.
23. Cobb, *Most Southern Place*, 213–15.
24. Daniel, *Lost Revolutions*, 197, 210–16.
25. Hansen, *Gaining Access*, 169.
26. See, e.g., Fite, *Cotton Fields*, 218 ("To help poor farmers throughout the region [South] meant assisting black producers" and "by the 1950s any program for

small, poverty-ridden farmers in the South became entangled with the civil rights movement").
27. Democrats held a 232–203 advantage over Republicans in the 84th Congress, with Sam Rayburn (D-TX) as Speaker. See U.S. House of Representatives, History, Art & Profile, Congress Profiles, https://history.house.gov/Congressional-Overview/Profiles/84th/. Harold D. Cooley (D-NC) was chair of the House Agriculture Committee, and W. R. Poage (D-TX) was his second-in-command. Allen J. Ellender (D-LA) was chair of the Senate Agriculture Committee, which included notorious segregationist Senator James O. Eastland (D-MS). See H. Rept. 84-203, at 5; Cong. Rec., May 3, 1955, at 5455–56; Cong. Rec., May 5, 1955, at 5804 (motion to recommit the bill defeated by a vote of 199–212 [20 not voting; 3 voting present]) and 5806 (final passage by a vote of 206–201 [22 not voting; 5 voting present]). See also CQ *Almanac*, "Farm Price Supports"; Schapsmeier and Schapsmeier, "Eisenhower"; Daniel, *Lost Revolutions*, 50.
28. See Eisenhower, "Special Message to Congress on Agriculture," January 9, 1956, 38; CQ *Almanac*, "President Outlines."
29. See S. Hrg. 84-1956, "Price-Support," part 8, at 3293 (Secretary Benson), 3344–45 (under Secretary True D. Morse), 3347–49 (Secretary Benson), and 3333–34 (Senator [and former Secretary of Ag] Anderson). See also Hansen, *Gaining Access*, 133–35; CQ *Almanac*, "Soil Bank Enacted."
30. S. Hrg. 84-1956, "Price-Support," part 8, at 3293–94 (Secretary Benson), 3312 (Secretary Benson), 3340–41 (Secretary Benson questioned by Senator Hubert H. Humphrey [D-MN]), and 3355 (Secretary Benson).
31. Cong. Rec., February 22, 1956, at 3123 (Senator Thye [R-MN]); Cong. Rec., February 22, 1956, at 3123 (Senator Aiken).
32. See S. Hrg. 84-1956, "Price-Support," part 8, at 3350 (text of S.1396), 3360, 3362–64, and 3371–73; S. Rept. 84-1484, at 8, 15; P.L. 84-540; 84 S.3183; Cong. Rec., February 24, 1956, at 3317 (Senator and ranking member George Aiken [R-VT]).
33. See S. Rept. 84-1484, at 36; CQ *Almanac*, "Soil Bank Enacted" (reporting that the committee reported it on a 12–3 vote but voted 8–7 on restoring the 90 percent price support); Cong. Rec., February 22, 1956, at 3117–18 (Chairman Ellender).
34. See Cong. Rec., February 22, 1956, at 3117–22 (Ellender), 3127 (Ellender), and 3181 (Senator Young [R-ND]); Cong. Rec., February 24, 1956, at 3317–18 (Senator and Ranking Member George Aiken [R-VT]); Cong. Rec., March 8, 1956, at 4288.
35. See Cong. Rec., February 22, 1956, at 3127 (Senator Holland [D-FL]); Cong. Rec., February 24, 1956, at 3320 (Aiken) and 3322–23 (Aiken); Cong. Rec., March 8, 1956, at 4287–88 (Senator Clinton Anderson [D-NM]). For veto threat, see Cong.

Rec., February 23, 1956, at 3180–81 (Senator Young); Cong. Rec., February 29, 1956, at 3564–65 (Senator McNamara [D-MI]); Cong. Rec., February 29, 1956, at 3572 (Senator Humphrey); Cong. Rec., March 5, 1956, at 3886 (Senator James Murray [D-MT]); Cong. Rec., March 7, 1956, at 4151 (Senator Olin Johnston [D-SC]); Cong. Rec., March 8, 1956, at 4290–91 (Senator Holland); Cong. Rec., March 9, 1956, at 4425 (Senator Welker). For the debate over striking price supports, see Cong. Rec., March 8, 1956, at 4288–89 (Ellender); Cong. Rec., March 8, 1956, at 4304 (Anderson amendment to strike 90 percent of parity agreed to by a vote of 54–21).

36. See S. Rept. 84-1484, at 8, 12; P.L. 75-430, at Sec. 301(b)(4)(A).
37. See, e.g., H. Rept. 84-1986, at 3–4, 6, 22, 46; Cong. Rec., March 8, 1956, at 4304–7 (Hickenlooper amendment and colloquy with Senator Mundt). See also Cong. Rec., March 8, 1956, at 4304–7 (Hickenlooper), 4308 (Ellender), and 4309 (Senator Humphrey). See also Cong. Rec., March 8, 1956, at 4316–17 (Senators Humphrey [modified amendment], Ellender, and Mundt); Cong. Rec., March 9, 1956, at 4379 (Ellender).
38. See S. Hrg. 84-1956, "Price-Support," part 8, at 3293–97 (Secretary Benson); S. Rept. 84-1484, at 68–70 (individual views of Senator Hickenlooper [R-IA]); Cong. Rec., February 24, 1956, at 3324–25, 4289–90; Cong. Rec., March 8, 1956, at 4305–6 (Senators Hickenlooper, Thye [R-MN], Senator Mundt [R-SD]) and 4307–8 (Chairman Ellender); Cong. Rec., March 9, 1956, at 4378, 4386 (Senator Anderson), 4379 (Chairman Ellender), 4380–81 (colloquy among Senators Young [R-ND], Holland [D-FL], Anderson [D-NM], and Humphrey [D-MN]), 4383–85 (Senators Young and Russell [D-GA]), 4386–87 (Senators Russell and Humphrey), and 4390 (compromise accepted by voice vote).
39. See Cong. Rec., March 9, 1956, at 4385. See also Cong. Rec., March 8, 1956, at 4289 (Senator Holland).
40. See, e.g., Cong. Rec., February 23, 1956, at 3128 (Ellender); Cong. Rec., March 8, 1956, at 4297 (Senator Barkley [D-KY] and Senator Young); Cong. Rec., March 8, 1956, at 4290–91 (Senator Holland); Cong. Rec., March 9, 1956, at 4386–87 (Senators Russell and Humphrey); Cong. Rec., March 19, 1956, at 5078 (by a vote of 93–2 [1 not voting]).
41. See CQ Almanac, "Soil Bank Enacted" (the conference agreement was agreed to by an 8–2 vote on April 6, 1956); H. Rept. 84-1986; Cong. Rec., April 11, 1956, at 6111–25, 6127 (Chairman Cooley).
42. H. Rept. 84-1986, at 36.
43. See Cong. Rec., April 11, 1956, at 6157–58 (in the House, the motion to recommit was defeated by a vote of 181–238 [14 not voting] and the conference report was passed 237–181 [15 not voting]) and 6108 (the Senate passed the confer-

ence report by a relatively close vote of 50–35 [11 not voting]); Schapsmeier and Schapsmeier, *Ezra Taft Benson*, 161 (quoting the president); CQ *Almanac*, "Soil Bank Enacted"; Cong. Rec., April 18, 1956, at 6541 (veto override failed by a vote of 202–211 [20 not voting]), 6532–33 (Chairman Cooley), and 6538 (Representative Halleck).

44. CQ *Almanac*, "Soil Bank Enacted" (Senate Ag Committee reported its revised version of the farm bill on May 11 by a vote of 13–2; Eisenhower's quote). See, e.g., H. Rept. 84-2197, at 3; P.L. 84-540; Cong. Rec., May 2, 1956, at 7344–47 (Chairman Cooley); Cong. Rec., May 3, 1956, at 7410–16 (text), 7417 (Representative Poage), and 7449–50 (House passed the revised farm bill on May 3, 1956, by a vote of 314–78 [41 not voting]); Cong. Rec., May 18, 1956, at 8472 (Senator Holland), 8474 (Senator Young), 8476–77 (Chairman Ellender; Holland amendments agreed to by the Senate on a vote of 73–14 [8 not voting]), and 8514 (passed by a voice vote in the Senate on May 18, 1956); Cong. Rec., May 23, 1956, at 8821, 8824–25 (Representative Poage on the 1956 implementation requirement), and 8832 (House passed the conference report on May 23, 1956, by a vote of 305–59 [69 not voting]).

45. See H. Rept. 84-1986, at 37–39, 54; Cong. Rec., April 11, 1956, at 6079, 6127 (Cooley).

46. H. Rept. 84-1986, at 13, 39–40. Compare S.3183, at Sec. 213 and Sec. 217.

47. P.L. 84-540.

48. See H. Rept. 84-1986, at 38, 53; S. Rept. 84-1484, at 11, 16; P.L. 84-540. See also Cong. Rec., April 11, 1956, at 6127 (Cooley); Cong. Rec., February 23, 1956, at 3152–53 (Chairman Ellender); H. Rept. 86-2219, at 69 (table 30); Eisenhower, "Special Message to Congress on Agriculture," January 9, 1956; Schapsmeier and Schapsmeier, "Eisenhower," 152.

6. Southern Sabotage

1. Eisenhower, "Statement by the President Upon Signing"; P.L. 84-540; H. Ex. Doc. 85-266, 1959 Budget; H. Hrg. 86-1959, Appropriations, part 3, at 2135–36 (statement of Assistant Secretary of Agriculture Marvin L. McLain); H. Rept. 86-2219. See also Schapsmeier and Schapsmeier, "Eisenhower."

2. See Kotz, *Let Them Eat Promises*, 84–102; Nick Kotz, "Jamie Whitten: The Permanent Secretary of Agriculture," *Washington Monthly*, October 1969, 8–19, https://www.unz.com/print/WashingtonMonthly-1969oct-00008/. See also J. Y. Smith, "Jamie Whitten Dies at 85," *Washington Post*, September 10, 1995, https://www.washingtonpost.com/archive/local/1995/09/10/jamie-whitten-dies-at-85/a12504d4-34ea-4cb2-8aca-c6ef20e43e4f/; David Binder, "Jamie Whitten, Who Served 53 Years in House, Dies at 85," *New York Times*, September 10, 1995,

https://www.nytimes.com/1995/09/10/obituaries/jamie-whitten-who-served-53-years-in-house-dies-at-85.html. See, e.g., Cong. Rec., May 4, 1955, at 5671 (Representative Jamie Whitten [D-MS]); Daniel, *Lost Revolutions*, 50; Cong. Rec., May 23, 1956, at 8823–24 (Representative Poage) and 8825 (Representative H. Carl Andersen [R-MN]).

3. See, e.g., Caro, *Master of the Senate*, 164–202; S. Hrg. 85-1957, Appropriations, at 24 (Senator Russell) and 26.
4. This would be fiscal year 1957. P.L. 84-540, at Sec. 120(a).
5. See, e.g., Cong. Rec., May 14, 1957, at 6878 (Whitten acknowledges that for the Soil Bank, his appropriation is "a reduction of $272 million in this bill from the budget"); Kotz, "Jamie Whitten." See also Knapp, "Congressional Control."
6. See P.L. 84-40 (covering the fiscal year ending June 30, 1956); P.L. 84-554 (covering the fiscal year ending June 30, 1957).
7. See, e.g., P.L. 84-540, at Sec. 103(a); P.L. 75-430, at Sec. 301(b)(7) (defining crop marketing years) and Sec. 508(a) (crop insurance).
8. See H. Hrg. 84-1956, H.R. 11958, at 2–5 (Poage); S. Hrg. 84-1956, S.3280, at 12 (comments of Mr. Robert P. Beach, Deputy Administrator of Operations, Office of Commodity Exchange Authority, U.S. Dept. of Agriculture). See also Hurt, *Dust Bowl*, 139–47, 152–55.
9. See, e.g., Bonnen, "Farm Policy Debate," 432.
10. See H. Hrg. 84-1956, H.R. 11958, at 7 (Rep. Poage), 13–14 (Assistant Secretary McLain), 18–20 (Assistant Secretary McLain), 23, 27 (Representative Poage), 38 (Poage and Chairman Cooley), 40–41 (Cooley).
11. See H. Hrg. 85-1957, "Soil Bank," at 1 (Chairman Cooley), 3–4 (Cooley), 7–8 (Cooley), 11–12 (Representative Poage), 14–15, 17 (Poage), 18–21 (Chairman Cooley), 30–32 (Representative Abernethy [D-MS]), 36 (Assistant Secretary McLain), 51–53 (Poage). The Soil Bank corn base was set at 51 million acres for the commercial corn region, whereas the allotment would revert to 37 million acres. Lacey, *Farm Bureau in Illinois*, 210–11. Illinois farmers voted overwhelmingly for the Soil Bank, but the referendum failed to achieve the two-thirds vote (only 61.2 percent), and the acreage allotments won.
12. See H. Hrg. 85-1957, "Soil Bank," at 33–34, 43–45; H. Hrg. 85-1957, "Acreage Reserve," at 2, 7–8 (Assistant Secretary Marvin L. McLain), 11, 44.
13. H. Hrg. 85-1957, "Present Conditions," at 1–2, 42–43 (Chairman Whitten), 136 (Representative H. Carl Andersen [R-MN]); S. Hrg. 85-1957, Appropriations, at 24, 26 (Senator Russell); H. Rept. 86-2219 (Gov't Ops, 1960) at 68 (table 29 reported obligations of $260 million in 1956 and $614 million in 1957; for the entire three years, the program obligated $1.57 billion); P.L. 84-540, at Sec. 120(a).

14. Note that Whitten sought to take credit for a reduction of $254 million that was sent to Congress and sought a total reduction of $403 million below what was available for fiscal year 1957 and $272.6 million below the original budget request for fiscal year 1958. See H. Rept. 85-438, at 1; Cong. Rec., May 14, 1957, at 6878 (Whitten).
15. See H. Hrg. 85-1957, "Present Conditions," at 3, 43, 46 (Benson); Cong. Rec., May 14, 1957, at 6887–88 (Secretary Benson sent a letter to Whitten).
16. H. Rept. 85-438, at 9; H. Hrg. 85-1958, Appropriations, part 4, at 1562–67 (Whitten and Benson) and 1624–26 (Whitten and McLain, in which McLain also blamed Poage and others on conference).
17. USDA was expecting 21 million in the acreage reserve in 1957, and as of April 12, 1957, USDA reported 21.4 million acres enrolled. It reported nearly the same acreage enrolled in 1956 (21.4 million) at a cost of $261 million (roughly $12.19 per acre). See S. Ex. Doc. 85-67, 1957, at 34; H. Ex. Doc. 85-149, 1957; H. Hrg. 85-1958, Appropriations, part 4, at 1551–52 (Secretary Benson), 1556–58 (recent recomputation), 1621 (McLain testimony), 1645 (USDA table), 1658 (more data on acreage reserve), 1665 (reporting funds obligated through April 1957).
18. See H. Rept. 85-438, at 1, 9–10, 25.
19. See Cong. Rec., May 15, 1957, at 7023–24 (Representative Burr Powell Harrison [D-VA] offered the amendment), 7024 (Chairman Cooley), 7026 (Representative Poage), 7027 (selective concerns about spending), 7029 (Representative Paul C. Jones [D-MO]), 7029–30 (Representative Howard Smith [D-VA]), 7030–31 (Representative George S. Long [D-LA]), 7031–32 (Representative William J. B. Dorn [D-SC]), 7032 (Representative E. C. Gathings [D-AR], Soil Bank was going to "liquidate our farmers"), 7032 (Representative Henderson Lovelace Lanham [D-GA] preferred General Lee over General Eisenhower); Cong. Rec., July 9, 1957, at 11107 (Representatives Whitten and Mahon [D-TX]). See also H. Rept. 85-438, at 2.
20. Cong. Rec., May 15, 1957, at 7025 (H. Carl Andersen [R-MN]), 7027–28 (Representative Page Belcher [R-OK] attacked Cooley and Poage, as did Representative Charles Halleck [R-IN]), 7028 (Representative Charles Vursell [R-IL] attacked Poage for sabotaging the Soil Bank), 7029 (Representative August Andresen quote), 7031 (Representative William Avery [R-KS]).
21. Cong. Rec., May 15, 1957, at 7033 (amendment was agreed to on tellers, 134–120) and 7038 (passage).
22. Cong. Rec., August 1, 1957, at 13280 (Senator Young), 13283 (Senator Russell), 13284 (Senators Aiken and Russell).
23. P.L. 85-118; Cong. Rec., July 9, 1957, at 11107 (Whitten quote); H. Rept. 85-682, at 7.

24. H. Hrg. 85-1957, Appropriations, part 4, at 1572.
25. See, e.g., H. Hrg. 85-1957, Appropriations, part 4, at 1556–58, 1629, 1648–51, 1658, 1679–80, 1687 (McLain and Rep. H. Carl Andersen). H. Rept. 85-438, at 1; Cong. Rec., May 14, 1957, at 6878 (Whitten). See also S. Hrg. 85-1957, Appropriations, at 6, 10.
26. See, e.g., S. Hrg. 85-1958, "Farm Program," part 2, at 304–5 (Senator Lister Hill [D-AL]), 320 (Senator Sparkman [D-AL]), 331–32 (Senator Yarborough [D-TX]), 366–69, and 373–74 (William Rhea Blake, Executive Vice President, National Cotton Council of America).
27. S. Hrg. 85-1958, "Farm Program," part 2, at 306, 322.
28. See H. Rept. 86-2219.
29. See, e.g., S. Hrg. 85-1958, "Farm Program," part 2, at 360–61 (Senator Symington quote), 373–78 (William Rhea Blake, Executive Vice President, National Cotton Council of America), 444–45 (Robert C. Jackson, Executive Vice President of the American Cotton Manufacturers Institute), 427 (Senator Albert Gore [D-TN]), 447–48, 451–52, 457, 460–63 (Charles R. Sayer, Chairman, Agriculture Committee, Delta Council).
30. See S. Hrg. 85-1958, "Farm Program," part 3, at 480 (Assistant Secretary McLain); S. Hrg. 85-1958, "Farm Program," part 2, at 350–51, 354 (McLain), and 359–60 (McLain and Senator Eastland [D-MS]).
31. H. Ex. Doc. 85-266, 1959 Budget, at M49; S. Hrg. 85-1958, "Farm Program," part 2, at 333 and 350–51 (Assistant Secretary Marvin L. McLain); S. Hrg. 95-158, "Farm Program," part 3, at 480 (McLain); H. Hrg. 85-1958, "Acreage Reserve," at 38. See also H. Hrg. 86-1959, Appropriations, part 3, at 2119 (USDA explained to Whitten and the House Agricultural Appropriations subcommittee on March 17, 1959), 2135–36 (McLain).
32. The program ended up costing $525 million in fiscal year 1957 with $600 million expected in fiscal year 1958. H. Rept. 85-1584, at 27 (quote); S. Rept. 85-1438, at 12–13; P.L. 85-459; H. Rept. 85-1776, at 5–6.
33. H. Rept. 85-438, at 26.
34. See H. Hrg. 85-1957, "Acreage Reserve," at 46 (Benson), 64.
35. See H. Hrg. 85-1957, Appropriations, part 6, at 138–40, 142–43, 165–68, 211.
36. See H. Hrg. 85-1957, Appropriations, part 6 (Soil Bank report), at 143, 148 (quote), 194–95.
37. See H. Hrg. 85-1957, Appropriations, part 6 (Soil Bank report), at 144, 196, 223.
38. See H. Hrg. 85-1957, Appropriations, part 6 (Soil Bank report), at 149–50, 231–33.
39. H. Hrg. 85-1958, Appropriations, part 1, at 239, 248, 299 (quotes).
40. H. Hrg. 85-1958, Appropriations, part 1, at 303–4, 324, 326.

41. H. Hrg. 85-1958, Appropriations, part 1, at 307, 326, 328.
42. H. Hrg. 85-1958, Appropriations, part 1, at 323, 325, 328, 329–30 (quote).
43. See, e.g., S. Hrg. 85-1958, "Farm Program," part 2, at 383–84 (Farm Bureau legislative counsel, Frank K. Woolley).
44. See, e.g., H. Hrg. 85-1957, Appropriations, part 6, at 36–39 (Whitten); S. Hrg. 85-1957, Appropriations, at 24 (Senator Russell), 26 (Russell), 26–27 (Senator Stennis [D-MS]), and 42 (Senator Milton Young [R-ND]); S. Hrg. 85-1958, "Farm Program," part 2, at 303–4 (Senator Lister Hill [D-AL]).
45. See, e.g., S. Hrg. 85-1957, Appropriations, at 24 (Senator Russell) and 27–28 (Senator Olin Johnston [D-SC]); H. Hrg. 85-1957, Appropriations, part 6, at 161–62 (Whitten); S. Hrg. 85-1958, "Farm Program," part 2, at 331–32 (Senator Yarborough); H. Rept. 85-438, at 25. See also Talbot, "Farm Legislation," 598–99.
46. See, e.g., Cochrane and Ryan, *American Farm Policy*, 178–79 (feed grains and corn surplus) and 225 (cotton surplus); H. Rept. 86-2219 (Gov't Ops, 1960 for payment rates).
47. See S. Hrg. 85-1958, "Farm Program," part 2, at 354 (McLain); "Farm Program," part 2, at 359–60 (Senator Eastland); H. Hrg. 85-1957, "Soil Bank," at 21 (USDA's Dr. Paarlberg); H. Hrg. 85-1957, "Acreage Reserve," at 86–87 (Representative August Andresen), 70, and 90 (Representative Abernethy).
48. See, e.g., H. Hrg. 85-1957, "Soil Bank," at 36 and 58 (Benson), 75–76 (Gehlbach), 136 (Representative H. Carl Andersen [R-MN] took credit, with Representative Fred Marshall [D-MN]); H. Hrg. 85-1957, "Acreage Reserve," at 14 (Chairman Cooley), 16 (Representative Abernethy), 52, 61 (McLain), 59 (Chairman Cooley and Secretary Benson), 80, 90–91 (Charles B. Shuman for AFBF), and 93 (Representative August Andresen).
49. See H. Hrg. 85-1957, Appropriations, part 6, at 42–43 (Whitten), 51 (Representative Poage), and 163–64 (Whitten submits for the record a news release about a speech to cotton farmers from March 18, 1957, in Lubbock, Texas).
50. See, e.g., Bridges, "Deserving Poor"; Kotz, "Jamie Whitten"; H. Hrg. 85-1957, "Soil Bank," at 42. See also Coppess, "High Cotton."
51. See, e.g., Cong. Rec., March 12, 1956, at 4460 (Senator Walter George [D-GA] introduced the Southern Manifesto in the Senate) and 4515–16 (Representative Howard Smith [D-VA] introduced it in the House); Brown v. Board of Education of Topeka, 347 U.S. 483 (1954) (decided May 17, 1954). See, generally, Daniel, *Lost Revolutions*; Zelizer, *On Capitol Hill*; Katznelson, *Affirmative Action*; MacLean, *Democracy in Chains*; Mayer, "Eisenhower"; Winquist, "Civil Rights"; Lane, "Civil Rights"; Brown, "Congressional"; Library of Congress, "The Civil Rights Act of 1964: A Long Struggle for Freedom," https://www.loc.gov/exhibits/civil-rights-act/civil-rights-era.html; U.S. House of Representatives,

History, Art & Archives, Historical Highlights: The Southern Manifesto of 1956 (March 12, 1956), https://history.house.gov/Historical-Highlights/1951-2000/The-Southern-Manifesto-of-1956/. McCarthyism was at its peak from 1950 to 1954. See, e.g., The Miller Center, University of Virginia, "McCarthyism and the Red Scare," https://millercenter.org/the-presidency/educational-resources/age-of-eisenhower/mcarthyism-red-scare; Dwight D. Eisenhower Presidential Library, "McCarthyism/The 'Red Scare,'" https://www.eisenhowerlibrary.gov/research/online-documents/mccarthyism-red-scare; U.S. Senate, "Have You No Sense of Decency?" https://www.senate.gov/about/powers-procedures/investigations/mccarthy-hearings/have-you-no-sense-of-decency.htm.

52. See, e.g., Katznelson, *Affirmative Action*, 30–31; Cobb, *Most Southern Place*, 207–29.
53. See, e.g., Fite, *Cotton Fields No More*, 218.
54. Daniel, *Lost Revolutions*, 217, 227, and 296; Cobb, *Most Southern Place*, 254–56.
55. See H. Rept. 84-1986, at 40 (dividing payments; landlords); H. Hrg. 85-1957, "Acreage Reserve," at 6, 11; H. Hrg. 85-1957, "Soil Bank," at 6 and 53 (Representative Jones [D-MO]); H. Hrg. 85-1957, Appropriations, part 4, at 111–12 (USDA's general counsel Farrington); H. Hrg. 85-1958, Appropriations, part 4, at 1979–91, 1995–96.
56. Daniel, *Lost Revolutions*, 8, 216–17, 296; Daniel, *Breaking the Land*, 168. See also Daniel, *Shadow of Slavery*; Daniel, *Dispossession*. See, e.g., Baldwin, *Poverty and Politics*; Conrad, *Forgotten Farmers*; Fite, *Cotton Fields*; Reynolds, "Black Farmers"; U.S. Commission on Civil Rights, "Equal Opportunity"; U.S. Commission on Civil Rights, "Decline of Black Farming"; Banfield, "Ten Years"; Browne, "Only Six Million Acres"; Mitchell, "From Reconstruction"; Mitchell, "Destabilizing"; Jones, "Negro Farmer"; Levine, "Economic"; V. R. Newkirk, "The Great Land Robbery," *The Atlantic*, September 2019, https://www.theatlantic.com/magazine/archive/2019/09/this-land-was-our-land/594742/; T. Philpott, "Black Farmers Have Been Robbed of Land. A New Bill Would Give Them a 'Quantum Leap' Toward Justice," *Mother Jones*, November 19, 2020, https://www.motherjones.com/food/2020/11/black-farmers-have-been-robbed-of-land-a-new-bill-would-give-them-a-quantum-leap-toward-justice/; See also J. Coppess, "The History and Development of USDA Farm Loan Programs, Part 3: 1946 to 1961," *Farmdoc Daily* 11, no. 46, Department of Agricultural and Consumer Economics, University of Illinois at Urbana-Champaign, March 25, 2021, https://farmdocdaily.illinois.edu/2021/03/the-history-and-development-of-usda-farm-loan-programs-part-3-1946-to-1961.html.
57. See, e.g., Bobby J. Smith II, "Mississippi's War against the War on Poverty: Food Power, Hunger, and White Supremacy," *Study the South*, Center for the Study of

Southern Culture, University of Mississippi, July 1, 2019, https://southernstudies.olemiss.edu/study-the-south/ms-war-against-war-on-poverty/.
58. See Madison, *Federalist*, no. 10. See also Cong. Rec., March 9, 1956, at 4385 (Senator Richard Russell [D-GA]).
59. See H. Hrg. 85-1957, Appropriations, part 4, at 44–45 (McLain) and 160; H. Hrg. 85-1958, Appropriations, part 1, at 329–30.
60. Cong. Rec., March 12, 1956, at 4515–16 (Representative Howard W. Smith's statement submitting the manifesto).

7. From Dust to Dust

1. See H. Ex. Doc. 87-15, 1962 Budget, at M55; H. Rept. 87-448, at 2. See also Coppess, *Fault Lines*, 112–14.
2. See S. Ex. Doc. 87-12, 1961.
3. See Cong. Rec., March 7, 1961, at 3397 (Cooley) and 3403 (Whitten opposed to payment-in-kind provision of an amendment and linked it to the Soil Bank); Cong. Rec., March 8, 1961, at 3474 (side-by-side comparison in Senate debate), 3481 (Senate Ag Chairman Allen Ellender [D-LA] blamed the Soil Bank for not cutting corn production and subtly threatened corn farmers to make this one work), 3651 (Representative James Battin [R-MT] called it "nothing more than another soil bank"), 3669 (Representative J. Ernest Wharton [R-NY]), 3670–71 (House passage by 209–202 [21 not voting]); Cong. Rec., March 10, 1961, at 3727 (Senator Everett Dirksen [R-IL] connects it to the Soil Bank), 3736 (Senator Roman Hruska [R-NE] connects it to the Soil Bank), 3768 (Senate passage of its version by 52–26 [22 not voting]); H. Rept. 87-170; Cong. Rec., March 21, 1961, at 4412 (House passage of conference report, 231–185 [15 not voting]); Cong. Rec., March 22, 1961, at 4508 (Chairman Ellender noted that there were no payment limits for the feed grains program as compared to the Soil Bank) and 4509 (Senate passage of the conference report, 58–31 [11 not voting]). See also Cochrane and Ryan, *American Farm Policy* (surplus situation); Bigelow and Borchers, "Major Land Uses."
4. See, e.g., P.L. 87-5 (1961 crop); P.L. 87-128 (1962 crop); P.L. 87-703 (1963 crop); P.L. 88-26 (applying to the 1964 and 1965 crops); P.L. 89-321 (applying to the 1966 through 1969 crops).
5. See Coppess, *Fault Lines*, 116–30; Coppess, "High Cotton."
6. Note that an average of 45 million acres were removed from production each year since the 1956 Soil Bank and nearly 55 million acres each year from 1961 to 1969. See P.L. 91-524; Cochrane and Ryan, *American Farm Policy*, 56–61; Coppess, *Fault Lines*, 136–38. See also Nixon, "Remarks," 1971; Nixon, "Letter Accepting," 1971 (Hardin letter of resignation was dated October 25, 1971);

Richard Goldstein, "Earl L. Butz, Secretary Felled by Racial Remark, Is Dead at 98," *New York Times*, February 4, 2008, https://www.nytimes.com/2008/02/04/washington/04butz.html; Purdue University, Archives and Special Collections, Butz, Earl L. (Earl Lauer), 1909–2008, Biographical Information, https://archives.lib.purdue.edu/agents/people/937. See also H. Rept. 91-1329; S. Rept. 91-1154; H. Rept. 91-1594.

7. See Christine Mai-Duc, "The 1969 Santa Barbara Oil Spill That Changed Oil and Gas Exploration Forever," *Los Angeles Times*, May 20, 2015, https://www.latimes.com/local/lanow/la-me-ln-santa-barbara-oil-spill-1969-20150520-htmlstory.html; Lila Thulin, "How an Oil Spill Inspired the First Earth Day," *Smithsonian Magazine*, April 22, 2019, https://www.smithsonianmag.com/history/how-oil-spill-50-years-ago-inspired-first-earth-day-180972007/; Jon Hamilton, "How California's Worst Oil Spill Turned Beaches Black and the Nation Green," NPR.org, January 28, 2019, https://www.npr.org/2019/01/28/688219307/how-californias-worst-oil-spill-turned-beaches-black-and-the-nation-green. See also Spezio, "Santa Barbara"; Clarke and Hemphill, "Santa Barbara."

8. See Lorraine Boissoneault, "The Cuyahoga River Caught Fire at Least a Dozen Times, but No One Cared Until 1969," *Smithsonian Magazine*, June 19, 2019, https://www.smithsonianmag.com/history/cuyahoga-river-caught-fire-least-dozen-times-no-one-cared-until-1969-180972444/; Cronin, "Cuyahoga Fire"; Stradling and Stradling, "Perceptions"; Adler, "Fables." See also Silverstein, *Law's Allure*, 128; Switzer, *Green Backlash*, 7; Kagan, *Adversarial Legalism*, 47; Kraft, "US Environmental"; EarthDay.org, https://www.earthday.org/history/.

9. The Senate agreed to the amended bill by unanimous consent (without a recorded vote) on July 10, 1969, and the House overwhelmingly agreed to its version on September 23, 1969. The Senate adopted the conference report on December 20, 1969, again by unanimous consent, and the House followed suit two days later, adopting the conference report without a recorded vote. Cong. Rec., July 10, 1969, at 19103 (Senate passage); Cong. Rec., September 23, 1969, at 26590 (by a vote of 372–15 [43 not voting]); Cong. Rec., December 20, 1969, at 40427 (Senate adopted the conference report); Cong. Rec., December 22, 1969, at 40928 (the House adopted the conference report without a recorded vote); P.L. 91-190; Jim Kershner, "NEPA, the National Environmental Policy Act," Historylink.org, August 27, 2011, https://www.historylink.org/File/9903; Lewis and Hensley, "May 4 Shootings."

10. Technically, the president had proposed Reorganization Plan No. 3 for the creation of the EPA, which the House allowed to go into effect when a majority voted against the resolution to disapprove the reorganization plan on September 28, 1970, without a recorded vote. See Cong. Rec., September 28, 1970, at 33884;

H. Ex. Doc. 91-364, 1970; H. Ex. Doc. 9-366, 1970; H. Rept. 91-1464; 91 H. Res. 1209; S. Rept. 91-1250; P.L. 91-604; Flippen, *Nixon and the Environment*, 80.

11. See Flippen, *Nixon and the Environment*, 87–88, 106–10, 113–16, 129–34.
12. See, e.g., Flippen, *Nixon and the Environment*; Bosso, *Pesticides and Politics*, 153–54, 166; Gottlieb, "Next Environmentalism," 302; Kraft, "US Environmental," 25; Silverstein, *Law's Allure*, 138. See also R. W. Apple Jr., "And Now, On to '72," *New York Times*, November 6, 1970, https://timesmachine.nytimes.com/timesmachine/1970/11/06/93891073.html?pageNumber=65; "The Nation," *New York Times*, November 8, 1970, https://www.nytimes.com/1970/11/08/archives/the-nation-what-vote-did-election-what-vote-meant-nixon-must-change.html; Fred P. Graham, "Environment Watchdog," *New York Times*, November 12, 1970, https://timesmachine.nytimes.com/timesmachine/1970/11/12/issue.html; "The Powell Memo: A Call-to-Arms for Corporations," BillMoyers.com (Moyers & Company), September 14, 2012, https://billmoyers.com/content/the-powell-memo-a-call-to-arms-for-corporations/ (link to memo: https://billmoyers.com/wp-content/uploads/2012/09/Lewis-Powell-Memo.pdf). Of note, President Nixon nominated Powell to the Supreme Court on October 22, 1971, and the Senate confirmed his appointment on December 6, 1971. See Supreme Court Historical Society, https://supremecourthistory.org/associate-justices/lewis-f-powell-jr-1972-1987/.
13. See Flippen, *Nixon and the Environment*, 16, 54, 102–3; Bosso, *Pesticides and Politics*, 58, 145–51, 168 (figure 3), 174, 176; Switzer, *Green Backlash*, 10; Farber, "Conservative," 1024; Kraft, "US Environmental," 23; Silverstein, *Law's Allure*, 130–32; P.L. 80-104; P.L., 80-845; Cong. Rec., October 4, 1972, at 33718 (passed Senate 74–0) and 33767 (passed the House by a vote of 366–11); Cong. Rec., October 17, 1972, at 36859–60 (Nixon veto message entered in the record of the Senate); Cong. Rec., October 17, 1972, at 36879 (Senate override); Cong. Rec., October 18, 1972, at 37060–61 (House override); Hines, "History," 84; Silverstein, *Law's Allure*, 141.
14. It passed the Senate on November 2, 1971, by a vote of 86–0 (14 not voting), and the House by a vote of 380–14 (37 not voting) on March 29, 1972. Both chambers passed the conference report overwhelmingly on October 4, 1972—the Senate by a vote of 74–0 (26 not voting), and the House by a vote of 366–11 (53 not voting). The Senate voted to override the president's veto on October 17, 1972, by a vote of 52–12 (36 not voting), and the next day the House also voted to override his veto by a vote of 247–23 (160 not voting, 1 voting present). See 92 S.2770; 92 H.R. 11896. See Cong. Rec., November 2, 1971, at 38865 (Senate passage); Cong. Rec., March 29, 1972, at 10803 (House passage); Cong. Rec., October 4, 1972, at 33718 (Senate) and 33767 (House); S. Doc. 92-93; Cong. Rec.,

October 17, 1972, at 36859–60 (veto message) and 36879 (Senate override); Cong. Rec., October 18, 1972, at 37060–61 (House override). See also Hines, "History," 81 (quoting and citing Richard J. Lazarus, *The Making of Environmental Law* [Chicago: University of Chicago Press, 2004], 59, 97); Bosso, *Pesticides and Politics*, 115–26, 129, 135–41, 150–56, 158–71, 174–77; Flippen, *Nixon and the Environment*, 156–57, 178–79.

15. See Flippen, *Nixon and the Environment*, 16, 54, 102–3; Bosso, *Pesticides and Politics*, 58, 145–51, 168 (figure 3), 174, 176; Switzer, *Green Backlash*, 10; Farber, "Conservative," 1024; Kraft, "US Environmental," 23; Silverstein, *Law's Allure*, 130–32; P.L. 80-104; P.L. 80-845; Cong. Rec., October 4, 1972, at 33718 (passed Senate 74–0) and 33767 (passed the House by a vote of 366–11); S. Ex. Doc. 92-93, 1972; Cong. Rec., October 17, 1972, at 36859–60 (Nixon veto message entered in the record of the Senate); Cong. Rec., October 17, 1972, at 36879 (Senate override); Cong. Rec., October 18, 1972, at 37060–61 (House override); Hines, "History," 84; Silverstein, *Law's Allure*, 141.

16. Hines, "History," 81.

17. See Gottlieb, "Next Environmentalism," 300–302; Turner, "Specter," 123–24, 129; Batie, *Soil Erosion*; Sampson, *Farmland or Wasteland*. See also S. Hrg. 97-1982, S.1825, at 21. According to Tom Barlow of the Natural Resources Defense Council, the "national environmental and conservation organizations have not been putting in the time on this issue . . . and that is a disgrace."

18. Batie, *Soil Erosion*, 5.

19. All calculations for acreage changes were made by the author using USDA publicly reported datasets. See Bigelow and Borchers, "Major Land Uses." See also J. Coppess, C. Zulauf, G. Schnitkey, N. Paulson, and K. Swanson, "Production Controls & Set Aside Acreage Discussion, Part IV: Reviewing PIK Policies," *Farmdoc Daily* 10, no. 141, Department of Agricultural and Consumer Economics, University of Illinois at Urbana-Champaign, July 30, 2020, https://farmdocdaily.illinois.edu/2020/07/production-controls-set-aside-acreage-discussion-part-iv-reviewing-pik-policies.html; J. Coppess, N. Paulson, G. Schnitkey, C. Zulauf, and K. Swanson, "Acres & Crisis: Prospective Plantings and Perspectives from the Past," *Farmdoc Daily* 10, no. 66, Department of Agricultural and Consumer Economics, University of Illinois at Urbana-Champaign, April 9, 2020, https://farmdocdaily.illinois.edu/2020/04/acres-crisis-prospective-plantings-and-perspectives-from-the-past.html.

20. See S. Rept. 93-1033.

21. See Barrett et al., "Population"; Ros-Tonen and van Boxel, "El Nino"; William Safire, "El Nino," *New York Times*, August 30, 1979, https://www.nytimes.com/1973/08/30/archives/el-nino-essay.html; Akihiko, "Formation"; Madigan,

"Adverse." See also J. Coppess, "A Brief Review of the Consequential Seventies," *Farmdoc Daily* 9, no. 99, Department of Agricultural and Consumer Economics, University of Illinois at Urbana-Champaign, May 30, 2019, https://farmdocdaily.illinois.edu/2019/05/a-brief-review-of-the-consequential-seventies.html; Coppess, *Fault Lines*, 314 (appendix 1, figure 1).

22. S. Rept. 93-173, at 11 (reprinting the secretary's speech before the National Agricultural Advertising and Marketing Association, Memphis, Tennessee, May 1, 1973); Sampson, *Farmland or Wasteland*, 9; Chen, "Get Green," 335; Rosenberg and Stucki, "Butz Stops Here," 17; P.L. 93-86; Coppess, *Fault Lines*, 138–46; Cochrane and Ryan, *American Farm Policy*.

23. GAO, "To Protect," 6.

24. GAO, "To Protect," 6; S. Com. Print., 94-1975, "Conservation," at 15; CRS Report 80-144S, "Soil Conservation," at 26; S. Hrg. 97-1981, "Recommendations," at 20.

25. Sampson, *Farmland or Wasteland*, 45; Batie, *Soil Erosion*, 5–7 (citing U.S. Water Resources Council, *Second National Water Assessment: The Nation's Water Resources, 1975–2000*, vol. 1, summary [Washington DC: U.S. Government Printing Office, 1978], 60) and 10–11 (quoting J. L. Simon, "Are We Losing Ground?" *Illinois Business Review* 37, no. 3 [1980]: 3).

26. See, generally, GAO, "To Protect"; Sampson, *Farmland or Wasteland*; Batie, *Soil Erosion*. See also P.L. 93-205; Nixon, "Endangered Species Act of 1973," Signing Statement (December 28, 1973); P.L. 93-523; Ford, "Safe Drinking Water Act," Signing Statement (December 17, 1974); P.L. 94-469; Ford, "Toxic Substances Control Act," Signing Statement (October 12, 1976).

27. See, e.g., Flippen, *Nixon and the Environment*, 106, 210–11, 214–18; Bosso, *Pesticides and Politics*, 151, 180; Sampson, *Farmland or Wasteland*, 9. See also Jones and Strahan, "Effect"; Mitchell, "Resources"; U.S. Department of State, "1973 Arab-Israeli War"; P.L. 93-159; P.L. 93-511; P.L. 93-319.

28. See Flippen, *Nixon and the Environment*, 214-19; Kraft, "US Environmental," 25; Silverstein, *Law's Allure*, 139; GAO, "Greater Conservation Benefits," 5; P.L. 93-344.

29. S. Hrg. 93-1974, "Implementation," at 6–7 (Senator Clark) and 41–42 (statement of Lyle Bauer, Vice President, National Association of Conservation Districts). See also GAO, "To Protect," 6.

30. See, e.g., Bosso, *Pesticides and Politics*, 166, 186–87; Rieselbach, *Congressional Reform*, 43–44; Ornstein, "Democrats Reform," 2–3, 35; Rieselbach and Unekis, "Ousting," 87–92. See also Paarlberg, *Farm*, 121–22.

31. S. Com. Print., 94-1975, "Conservation." See also Heady, "Externalities."

32. S. Hrg. 94-1976, "Cost-Sharing," at 1 (Senator Clark).

33. H. Hrg. 94-1976, "Resource Conservation," at 20, 23, 29 (Barlow).
34. See, e.g., Rodriguez, Bullock, and Boerngen, "Origins," 725; Kramer and Batie, "Cross Compliance," 314.
35. See Cong. Rec., October 1, 1976, at 35081 (inserted into the record, Ford's statement of disapproval [pocket veto] received by the Senate on October 26, 1976).
36. See S. Rept. 94-895, at 1; H. Rept. 94-1744, at 4; Cong. Rec., May 25, 1976, at 15334; Cong. Rec., October 1, 1976, at 35148 (de la Garza called it up) and 35152 (House passage); Cong. Rec., October 1, 1976, at 34440 (Senate passage). See also Leman, "Political Dilemmas," 58–59.
37. See GAO, "To Protect," 6, 30, 41, 45; S. Hrg. 95-1977, "Protection," at 29 (comment by R. M. Davis, administrator of SCS) and 65 (Marion Edey from the Council on Environmental Quality).
38. S. Hrg. 95-1977, "Protection," at 902–3 (Senator Melcher [D-MT] and Barlow); H. Hrg. 95-1977, "General Farm Bill," at 332–33 (Barlow).
39. S. Hrg. 95-1977, "Protection," at 902–3 (Senator Melcher [D-MT] and Barlow); H. Hrg. 95-1977, "General Farm Bill," at 332–33 (Barlow); S. Rept. 95-180, at 186–88, 190; H. Rept. 95-348, at 155 (Representative Harkin amendment); S. Rept. 95-418, at 238–39; H. Rept. 95-599, at 238–39.
40. P.L. 95-113. See, e.g., H. Rept. 95-348, at 9, 36–37, 156 (Poage amendment), 241; S. Rept. 95-418, at 241; P.L. 95-192; 94 S.2081; 94 H.R. 10456; 94 H.R. 14912; 94 H.R. 75; 95 S.106; CRS Report 80-144S, "Soil Conservation." See also Kramer and Batie, "Cross Compliance," 314 (citing W. D. Seitz and R. G. F. Spitze, "Environmentalizing Agricultural Production Control Policies," *Journal of Soil and Water Conservation* 28 [March 1973]).

8. A New Foundation

1. P.L., 99-198, at Sec. 1211, Sec. 1221; Malone, "Historical Essay," 577.
2. See Cong. Rec., February 7, 1980, at 2260; 96 H.R. 3681; H. Hrg. 96-1979, "Conservation," at 8–10, 12, 14, 15 (statement of M. Rupert Cutler, Assistant Secretary of Agriculture for Natural Resources and Environment).
3. See H. Hrg. 96-1979, "Conservation," at 211, 216 (subcommittee chair Ed Jones [D-TN] and Richard Ekstrum, president, South Dakota Farm Bureau); H. Hrg. 96-1979, "Conservation," at 794–95 (Harold B. Steele, Illinois Farm Bureau President), 854 (Representative Ed Madigan [R-IL] and witness Ed Bowman), 855 (Gilbert Fricke with the Illinois Department of Agriculture). Compare H. Hrg. 96-1979, "Conservation," at 190 (South Dakota State Representative R. Lars Herseth), 348–49 (comments of John Ware [Buddy] Moses, farmer, Brownsville, Tennessee), 839 (Harold Parett, farmer, Mahomet, Illinois), and 856 (Charles Guthrie [Taylorville, Illinois]).

4. See H. Hrg. 96-1979, "Conservation," at 65 (statement of Neil Sampson, Executive Vice President, National Association of Conservation Districts); H. Hrg. 96-1979, "Conservation," at 21 (Tennessee NACD witness).
5. See 96 H.R. 3681; Cong. Rec., February 6, 1980, at 2150–51 (Representatives Beilenson [D-CA], Latta [R-OH], and Grassley [R-IA]), 2152 (Representatives Jeffords, Fithian [D-IN], and Glickman [D-KS]), 2153–54 (Representative Tauke [R-IA] and House Ag Chairman Tom Foley [D-WA]), 2151–52 (Representatives Marlanee [R-MT] and Symms [R-ID]), 2155 (Representative Wampler [R-VA]), 2159 (Grassley), and 2162 (Representative Charlie Stenholm [D-TX]); Cong. Rec., February 7, 1980, at 2266, 2268 (Stenholm), 2164 (Stockman), and 2270–72 (Grassley offered the motion to recommit, passed 227–163 [43 not voting]; and then on final passage with the Grassley amendments, it was defeated 177–210 [46 not voting]). See also GAO, "Preserving America's Farmland"; Benbrook, "Integrating"; Adler, "Back to the Future"; Switzer, *Green Backlash*, 247–76.
6. See, generally, Harl, *Farm Debt Crisis*; Sunbury, *Farm Credit Empire*; Barnett, "U.S. Farm Financial Crisis"; Smith, "Social and Ecological"; Sigelman, "Politics"; Dinterman, Katchova, and Harris, "Financial Stress"; Stam and Dixon, "Farmer Bankruptcies." See also USDA, ERS, *Data Files*, "Net Cash Income."
7. See 97 S.1825; S. Hrg. 97-1982, S.1825, at 1–4 (Senator Armstrong), 7–10 (ASCS administrator Everett Rank), 13 (Senator Thad Cochran [R-MS] and Rank), and 21 (Thomas Barlow). See also Joey Bunch and K. Mitche, "Former U.S. Senator, University President Bill Armstrong Has Died," *Denver Post*, July 6, 2016, https://www.denverpost.com/2016/07/06/bill-armstrong-colorado-dies/; Dan Elliott, "William L. Armstrong, Republican Senator from Colorado, Dies at 79," *Washington Post*, July 7, 2016, https://www.washingtonpost.com/politics/william-l-armstrong-republican-senator-from-colorado-dies-at-79/2016/07/07/5a4fa18a-4458-11e6-8856-f26de2537a9d_story.html; Associated Press, "William Armstrong, Conservative Voice in the Senate, Dies at 79," *New York Times*, July 6, 2016, https://www.nytimes.com/2016/07/07/us/politics/william-armstrong-two-term-colorado-senator-dies-at-79.html.
8. P.L. 95-192; S. Rept. 97-126, at 25–27, 225; H. Rept. 97-106, at 287.
9. See P.L. 97-98, at Title XV ("Resource Conservation") and Sec. 1502; H. Rept. 97-377, at 251; S. Rept. 97-290, at 251. See also S. Rept. 97-126, at 2–3, 25–27; H. Rept. 97-106, at 81.
10. P.L. 97-98, at Sec. 1503; S. Rept. 97-126, at 2–3, 37–38, 145–44; H. Rept. 97-106, at 81–84. See also Braden and Uchtmann, "Soil Conservation."
11. See P.L. 97-98, at Title XV; H. Rept. 97-337, at 25; H. Rept. 97-106, at 73, 81, 90; S. Rept. 97-126, at 151, 215–16; Cong. Rec., October 22, 1981, at 24903–4.

12. See CQ *Almanac*, "$34 Billion"; S. Hrg. 97-1982, "Oversight," at 14 (Milton Mekelburg, President of the National Association of Conservation Districts), 19–20 (Sampson), 24–25 (Sampson); S. Hrg. 97-1982, S.3074, at 42–43; S. Hrg. 97-1982, S.1825, at 15–17 (Sampson).
13. S. Hrg. 97-1982, "Oversight," at 3 (Secretary Block), 4 (Senator John Melcher [D-MT]), 6 (Senator Ed Zorinksy [D-NE]), 10–13 (Richard D. Siegel, Deputy Assistant Secretary for Natural Resources and Environment), 21–23 (Senator Jepsen and Sampson), and 29–30 (Mary Kay Thatcher, AFBF); S. Hrg. 97-1982, S.1825, at 15–17, 19–20 (Sampson). See also Prentiss, "Sodbusting"; Ogg, "Soil Conservation"; McSweeny and Kramer, "Farm Level"; Ogg, Johnson, and Clayton, "Policy Option"; Braden, "Some Emerging"; Ervin, Heffernan, and Green, "Cross-Compliance"; Braden and Uchtmann, "Soil Conservation," 665, 670, 686–87.
14. See S. Hrg. 98-254, "Oversight," at 1–2 (Senator Jepsen [R-IA]), 5 (Senator Melcher), 7–8, 11 (Block), 13, 15 (response of John B. Crowell, Assistant Secretary for Natural Resources and Environment), and 19 (Sampson). See also Coppess, *Fault Lines*, 158–59; J. Coppess, C. Zulauf, G. Schnitkey, N. Paulson, and K. Swanson, "Production Controls & Set Aside Acreage Discussion, Part IV: Reviewing PIK Policies," *Farmdoc Daily* 10, no. 141, Department of Agricultural and Consumer Economics, University of Illinois at Urbana-Champaign, July 30, 2020, https://farmdocdaily.illinois.edu/2020/07/production-controls-set-aside-acreage-discussion-part-iv-reviewing-pik-policies.html.
15. See S. Hrg. 98-254, "Oversight," at 12 (Block). See, e.g., Pigford v. Glickman, 185 F.R.D. 82 (D.D.C. 1999); Daniel, *Dispossession*; Reynolds, "Black Farmers"; U.S. Commission on Civil Rights, "Equal Opportunity"; U.S. Commission on Civil Rights, "Decline of Black Farming"; V. R. Newkirk, "The Great Land Robbery," *The Atlantic*, September 2019, https://www.theatlantic.com/magazine/archive/2019/09/this-land-was-our-land/594742/; T. Philpott, "Black Farmers Have Been Robbed of Land. A New Bill Would Give Them a 'Quantum Leap' Toward Justice," *Mother Jones*, November 19, 2020, https://www.motherjones.com/food/2020/11/black-farmers-have-been-robbed-of-land-a-new-bill-would-give-them-a-quantum-leap-toward-justice/.
16. See 98 S.663; S. Rept. 98-296, at 1, 5, 8–10, 11 (reported out by unanimous vote, as amended on October 19, 1983), 13 (Block's letter), and 15 (Huddleston letter).
17. See, e.g., S. Hrg. 98-1983, S.663, at 26–27 (Norman A. Berg, Washington Representative for Soil Conservation Society of America and Senior Advisor to American Farmland Trust); Cong. Rec., November 18, 1983, at 34403–4 (Majority Leader Baker and Senator Armstrong), 34406, 34409 (Senator Jepsen), 34411 (Senator Baucus); H. Hrg. 98-1983, "Conservation," at 63 (Representative

Charlie Stenholm [D-TX]), 73–74 (Ronald A. Michielli, Vice President, Natural Resources, National Cattlemen's Association), 320 (Representative Cooper Evans [R-IA]), and 372 (John Tarburton, President, Delaware Farm Bureau, on behalf of the American Farm Bureau Federation).

18. See H. Rept. 98-696, at 1, 6–9, 11, 28–29, 32–44 (text of substitute); Cong. Rec., May 8, 1984, at 11362. See also Jay Kohn and Mike Dennison, "Longtime Montana GOP Congressman Ron Marlanee Dead at 84," *Missoula Current/MTN News*, April 29, 2020, https://missoulacurrent.com/government/2020/04/ron-marlenee/.

19. Cong. Rec., May 8, 1984, at 11367 (Representative Marlanee).

20. Cong. Rec., May 8, 1984, at 11368 (Chairman de la Garza), 11369–71 (Representative Jeffords), and 11374 (passage).

21. See S. Rept. 98-1984, S.663, unpublished conference transcript; H. Hrg. 99-5, 1985, "General Farm Bill," at 655 (April 4, 1985; Representative Ed Jones, subcommittee chair); Cong. Rec., August 9, 1984, at 23292–93 (Senator Armstrong appropriations amendment), 23294–96 (Senator Melcher substitute), and 23299 (Melcher's substitute passed 62–34). The sodbuster provision was dropped in conference with the House. See CQ *Almanac*, "Sodbuster"; CQ *Almanac*, "Farm Program Funds Included in Stopgap Bill."

22. See H. Hrg. 99-2, 1985, "Economic Conditions," at 14–15 (Secretary Block); H. Hrg. 99-5, 1985, "General Farm Bill," at 428–30 (Robert Gray, Director of Policy Development, American Farmland Trust), 432–33 (Norman Berg, Soil Conservation Society), 434–38 (Ken Cook); H. Hrg. 99-5, 1985, "General Farm Bill," at 663–64 (USDA), 736 (Representative Brown quote), 740 (Robert J. Gray, Director of Policy Development for American Farmland Trust), 742–43 (Ken Cook), 743–44 (David Senter, National Director, American Agricultural Movement), 744–45 (John Tarburton, President, Delaware Farm Bureau Federation), 752–53, 757 (NACD's Charles Boothby and Representative Thomas; AFT's Robert Gray), 763–64 (Norman Berg, Washington Representative, Soil Conservation Society of America), 765–67 (Justin Ward, NRDC), 768–69 (Marjorie Raines, Chairman, West Tennessee Sierra Club), 770 (Michael Heller, Chesapeake Bay Foundation), 772 (Mr. Weiss), 773 (Representative Thomas), and 777–78 (Neil Sampson, Executive Vice President, American Forestry Association); S. Hrg. 99-55, 1985, "Framework"; Cong. Rec., April 4, 1985, at 7598 (Representatives Robert Smith [R-OR] and Coleman [R-MO]).

23. H. Hrg. 99-5, 1985, "General Farm Bill," at 738–39 (Representative Stenholm).

24. H. Hrg. 99-5, 1985, "General Farm Bill," at 737–38 (Representative Wolpe) and 746–47 (Laverne Lemen, National Cotton Council). See also H. Hrg. 99-5, 1985, "General Farm Bill," at 655 (Chairman Ed Jones), 657–58 (ranking Republican

Coleman), 659–60 (Representative Pat Roberts [R-KS]), 735 (Representative Hank Brown [R-CO]), 750–52 (Representative Dan Glickman [D-KS]).

25. H. Hrg. 99-5, 1985, "General Farm Bill," at 736 (Daschle quote); H. Rept. 99-271, at 410–11, 413–17. See also 99 H.R. 1656; 99 H.R. 2108; 99 H.R. 2112; 99 H.R. 2100-IH; Carroll, "Soil Conservation"; Cook, "1985 Farm Bill"; McCullough and Weiss, "Environmental Look"; Sara Wyant, "The Man behind the Chairman," *High Plains Journal*, September 4, 2006, https://m.hpj.com/archives/the-man-behind-the-chairman/article_7a9bc83e-01c3-523c-852c-ea7b616f002f.html.

26. Cong. Rec., September 20, 1985, at 24533 (Chairman de la Garza), 24539 (Jones quote), 24552 (Coleman quote); Cong. Rec., October 3, 1985, at 26012 (Wolpe quote), 26011–13 (Glickman), 26017 (the recorded vote was 313–99 [31 not voting]). See also H. Rept. 99-271, at 417–19; CQ *Almanac*, "Farm Bill Granted."

27. Cong. Rec., October 3, 1985, at 26012–14 (Marlanee quote and Representatives Coleman and Wolpe) and 26016 (Glickman quote); Cong. Rec., October 8, 1985, at 26667 (passed by a vote of 282–141 [11 not voting]). See also Wickard v. Filburn, 317 U.S. 111, 131 (1942).

28. See 99 H.R. 2100-PCS.

29. See 99 S.1050; S. Rept. 99-145, at 300–304, 306–8, 354–55, 360–62, 463–69; Cong. Rec., October 23, 1985, at 28578.

30. See, e.g., Cong. Rec., October 25, 1985, at 29124 (Senator Zorinksy [D-NE], Senator Bob Kasten [R-WI]); Cong. Rec., November 22, 1985, at 33240–41 (Senator Kasten technical amendment); Cong. Rec., November 19, 1985, at 32313 (Senator Leahy [D-VT]). See also Coppess, *Fault Lines*, 162–64.

31. H. Rept. 99-447, at 454–55, 458–60, 461–65; P.L. 99-198; Reagan, "Statement on Signing," H.R. 2100 (December 23, 1985). See also Malone, "Historical Essay."

32. See, e.g., Malone, "Historical Essay"; Cook, "Environmental Era"; Hamilton, "Legal Issues"; Malone, "Conservation at the Crossroads."

33. See P.L. 99-198; H. Rept. 99-447, at 158, 461 (environmental quote), 462, 464 (priority quote), 465, 468.

34. P.L. 99-128; H. Rept. 99-447, at 462 (local economy quote), 464 (practicable quote), 467.

35. H. Hrg. 99-5, 1985, "General Farm Bill," at 736 (Representative Brown) and 764 (Berg).

36. P.L. 99-128, at Sec. 1211, 1212 (HEL compliance), 1221, and 1222 (wetlands compliance); H. Rept. 99-447, at 160–61, 454, 456, 460.

37. See Lou Cannon, "Stockman Quits OMB to Join Banking Firm," *Washington Post*, July 10, 1985, https://www.washingtonpost.com/archive/politics/1985/07/10/stockman-quits-omb-to-join-banking-firm/ba8add32-5f1f-4736-b302

-58894b82fba9/; Bernard Weinraub, "Stockman Resigns Top Budget Post; Boing to Wall St.," *New York Times*, July 10, 1985, https://www.nytimes.com/1985/07/10/us/stockman-resigns-top-budget-post-going-to-wall-st.html.

38. See, e.g., de la Garza, "Linking." See also U.S. House of Representatives, "de la Garza, Eligio (Kika), II," https://history.house.gov/People/Detail/12083; Library of Congress, "Hispanic Americans in Congress, 1822–1995: Eligio 'Kika' de la Garza II," https://www.loc.gov/rr/hispanic/congress/delagarza.html; Sam Roberts, "Kika de la Garza, Texas Congressman and Farmers' Ally, Dies at 89," *New York Times*, March 15, 2017, https://www.nytimes.com/2017/03/15/us/politics/kika-de-la-garza-dead.html.

39. Waldron, *Dignity*, 155–56.

40. See P.L. 99-128, at Sec. 1233, 1234, 1235; H. Rept. 99-447, at 465, 468.

9. Modern Developments

1. See USDA, ERS, *Data Files*, "Federal Government Direct Farm Payments"; P.L. 115-334, Title II; CBO, "May 2019 Baseline," 26; CBO, "Baseline as of March 6, 2020," 26.

2. See, e.g., Doering and Smith, "Examining"; Claassen et al., "Environmental Compliance"; GAO, "Farm Programs: Conservation Compliance"; Holland, Bennett, and Secchi, "Complying"; Claassen et al., "Environmental Compliance," AER-832. See also P.L. 104-127, at Sec. 311; P.L. 113-79, at Sec. 2611; S. Rept. 113-88, at 101–6; Cong. Rec., June 20, 2012, at s4353–54 (Chambliss amendment agreed to by a vote of 52–47 [1 not voting]); Daniel Looker, "Crop Insurance Linked to Income, Conservation," *Successful Farming*, June 19, 2012, http://www.agriculture.com/news/policy/crop-insurce-linked-to-income_4-ar24816; David Rogers, "House Agriculture Committee Pulls Back on Farm Bill Markup," *Politico*, June 20, 2012, http://www.politico.com/story/2012/06/cantor-hits-pause-on-farm-bill-action-077651; NACD, "Conservation Compliance."

3. H. Rept. 101-916, at 909–11; P.L. 101-624, at Sec. 1421, 1422.

4. See, e.g., S. Hrg. 104-359, 1995, "Conservation," at 16 (Senator Kent Conrad [D-ND]) and 28–29 (John Laurie, President, Michigan Farm Bureau).

5. See, e.g., McCorvie and Lant, "Drainage District"; Urban, "Uninhabited Waste"; J. Coppess, "Dead Zones & Drinking Water, Part 2: Why Tile Matters," *Farmdoc Daily* 6, no. 47, Department of Agricultural and Consumer Economics, University of Illinois at Urbana-Champaign, March 10, 2016, https://farmdocdaily.illinois.edu/2016/03/dead-zones-drinking-water-part2.html; Jonathan Coppess, "Thinking About the Des Moines Water Works Lawsuit and the History of Drainage," Policy Matters, Department of Agricultural and Consumer Econom-

ics, University of Illinois at Urbana-Champaign, September 30, 2015, https://policymatters.illinois.edu/thinking-about-the-des-moines-water-works-lawsuit-and-the-history-of-drainage/.
6. S. Hrg. 104-359, 1995, "Conservation," at 16–17; GAO, "To Protect."
7. P.L. 104-127, at Sec. 321, 322; H. Rept. 104-494, at 379–81.
8. See, e.g., P.L. 107-171, at Sec. 2002; P.L. 110-234, at Sec. 2003; P.L. 113-79, at Sec. 2609, 2611; H. Rept. 113-333, at 419–20.
9. See, e.g., CRS Report R42459, "Conservation Compliance," at 8–9; Mikkel Pates, "ND Senators Decry Crop Insurance, Conservation Compliance Link," *Agweek*, May 16, 2013, http://www.agweek.com/event/article/id/20915/; Daniel Looker, "Senate Tacks On Conservation Compliance," *Successful Farming*, May 14, 2013, http://www.agriculture.com/news/policy/senate-tacks-on-conservation-complice_4-ar31526; S. Rept. 113-88, at 52–53, 101–6.
10. See P.L. 115-334, at Sec. 2101. See also GAO, "Farm Programs: USDA Should Take"; CRS Report R45698, "Agricultural Conservation," at 10. See also Claassen et al., "Conservation Compliance"; Claassen and Bowman, "Conservation Compliance."
11. P.L. 101-624, at Sec. 1431; H. Rept. 101-916, at 916–17 (quote) and 925–26. Acreage data can be found on the Farm Service Agency website. USDA, FSA, "Reports and Statistics," Conservation Reserve Program.
12. See P.L. 104-127; H. Rept. 104-494, at 113, 382–83. See also Schertz and Doering, *1996 Farm Act*; Coppess, *Fault Lines*, 181–94.
13. See H. Hrg. 103-91, 1994, "Review," at 7–8, 9–11 (Dr. Keith Collins, Acting Assistant Secretary for Economics), 37–38 (quote from the statement of Kenneth Cook, President, Environmental Working Group), and 43; H. Hrg. 103-92, 1994, "Future," at 9–10 (Secretary Mike Espy).
14. H. Hrg. 103-91, 1994, "Review," at 1 (Representative Johnson quote); H. Hrg. 103-92, 1994, "Future," at 2 (Senator Tom Daschle [D-SD]), 19–20 (testimony of Carl G. Anderson, Executive Secretary, South Dakota Grain and Feed Association), and 29 (Representative Johnson [D-SD]); S. Hrg. 104-359, 1995, "Conservation," at 2 (Chairman Richard Lugar [R-IN]), 3 (Senator Leahy [D-VT], ranking member), 11–12 (statement of Paul Johnson, Chief, Natural Resources Conservation Service, U.S. Department of Agriculture), 43 (Dr. Otto Doering of Purdue University), 44–45 (Dr. Bruce Babcock of Iowa State).
15. See H. Hrg. 103-91, 1994, "Review," at 33–34 (testimony of Theresa Gullo, Principal Analyst, Budget Analysis Division, Congressional Budget Office); H. Hrg. 103-92, 1994, "Future," at 8 (Secretary Espy), 9–10 (Senator Daschle), 12 (Representative Johnson).

16. H. Rept. 104-462, at 50; H. Rept. 102-617, at 80; S. Rept. 102-334, at 78; H. Rept. 103-153, at 76–77; S. Rept. 103-102, at 76; H. Rept. 103-542, at 70–71; S. Rept. 103-290, at 71; H. Rept. 104-172, at 61; S. Rept. 104-142, at 73.
17. See, e.g., P.L. 104-127, at Sec. 341; S. Rept. 104-317, at 67, 80–81; H. Rept. 104-726, at 26; H. Rept. 104-494, at 395; H. Rept. 104-462, at 32, 50.
18. See P.L. 107-171, at Sec. 2101. See also H. Rept. 107-424, at 482, 484; H. Rept. 107-191, at 128–29; S. Rept. 107-117, at 109.
19. See, e.g., Ifft, Rajagopal, and Weldzuis, "Ethanol"; Lark, Salmon, and Gibbs, "Cropland."
20. P.L. 110-234, at Sec. 2103; H. Rept. 110-627, at 710.
21. See P.L. 113-79, at Sec. 2001.
22. See P.L. 115-334, at Sec. 2201; USDA, FSA, "Reports and Statistics," Conservation Reserve Program.
23. H. Rept. 101-916, at 928–31.
24. P.L. 104-127, at Sec. 388; H. Rept. 104-494, at 381.
25. See P.L. 107-171, at Sec. 2201, Sec. 2401; H. Rept. 107-424, at 488–89, 502–3; P.L. 110-234, at Sec. 2201; H. Rept. 110-627, at 715–16.
26. Under budget rules, this helped all three achieve permanent CBO baseline status and thus helped secure funding going forward. See P.L. 113-79, at Sec. 2301; H. Rept. 113-333, at 408–10.
27. P.L. 115-334, at Sec. 2601; H. Rept. 115-1072, at 582–87; USDA, NRCS, "USDA Celebrates."
28. S. Hrg. 104-359, 1995, "Conservation," at 2.
29. P.L. 101-624, at Sec. 1439; H. Rept. 101-916, at 936–42.
30. P.L. 104-127, at Sec. 334, Sec. 341 (CCC funding); H. Rept. 104-494, at 385–89; P.L. 107-171, at Sec. 2701.
31. P.L. 110-234, at Sec. 2502–10, Sec. 2701.
32. P.L. 113-79, at Sec. 2201–8; H. Rept. 110-627, at 725–29; H. Rept. 113-333, at 405–8.
33. See P.L. 107-171, at Sec. 2001, Sec. 2701 (CCC funding); H. Rept. 107-424, at 477–82.
34. S. Hrg. 107-225, 2001, "Conservation," at 3 (Katherine R. Smith, Director, Resource Economics Division, Economic Research Service, U.S. Dept. of Agriculture), 5 (Jeffrey A. Zinn, Senior Analyst, Natural Resources Policy, Congressional Research Service), 8 (Chairman Lugar), 9–10 (Ranking Member Harkin), 92–93 (March 1, 2001; quote by Craig Cox with the Soil and Water Conservation Society), 98–99 (Paul Johnson, formerly NRCS Chief), 94–95 (John Hassell of the Conservation Technology Information Center at Purdue University), 97 (Nathan Rudgers, Commissioner, New York State Department of Agriculture

and Markets, National Association of State Departments of Agriculture), and 115 (Tom Buis, Executive Director, National Farmers Union).
35. 106 S.1426; 106 S.3260. See also 106 H.R. 5511; 107 H.R. 1949; Cong. Rec., October 27, 2000, at 25153 (quotes). See also Even, "Green Payments."
36. See, e.g., S. Hrg. 107-225, 2001, "Conservation," at 10 (Gary Mast, National Association of Conservation Districts), 112 (Dan Specht, Sustainable Agriculture Coalition and an Iowa farmer), 118–19 (Gerald Cohn, Southeast Regional Director, American Farmland), and 121 (Chairman Lugar and Specht).
37. See S. Hrg. 107-225, 2001, "Conservation," at 100–104 (Chairman Lugar), 105–9 (Harkin), 110–11 (Bob Stallman, President, American Farm Bureau Federation); S. Hrg. 107-828, 2001, "Working Lands," at 5 (Lee Klein, National Corn Growers Association and American Soybean Association) and 8–9 (George Dunklin Jr., USA Rice Federation); Cong. Rec., February 13, 2003, at 4026–28 (Harkin speech for the record). See, e.g., Philip Brasher, "Weekly Farm: Harkin Pushing Farm Conservation Program That Splits Environmentalists," Associated Press, March 23, 2002; Philip Brasher, "Proposal Pays Farmers to Conserve," Associated Press, *Bismarck Tribune*, March 24, 2002; Philip Brasher, "Farm Bill Agreement Waiting for a Final Budget Scoring," *National Journal's Congress Daily*, April 26, 2002.
38. See, e.g., CQ *Almanac*, "Farm Bill Delayed a Year"; CQ *Almanac*, "'02 Farm Bill Revives Subsidies"; Forrest Laws, "Harkin Going to the Mat for CSP in New Farm Bill," *Farm Progress*, September 1, 2007; Ron Laird, "Harkin Looks Back as He Says Farewell," *Des Moines Register*, October 18, 2014, https://www.desmoinesregister.com/story/opinion/columnists/2014/10/18/tom-harkin-qa-interview-senator-says-farewell/17522089/; Cong. Rec., December 10, 2001, at s12765 (Chairman Harkin).
39. See, e.g., GAO, "Conservation Security Program," 5; Lehrer, "Negotiating," 106–8; Lenihan and Brasier, "Ecological Modernization"; CRS Report RS21740, "Conservation Security Program"; CBO, "H.R. 2646."
40. See GAO, "Conservation Security Program," 2, fn4, 5, 6–7, 23–24.
41. See GAO, "Conservation Security Program," 2, fn4, 21–22; S. Hrg. 108-564, 2004, "Oversight," at 10 (Bruce Knight, Chief of the NRCS), 12–14, 16–19 (Harkin), 21 (Al Christopherson, Minnesota Farm Bureau), 24 (John Hansen, Nebraska Farmers Union). See also Ted Monoson, "House Appropriators Rewrite Harkin's Signature Conservation Program," *Congressional Quarterly Today*, July 5, 2002, https://advance-lexis-com.proxy2.library.illinois.edu/api/document?collection=news&id=urn:contentItem:4BK8-MH50-01CP-F1GT-00000-00&context=1516831.

42. See Cong. Rec., February 13, 2003, at 4026–28 (Harkin) and 4012 (Senator Tim Johnson [D-SD]). See also Cong. Rec., November 3, 2005, at 24657 (Senator Kerry [D-MA]) and 24658 (Senator Clinton [D-NY]).
43. S. Rept. 110-220, at 28 (Harkin); CBO, "March 2007 CBO Baseline"; H. Hrg. 110-12, 2007, "Review," at 4 (House Ag Committee Chairman Collin Peterson [D-MN]).
44. See, e.g., S. Hrg. 110-18, 2007, "Working Lands," at 13 (NRCS Chief Arlen Lancaster and Ranking Member Saxby Chambliss [R-GA]) and 26–28 (Chairman Harkin and Lancaster); H. Hrg. 110-12, 2007, "Review," at 7 (Charles "Jamie" Jamison, National Corn Growers Association); S. Hrg. 110-155, 2007, "Recommendations," at 31–32 (NFU witness was John Hansen, President, Nebraska Farmers Union); H. Hrg. 109-31, 2006, "Review," at 141 (statement of Danny Robbins, cotton producer, Altus, Oklahoma).
45. H. Hrg. 110-12, 2007, "Review," at 20 (Representative Frank Lucas [R-OK]); S. Hrg. 110-18, 2007, "Working Lands," at 52 (Senator Thad Cochran [R-MS]); *CQ Almanac*, "Work Begins"; Morgan, "Farm Bill," 26–27; H. Rept. 110-256, at 216.
46. S. Rept. 110-220, at 28 (Harkin); CBO, "March 2007 CBO Baseline"; P.L. 110-234, at Sec. 2301, Sec. 2310; H. Rept. 110-627, at 718–22; CBO, "H.R. 2419."
47. See CBO, "March 2011 Baseline"; P.L. 113-79, at Sec. 2101 (acreage enrollment cap provision); H. Rept. 113-333, at 402–4; CBO, "H.R. 2642."
48. See, e.g., Jeff Zeleny, "Tom Harkin of Iowa Won't Seek Re-election to the Senate," *New York Times*, January 26, 2013, https://www.nytimes.com/2013/01/27/us/politics/tom-harkin-of-iowa-wont-seek-re-election-to-senate.html; James Hohmann, "Harkin Retirement: Jump Ball in Iowa," *Politico*, January 26, 2013, https://www.politico.com/story/2013/01/tom-harkin-wont-seek-sixth-senate-term-086759; S. Hrg. 115-592, 2017, "Perspectives," at 6 (quote from testimony of Jimmy Bramblett, Deputy Chief, Programs, Natural Resources Conservation Service) and 11; H. Rept. 115-661, at 174, 189, 191, 1477–78 (dissenting views of Representative Walz). See also J. Coppess, C. Zulauf, B. Gramig, G. Schnitkey, and N. Paulson, "Conferencing Conservation: Reviewing Title II of the House and Senate Farm Bills," *Farmdoc Daily* 8, no. 133, Department of Agricultural and Consumer Economics, University of Illinois at Urbana-Champaign, July 19, 2018, https://farmdocdaily.illinois.edu/2018/07/conferencing-conservation-reviewing-title-ii-of-the-house-and-senate-farm-bills.html; J. Coppess and B. Gramig, "Reviewing Directions in Conservation Policy: CSP and EQIP in the House Farm Bill," *Farmdoc Daily* 8, no. 104, Department of Agricultural and Consumer Economics, University of Illinois at Urbana-Champaign, June 7, 2018, https://farmdocdaily.illinois.edu/2018/06/reviewing-directions-in-conservation-policy-html.html. See also Chuck Abbott, "House Ag Committee Passes 2018

Farm Bill: The Bill Heads to the Floor for Full House Vote," *Successful Farming*, April 18, 2018, https://www.agriculture.com/crops/xtremeag-what-a-difference-the-rain-makes; Emily Anderson, "A Long Road for CSP: Reflections on the 2018 Farm Bill," Farm Bill Law Enterprise, February 12, 2019, http://www.farmbilllaw.org/2019/02/12/a-long-road-for-csp/.

49. See Cong. Rec., May 18, 2018, at H4230–32 (Representative Sean Patrick Maloney [D-NY]); Cong. Rec., May 18, 2018, at H4232–33 (farm bill initially defeated 198–213 [17 not voting]); Cong. Rec., June 21, 2018, at H5448–49 (motion to reconsider the May 18 vote was agreed to by a vote of 233–191 [3 not voting], and the bill passed 213–211 [4 not voting]). See also Philip Brasher, "CBO: Farm Bill Taps CSP, Reworks SNAP to Fund Priorities," Agri-Pulse.com, April 13, 2018, https://www.agri-pulse.com/articles/10847-new-farm-bill-taps-csp-shifts-snap-to-fund-priorities.

50. See, e.g., Cong. Rec., June 26, 2018, at S4383 (Ranking Member Stabenow [D-MI]); Cong. Rec., June 27, 2018, at S4480 (Senator Jon Tester [D-MT]); Cong. Rec., June 25, 2018, at S4368 (motion to proceed to the farm bill passed easily, 89–3); Cong. Rec., June 28, 2018, at S4717 (Senate passage of bill on June 28, 2018, by 86–11); Adam Wernick, "The 2018 Farm Bill Stirs Conflict and Controversy," PRI.org, July 14, 2018, https://www.pri.org/stories/2018-07-14/2018-farm-bill-stirs-conflict-and-controversy; Marc Heller, "Advocates Welcome Senate Farm Bill—with Caveats," *E&E Daily*, June 11, 2018, https://www.eenews.net/stories/1060083995; Paul Huttner and Cody Nelson, "Major Conservation Program at Risk in Farm Bill Draft," MPR News, October 3, 2018, https://www.mprnews.org/story/2018/10/03/climate-change-farm-bill; H. Rept. 115-1072, at 572.

51. See, e.g., S. Rept. 112-203, at 9; S. Rept. 113-88, at 9, 95–98; H. Rept. 113-333, at 408–10. See also J. Coppess and T. Kuethe. "The Regional Conservation Partnership Program in the Farm Bill," *Farmdoc Daily* 4, no. 99, Department of Agricultural and Consumer Economics, University of Illinois at Urbana-Champaign, May 29, 2014, https://farmdocdaily.illinois.edu/2014/05/the-regional-conservation-partnership-program-in-the-farm-bill.html.

52. H. Rept. 115-661, at 174 (summary), 196; H. Rept. 115-1072, at 587–88 ("The Managers streamline the operation. . . . The Managers . . . do not intend for USDA to continue to implement RCPP through the rules and regulations of the covered programs").

53. S. Hrg. 115-592, 2017, "Perspectives," at 3 (Stabenow quotes), 11 (backlogs), 21 (Senator Casey [D-PA] raising issues with implementation of RCPP), 40–41 (Adam Sharp, Ohio Farm Bureau, discussing RCPP in Western Lake Erie Watershed); H. Rept. 115-1072, at 59, 589–90.

54. See CBO, "Estimated Budgetary Effects of Public Law 117-169"; P.L. 117-169. See also J. Coppess, K. Swanson, N. Paulson, C. Zulauf, and G. Schnitkey, "Reviewing the Inflation Reduction Act of 2022; Part 1," *Farmdoc Daily* 12, no. 119, Department of Agricultural and Consumer Economics, University of Illinois at Urbana-Champaign, August 11, 2022, https://farmdocdaily.illinois.edu/2022/08/reviewing-the-inflation-reduction-act-of-2022-part-1.html.

55. CBO, "Estimated Budgetary Effects of Public Law 117-169."

56. J. Coppess, "Climate Change and the Farm Bill: A Brief History," *Farmdoc Daily* 12, no. 149, Department of Agricultural and Consumer Economics, University of Illinois at Urbana-Champaign, September 29, 2022, https://farmdocdaily.illinois.edu/2022/09/climate-change-and-the-farm-bill-a-brief-history.html.

10. Of Congress: Closing Argument

1. See USDA, ERS, *Data Files*, "Federal Government Direct Farm Payments."
2. See USDA, AAA, "Census of Agriculture Historical Archive: 1930"; USDA, AAA, "Census of Agriculture: 2017"; U.S. Census Bureau, "Historical Population Change Data (1910–2020)."
3. Hamilton, *Federalist*, no. 9; Madison, *Federalist*, no. 10.
4. Madison, *Federalist*, no. 10.
5. Olson, *Logic of Collective Action*.
6. Hovenkamp, "Appraising."
7. See Waldron, *Dignity*; Waldron, *Law and Disagreement*; Waldron, "Legislation."
8. Madison, *Federalist*, no. 42.
9. Madison, *Federalist*, no. 51.
10. Madison, *Federalist*, no. 10.
11. See, e.g., Bowman, "Antebellum Planters"; Richards, *Slave Power*; Freeman, *Field of Blood*; Key, *Southern Politics*; Bateman, Katznelson, and Lapinski, *Southern Nation*; Mickey, *Paths Out of Dixie*.
12. See, e.g., Katznelson, *Fear Itself*.
13. Bosso, *Pesticides and Politics*, 166; Bosso, *Pesticides and Politics*, 150–51 (quoting Schattschneider that "procedures for the control of the expansive power of conflict determine the shape of the political system").
14. See Brown v. Board of Education, 347 U.S. 483 (1954); Baker v. Carr, 369 U.S. 186 (1962); Reynolds v. Sims, 377 U.S. 533 (1964); Wesberry v. Sanders, 376 U.S. 1 (1964). See also Tolson, "Law of Democracy"; Mikva, "Justice Brennan"; U.S. House of Representatives, "Constitutional Amendments and Major Civil Rights Acts."
15. P.L. 101-624; J. Coppess, "Climate Change and the Farm Bill: A Brief History," *Farmdoc Daily* 12, no. 149, Department of Agricultural and Consumer Econom-

ics, University of Illinois at Urbana-Champaign, September 29, 2022, https://farmdocdaily.illinois.edu/2022/09/climate-change-and-the-farm-bill-a-brief-history.html.

16. See, e.g., Scott Waldman, "Bush Had a Lasting Impact on Climate and Air Policy," E&E News, Policy, December 3, 2018, https://www.scientificamerican.com/article/bush-had-a-lasting-impact-on-climate-and-air-policy/; Ariel Wittenberg. "How George H. W. Bush (Eventually) Rescued U.S. Wetlands," E&E News, GreenWire, December 3, 2018, https://www.eenews.net/articles/how-george-h-w-bush-eventually-rescued-u-s-wetlands/; Scott Waldman and Benjamin Hulac, "This Is When the GOP Turned Away from Climate Policy," E&E News, ClimateWire, December 5, 2018, https://www.eenews.net/articles/this-is-when-the-gop-turned-away-from-climate-policy/.

17. P.L. 93-344; Zelizer, *On Capitol Hill*, 150–54 (citing Allen Schick, *Congress and Money: Budgeting, Spending and Taxing* [Washington DC: Urban Institute, 1980], 66), 212, 234–36; Wildavsky, *New Politics*, 135, 141; Garrett, "Congressional Budget Process," 713. See also Ornstein, "The Politics of the Deficit," 312.

18. Stockman, *Triumph of Politics*, 159–60.

19. Wildavsky, *New Politics*, 160, 212.

20. See P.L. 99-177; Stith, "Rewriting"; Kahn, "Gramm-Rudman"; Hoadley, "Easy Riders"; West, "Gramm-Rudman-Hollings."

21. See P.L. 101-508; Garrett, "Harnessing Politics," 503.

22. See P.L. 105-33.

23. See P.L. 112-25. See also CRS Report IF10657, "Budgetary Effects."

24. CBO reports make this abundantly clear. Federal deficits are currently projected to average $1.6 trillion from 2023 to 2032 with the federal debt projected to reach 110 percent of gross domestic product in 2023. See CBO, "Budget and Economic Outlook: 2022 to 2023."

25. Garrett, "Congressional Budget Process," 705–7, 711, 715; Garrett, "Harnessing Politics," 502.

26. Wildavsky, *New Politics*, 8, 32, 142, 397, 439; Dam, "American Fiscal Constitution"; Stith, "Power of the Purse"; Garrett, "Harnessing Politics," 502.

27. Garrett, "Harnessing Politics," 503–4. See, e.g., King, *Budgeting Entitlements*, 168.

28. King, *Budgeting Entitlements*, 170.

29. See, e.g., CBO, "May 2022 Baseline"; J. Coppess, K. Swanson, N. Paulson, G. Schnitkey, and C. Zulauf, "Reviewing the Latest CBO Farm Bill Baseline," *Farmdoc Daily* 12, no. 80, Department of Agricultural and Consumer Economics, University of Illinois at Urbana-Champaign, June 1, 2022, https://farmdocdaily.illinois.edu/2022/06/reviewing-the-latest-cbo-farm-bill-baseline.html.

30. Garrett, "Congressional Budget Process," 727.
31. See, e.g., Wildavsky, *New Politics*, 251, 439; White and Wildavsky, *Deficit*, 589; Zelizer, *On Capitol Hill*, 246–47, 256; King, *Budgeting Entitlements*, 168. See, generally, Katznelson, *Fear Itself*.
32. Wildavsky, *New Politics*, 216–18, 240, 250. See also King, *Budgeting Entitlements*, 166.
33. Wildavsky, *New Politics*, 160–62, 197–98, 203–5, 439.
34. White and Wildavsky, *Deficit*, 504, 505.
35. Wildavsky, *New Politics*, 69, 142–49, 204, 207–9. See also White and Wildavsky, *Deficit*, 19; Garrett, "Congressional Budget Process," 718. See also King, *Budgeting Entitlements*, 165–66.
36. West, "Gramm-Rudman-Hollings," 97.
37. See White and Wildavsky, *Deficit*, 208; Zelizer, *On Capitol Hill*, 152; Wildavsky, *New Politics*, 69; MacLean, *Democracy in Chains*; Grofman, "Public Choice," 1542n1 (among the foundational texts are Kenneth J. Arrow, *Social Choice and Individual Values*, 2nd ed. [New York: John Wiley & Sons, 1963] and James M. Buchanan and Gordon Tullock, *The Calculus of Consent: Logical Foundations of Constitutional Democracy* [Ann Arbor: University of Michigan Press, 1962]), 1571, 1581, 1585; Rubin, "Beyond," 1, 5–6, 6–8, 10, 20; Bessette, *Mild Voice*, 56–58; Kelman, "Public Choice," 87, 94.
38. See, e.g., Ornstein, "The Politics of the Deficit," 312; Zelizer, *On Capitol Hill*, 153–54 (citing Schick, *Congress and Money*, 66), 212, 234–37, 246; White and Wildavsky, *Deficit*, 425; King, *Budgeting Entitlements*, 149.
39. See, e.g., William Greider, "The Education of David Stockman," *The Atlantic*, December 1981, https://www.theatlantic.com/magazine/archive/1981/12/the-education-of-david-stockman/305760/; J. Coppess, "Reviewing Farm Bill History: Budgets, Boll Weevils and the 1981 Farm Bill," *Farmdoc Daily* 7, no. 49, Department of Agricultural and Consumer Economics, University of Illinois at Urbana-Champaign, March 16, 2017, https://farmdocdaily.illinois.edu/2017/03/reviewing-farm-bill-history-budgets-boll-weevils.html.
40. See, e.g., Coppess, "High Cotton."
41. See CRS Report IF12024, "Farm Bill Primer." See also Scott Faber and Jared Hayes, "Growing Farm Conservation Backlog Shows Need for Congress to Spend Smarter," Environmental Working Group, News & Insights, August 18, 2021, https://www.ewg.org/news-insights/news/2021/08/growing-farm-conservation-backlog-shows-need-congress-spend-smarter.
42. See, e.g., West Virginia v. EPA, 142 S.Ct. 2587 (2022); J. Coppess, "Major Questions: The Supreme Court and the Decline of Textualism," *Farmdoc Daily* 12, no. 108, Department of Agricultural and Consumer Economics, University of

Illinois at Urbana-Champaign, July 22, 2022, https://farmdocdaily.illinois.edu/2022/07/major-questions-the-supreme-court-and-the-decline-of-textualism.html; Alex Guillen, "Supreme Court Handcuffs Biden's Climate Efforts," *Politico*, June 30, 2022, https://www.politico.com/news/2022/06/30/supreme-court-handcuffs-biden-on-major-climate-rule-00043423.

43. See, e.g., Jentleson, *Kill Switch*.
44. See, e.g., Berman, *Thinking Like an Economist*; Timothy Noah, "May God Save Us from Economists," *New Republic*, October 25, 2022, https://newrepublic.com/article/168049/tyranny-economists-government.
45. See, e.g., P.L. 117-169; CBO, "Estimated Budgetary Effects of Public Law 117-169"; J. Coppess, K. Swanson, N. Paulson, C. Zulauf, and G. Schnitkey, "Reviewing the Inflation Reduction Act of 2022; Part 1," *Farmdoc Daily* 12, no. 119, Department of Agricultural and Consumer Economics, University of Illinois at Urbana-Champaign, August 11, 2022, https://farmdocdaily.illinois.edu/2022/08/reviewing-the-inflation-reduction-act-of-2022-part-1.html.
46. See, e.g., J. Larsen et al., "A Turning Point for US Climate Progress: Assessing the Climate and Clean Energy Provisions in the Inflation Reduction Act," Rhodium Group, August 12, 2022, https://rhg.com/research/climate-clean-energy-inflation-reduction-act/; Robinson Meyer, "America Just Shrugged Off Biden's Big Climate Law," *The Atlantic*, November 11, 2022, https://www.theatlantic.com/science/archive/2022/11/midterm-elections-inflation-reduction-act-climate-change/672091/; Michael Tomasky, "Yes, the Inflation Reduction Act Is a Big Effing Deal," *New Republic*, August 8, 2022, https://newrepublic.com/article/167332/senate-passes-inflation-reduction-act.
47. See P.L. 117-169, at Sec. 21001, Sec. 21002; CBO, "Estimated Budgetary Effects of Public Law 117-169."
48. See, e.g., Elshtain, "Alchemy"; Kauffman, "Role"; Newman and Principe, "Alchemy"; Newman, "What Have," 314.
49. White, "Historical Roots."
50. Leopold, "Land Ethic," 208.
51. See, e.g., Bradshaw et al., "Underestimating"; Ben Ehrenreich, "We're Hurtling Toward Global Suicide," *New Republic*, March 18, 2021, https://newrepublic.com/article/161575/climate-change-effects-hurtling-toward-global-suicide.

BIBLIOGRAPHY

A note on the organization of the bibliography: due to the large volume, all congressional materials are in a separate section at the end of the bibliography.

Adler, Jonathan H. "Back to the Future of Conservation: Changing Perceptions of Property Rights & Environmental Protection." NYU Journal of Law & Liberty 1 (2005): 987–1022.

———. "Fables of the Cuyahoga: Reconstructing a History of Environmental Protection." Fordham Environmental Law Journal 14 (2002): 89–146.

Akihiko, H. "Formation of Japan's Food Security Policy: Relations with Food Situation and Evolution of Agricultural Policies." Norinchukin Research Institute, 2017. https://www.nochuri.co.jp/english/pdf/rpt_20180731-1.pdf.

Alston, Lee J., and Joseph P. Ferrie. "Labor Costs, Paternalism, and Loyalty in Southern Agriculture: A Constraint on the Growth of the Welfare State." Journal of Economic History 45, no. 1 (1985): 95–117.

———. "Paternalism in Agricultural Labor Contracts in the US South: Implications for the Growth of the Welfare State." American Economic Review 83, no. 4 (1993): 852–76.

———. "Social Control and Labor Relations in the American South Before the Mechanization of the Cotton Harvest in the 1950s." Journal of Institutional and Theoretical Economics (JITE) 141, no. 1 (1989): 133–57.

Alston, Lee J., Edwyna Harris, and Bernardo Mueller. "The Development of Property Rights on Frontiers: Endowments, Norms, and Politics." Journal of Economic History 72, no. 3 (2012): 741–70.

Amundson, Ronald, Asmeret Asefaw Berhe, Jan W. Hopmans, Carolyn Olson, A. Ester Sztein, and Donald L. Sparks. "Soil and Human Security in the 21st Century." Science 348, no. 6235 (2015): 1261071. https://www.science.org/doi/full/10.1126/science.1261071.

Anderson, Margot. "Conservation, the Environment, and the Farm Bill." Journal of Contemporary Water Research and Education 101, no. 1 (1995): 2–12.

Angelo, Mary Jane. "Corn, Carbon and Conservation: Rethinking US Agricultural Policy in a Changing Global Environment." *George Mason Law Review* 17, no. 3 (2010): 593–660.

Angelo, Mary Jane, and Jon Morris. "Maintaining a Healthy Water Supply While Growing a Healthy Food Supply: Legal Tools for Cleaning Up Agricultural Water Pollution." *University of Kansas Law Review* 62 (2013): 1003–41.

Baldwin, Sidney. *Poverty and Politics: The Rise and Decline of the Farm Security Administration.* Chapel Hill: University of North Carolina Press, 1968.

Banfield, E. C. "Ten Years of the Farm Tenant Purchase Program." *Journal of Farm Economics* 31, no. 3 (1949): 469–86.

Banker, David, J. M. MacDonald, and R. A. Hoppe. "Growing Farm Size and the Distribution of Farm Payments." Economic Brief No. (EB-6). U.S. Department of Agriculture, Economic Research Service, March 2006. https://www.ers.usda.gov/publications/pub-details/?pubid=42917.

Bardgett, Richard. *Earth Matters: How Soil Underlies Civilization.* Oxford University Press, 2016.

Barnett, Barry J. "The U.S. Farm Financial Crisis of the 1980s." *Agricultural History* 74, no. 2 (Spring 2000): 366–80.

Barrett, Robert C., Jonathan P. Caulkins, Andrew J. Yates, and D. L. Elliott. "Population Dynamics of the Peruvian Anchovy." *Mathematical Modelling* 6, no. 6 (1985): 525–48.

Bateman, David A., Ira Katznelson, and John Lapinski. "*Southern Politics* Revisited: On V. O. Key's 'South in the House.'" *Studies in American Political Development* 29, no. 2 (2015): 154–84.

Bateman, David A., Ira Katznelson, and John S. Lapinski. *Southern Nation: Congress and White Supremacy after Reconstruction.* Princeton: Princeton University Press, 2018.

Bates, J. Leonard. "Fulfilling American Democracy: The Conservation Movement, 1907 to 1921." *Mississippi Valley Historical Review* 44, no. 1 (1957): 29–57.

———. "The Teapot Dome Scandal and the Election of 1924." *American Historical Review* 60, no. 2 (1955): 303–22.

Batie, Sandra S. *Soil Erosion: Crisis in America's Croplands?* Conservation Foundation, 1983.

Beatty, Robert O. "The Conservation Movement." *Annals of the American Academy of Political and Social Science* 281, no. 1 (1952): 10–19.

Beckert, Sven. *Empire of Cotton: A Global History.* New York: Vintage Books, 2014.

Bell, John. "Path Dependence and Legal Development." *Tulane Law Review* 87 (2012): 787–810.

Benbrook, Charles. "Integrating Soil Conservation and Commodity Programs: A Policy Proposal." *Journal of Soil and Water Conservation* (1979): 160–67.

Benedict, Murray R. *Farm Policies of the United States, 1790–1950.* New York: Twentieth Century Fund, 1953.

Bennett, Hugh H. *Soil Conservation.* York PA: Maple Press; New York: McGraw-Hill, 1939. Reprint, Arno Press, 1970.

Bensel, Richard. *Sectionalism and American Political Development, 1880–1980.* Madison: University of Wisconsin Press, 1984.

Berardo, Ramiro, V. Kelly Turner, and Stian Rice. "Systemic Coordination and the Problem of Seasonal Harmful Algal Blooms in Lake Erie." *Ecology and Society* 24, no. 3 (2019).

Bergh, Emma L., Francisco J. Calderon, Andrea K. Clemensen, Lisa Durso, Jed O. Eberly, Jonathan J. Halvorson, Virginia L. Jin, et al. "Time in a Bottle: Use of Soil Archives for Understanding Long-Term Soil Change." *Soil Science Society of America Journal* (2022): 520–27. https://acsess.onlinelibrary.wiley.com/doi/full/10.1002/saj2.20372.

Berman, Elizabeth Popp. *Thinking Like an Economist: How Efficiency Replaced Equality in U.S. Public Policy.* Princeton: Princeton University Press, 2022.

Berry, Christopher R., and Anthony Fowler. "Congressional Committees, Legislative Influence, and the Hegemony of Chairs." *Journal of Public Economics* 158 (2018): 1–11.

Bessette, Joseph. *The Mild Voice of Reason: Deliberative Democracy and American National Government.* Chicago: University of Chicago Press, 1997.

Bickel, Alexander M. *The Least Dangerous Branch: The Supreme Court at the Bar of Politics.* New Haven: Yale University Press, 1986.

Bigelow, D., and A. Borchers. "Major Land Uses in the United States, 2012." Economic Information Bulletin No. (EIB-178). U.S. Department of Agriculture, Economic Research Service, August 2017. https://www.ers.usda.gov/publications/pub-details/?pubid=84879.

Biles, Amanda B. "Politics and Production Control: American Farmers and the Agricultural Adjustment Act of 1938." Master's thesis, University of Central Oklahoma, 2011.

Billings, Dwight B. "Class Origins of the 'New South': Planter Persistence and Industry in North Carolina." *American Journal of Sociology* 88 (1982): s52–85. http://www.jstor.org/stable/3083239.

Blakley, Leo V. "The Effects of Government Programs on Southern Cotton Production since 1945." *Journal of Farm Economics* 47, no. 5 (1965): 1160–71.

Bonnen, James T. "The Farm Policy Debate: Discussion." *Journal of Farm Economics* 42, no. 2 (1960): 429–34.

Bosso, Christopher J. *Pesticides and Politics: The Life Cycle of a Public Issue.* Pittsburgh: University of Pittsburgh Press, 1987.

Bowman, Shearer Davis. "Antebellum Planters and *Vormärz* Junkers in Comparative Perspective." *American Historical Review* 85, no. 4 (October 1980): 779–808. https://www.jstor.org/stable/pdf/1868872.pdf.

Braden, John B. "Some Emerging Rights in Agricultural Land." *American Journal of Agricultural Economics* 64, no. 1 (1982): 19–27.

Braden, John B., and Donald L. Uchtmann. "Soil Conservation Programs Amidst Faltering Environmental Commitments and the New Federalism." *Boston College Environmental Affairs Law Review* 10 (1982): 639–96.

Bradshaw, Corey J. A., Paul R. Ehrlich, Andrew Beattie, Gerardo Ceballos, Eileen Crist, Joan Diamond, Rodolfo Dirzo, et al. "Underestimating the Challenges of Avoiding a Ghastly Future." *Frontiers in Conservation Science* 1, no. 615419 (2021): 9. https://www.frontiersin.org/articles/10.3389/fcosc.2020.615419/full.

Braun, Artur, Debajeet K. Bora, Lars Lauterbach, Elisabeth Lettau, Hongxin Wang, Stephen P. Cramer, Feipeng Yang, and Jinghua Guo. "From Inert Gas to Fertilizer, Fuel and Fine Chemicals: N2 Reduction and Fixation." *Catalysis Today* 387 (2022): 186–96. https://www.sciencedirect.com/science/article/pii/S0920586121001863.

Breggin, Linda, and D. Bruce Myers Jr. "Subsidies with Responsibilities: Placing Stewardship and Disclosure Conditions on Government Payments to Large-Scale Commodity Crop Operations." *Harvard Environmental Law Review* 37 (2013): 487–538.

Breimyer, Harold F. "Conceptualization and Climate for New Deal Farm Laws of the 1930s." *American Journal of Agricultural Economics* 65, no. 5 (1983): 1153–57.

Breimyer, H. F. "Agricultural Philosophies and Policies in the New Deal." *Minnesota Law Review* 68 (1983): 333–34.

Brevik, Eric C., Jeffrey A. Homburg, and Jonathan A. Sandor. "Soils, Climate, and Ancient Civilizations." In *Developments in Soil Science*, 35:1–28. Elsevier, 2018. https://www.sciencedirect.com/science/article/pii/B9780444638656000016.

Bridges, Khiara M. "The Deserving Poor, the Undeserving Poor, and Class-Based Affirmative Action." *Emory Law Journal* 66 (2016): 1049–114.

Brown, Sarah Hart. "Congressional Anti-Communism and the Segregationist South: From New Orleans to Atlanta, 1954–1958." *Georgia Historical Quarterly* 80, no. 4 (1996): 785–816.

Browne, R. S. *Only Six Million Acres: The Decline of Black Owned Land in the Rural South.* New York: Black Economic Research Center, 1973.

Bruhl, Aaron-Andrew P. "Using Statutes to Set Legislative Rules: Entrenchment, Separation of Powers, and the Rules of Proceedings Clause." *Journal of Law and Politics* 19 (2003): 345–416.

Burford, Roger L. "The Federal Cotton Programs and Farm Labor Force Adjustments." *Southern Economic Journal* 33, no. 2 (October 1966): 223–36.

Burns, K. "The Dust Bowl." PBS.org, 2012. https://www.pbs.org/kenburns/dustbowl/.

Buttel, Frederick H. "The Environmental Movement: Consensus, Conflict, and Change." *Journal of Environmental Education* 7, no. 1 (1975): 53–63.

Caro, Robert A. *The Years of Lyndon Johnson: Master of the Senate.* New York: Vintage Books, 2002.

Carolan, Michael S. "Do You See What I See? Examining the Epistemic Barriers to Sustainable Agriculture." *Rural Sociology* 71, no. 2 (2006): 232–60.

Carroll, Curtis. "Soil Conservation: Proposed Legislation in the 1985 Congress." *South Dakota Law Review* 31 (1985): 453–61.

Carson, Rachel. *Silent Spring.* New York: Houghton Mifflin, 1962.

Case, H. C. M. "A Proposed Price Support Program Based on Soil Conservation and Production Control." University of Illinois, Department of Agricultural Economics, 630.7, IL6ae, No. 3086-3147 (December 15, 1955). On file with the University of Illinois ACES Library.

Chen, James Ming. "After Agrarian Virtue." *Indiana Law Review* 53 (2020): 1–34.

Chen, Jim. "Get Green or Get Out: Decoupling Environmental from Economic Objectives in Agricultural Regulation." *Oklahoma Law Review* 48 (1995): 333–52.

Chernow, Ron. *Alexander Hamilton.* New York: Penguin Press, 2004.

Claassen, Roger. "Have Conservation Compliance Incentives Reduced Soil Erosion?" *Amber Waves*, June 1, 2004. U.S. Department of Agriculture, Economic Research Service. https://www.ers.usda.gov/amber-waves/2004/june/have-conservation-compliance-incentives-reduced-soil-erosion/.

Claassen, Roger, and Maria Bowman. "Conservation Compliance in the Crop Insurance Era." *Amber Waves*, July 27, 2017. U.S. Department of Agriculture, Economic Research Service. https://www.ers.usda.gov/amber-waves/2017/july/conservation-compliance-in-the-crop-insurance-era/.

Claassen, Roger, M. Bowman, V. Breneman, T. Wade, R. Williams, J. R. Fooks, L. Hansen, R. Iovann, and C. Loesch. "Conservation Compliance: How Farmer Incentives are Changing in the Crop Insurance Era." Economic Research Report No. (ERR-234). U.S. Department of Agriculture, Economic Research Service, July 2017. https://www.ers.usda.gov/publications/pub-details/?pubid=84456.

Claassen, Roger, Vincent E. Breneman, Shawn Bucholtz, Andrea Cattaneo, Robert C. Johansson, and Mitchell J. Morehart. "Environmental Compliance in U.S. Agricultural Policy: Past Performance and Future Potential." Agricultural Economic

Report No. (AER-832). U.S. Department of Agriculture, Economic Research Service, May 2004. https://www.ers.usda.gov/publications/pub-details/?pubid =41660.

Clarke, Keith C., and Jeffrey J. Hemphill. "The Santa Barbara Oil Spill: A Retrospective." *Yearbook of the Association of Pacific Coast Geographers* 64 (2002): 157–62.

Cobb, James C. *The Most Southern Place on Earth: The Mississippi Delta and the Roots of Regional Identity*. Oxford University Press, 1992.

Cochrane, Willard W. *The Development of American Agriculture: A Historical Analysis*. Minneapolis: University of Minnesota Press, 1979.

Cochrane, Willard W., and Mary E. Ryan. *American Farm Policy, 1948–1973*. Minneapolis: University of Minnesota Press, 1976.

Comerford, Robert J. "The American Liberty League." PhD diss., St. Johns University, 1967. ProQuest, 6803816.

Comito, Jacqueline, Jon Wolseth, and Lois Morton. "The State's Role in Water Quality: Soil and Water Conservation District Commissioners and the Agricultural Status Quo." *Human Organization* 72, no. 1 (2013): 44–54.

Conrad, Eugene. *The Forgotten Farmers: The Story of the Sharecropper in the New Deal*. Urbana: University of Illinois Press, 1965.

Cook, Benjamin I., Ron L. Miller, and Richard Seager. "Amplification of the North American 'Dust Bowl' Drought through Human-Induced Land Degradation." *Proceedings of the National Academy of Sciences* 106, no. 13 (2009): 4997–5001.

Cook, Ken. "The 1985 Farm Bill: A Turning Point for Soil Conservation." *Journal of Soil and Water Conservation* 40, no. 2 (1985): 218–20.

Cook, Kenneth A. "The Environmental Era of US Agricultural Policy." *Journal of Soil and Water Conservation* 44, no. 5 (1989): 362–66.

Cook, Maurice G. "Hugh Hammond Bennett: The Father of Soil Conservation." In *Century of Soil Science*, 69. Soil Science Society of North Carolina, 2003.

Coppess, Jonathan. *The Fault Lines of Farm Policy: A Legislative and Political History of the Farm Bill*. Lincoln: University of Nebraska Press, 2018.

———. "High Cotton and the Low Road: An Unraveling Farm Bill Coalition and Its Implications." *Drake Journal of Agricultural Law* 23 (2018): 353–412.

———. "A Return to the Crossroads: Farming, Nutrient Loss, and Conservation." *University of Arkansas at Little Rock Law Review* 39 (2016): 351–88.

Coppess, Jonathan W. "A Perspective on Agricultural Policy in the Age of Nutrient Loss." *Drake Journal of Agricultural Law* 23 (2018): 29–44.

Cordova, Carlos, and Jess C. Porter. "The 1930s Dust Bowl: Geoarcheological Lessons from a 20th Century Environmental Crisis." *The Holocene* 25, no. 10 (2015): 1707–20.

Coulborn, Rushton. *Origin of Civilized Societies*. Princeton: Princeton University Press, Princeton Legacy Library, 1959. Online edition available via JSTOR at https://www.jstor.org/stable/j.ctt183pmpj.

Crawford, Amanda L. "Nutrient Pollution and the Gulf of Mexico Dead Zone: Will Des Moines Water Works be a Turning Point." *Tulane Law Review* 91 (2016): 157–87.

Cronin, John. "The Cuyahoga Fire at Fifty: A False History Obscures the Real Water Crisis That Never Ceased." *Journal of Environmental Studies and Sciences* 9, no. 3 (2019): 340–51.

Cronon, William. "A Place for Stories: Nature, History, and Narrative." *Journal of American History* 78, no. 4 (1992): 1347–76.

Cunfer, Geoff. "Introduction to Soils and Sustainability Special Issue." *Social Science History* 45, no. 4 (2021): 617–23.

———. "Scaling the Dust Bowl." In *Placing History: How Maps, Spatial Data, and GIS Are Changing Historical Scholarship*, edited by Anne Kelly Knowles and Amy Hillier, 95–121. Redlands CA: ESRI Press, 2008. https://downloads2.esri.com/ESRIpress/images/133/knowles.pdf.

———. "Soil Fertility on an Agricultural Frontier: The US Great Plains, 1880–2000." *Social Science History* 45, no. 4 (2021): 733–62.

Cunfer, Geoff, and Fridolin Krausmann. "Sustaining Soil Fertility: Agricultural Practice in the Old and New Worlds." *Global Environment* 2, no. 4 (2009): 8–47.

Currie, David P. "The Constitution in the Supreme Court: The New Deal, 1931–1940." *University of Chicago Law Review* 54, no. 2 (1987): 504–55.

Dahl, Robert A. *How Democratic Is the American Constitution?* New Haven: Yale University Press, 2003.

Dam, Kenneth W. "The American Fiscal Constitution." *University of Chicago Law Review* 44, no. 2 (1977): 272–320.

Daniel, Pete. *Breaking the Land: The Transformation of Cotton, Tobacco, and Rice Cultures since 1880*. Urbana: University of Illinois Press, 1985.

———. "The Crossroads of Change: Cotton, Tobacco, and Rice Cultures in the Twentieth-Century South." *Journal of Southern History* 50, no. 3 (1984): 429–56.

———. *Dispossession: Discrimination Against African American Farmers in the Age of Civil Rights*. Chapel Hill: University of North Carolina Press, 2013.

———. *Lost Revolutions: The South in the 1950s*. Chapel Hill: University of North Carolina Press, 2000.

———. *The Shadow of Slavery: Peonage in the South, 1901–1969*. Urbana: University of Illinois Press, 1972.

Davidson, Roger H. "The Advent of the Modern Congress: The Legislative Reorganization Act of 1946." *Legislative Studies Quarterly* 15, no. 3 (August 1990): 357–73.

Day, Richard H. "The Economics of Technological Change and the Demise of the Sharecropper." *American Economic Review* 57, no. 3 (1967): 427–49.

Deaton, B. James, and John P. Hoehn. "The Social Construction of Production Externalities in Contemporary Agriculture: Process Versus Product Standards as the Basis for Defining 'Organic.'" *Agriculture and Human Values* 22, no. 1 (2005): 31–38.

De la Garza, Kika. "Linking Trade Growth and the Environment: One Lawmaker's View." *Environmental Law* 23 (1993): 701–2.

Diamond, Jared. *Collapse: How Societies Choose to Fail or Succeed*. New York: Viking, 2005.

Dinterman, Robert, Ani L. Katchova, and James Michael Harris. "Financial Stress and Farm Bankruptcies in U.S. Agriculture." *Agricultural Finance Review* 78, no. 4 (2018): 441–56.

Doering, Otto, and Katherine R. Smith. "Examining the Relationship of Conservation Compliance and Farm Program Incentives." C-FARE Reports, no. 156624. Council on Food, Agricultural, and Resource Economics, July 2012. https://ageconsearch.umn.edu/record/156624.

Dotterweich, Markus. "The History of Human-Induced Soil Erosion: Geomorphic Legacies, Early Descriptions and Research, and the Development of Soil Conservation—A Global Synopsis." *Geomorphology* 201 (2013): 1–34.

Edwards, R. "Changing Perceptions of Homesteading as a Policy of Public Domain Disposal." *Great Plains Quarterly* 29, no. 3 (2009): 179–202.

Egan, Timothy. *The Worst Hard Time: The Untold Story of Those Who Survived the Great American Dust Bowl*. Boston: Houghton Mifflin, 2006.

Eisenhower, Dwight D. "Special Message to Congress on Agriculture, January 9, 1956, in the Public Papers of the Presidents of the United States, Federal Register Division, National Archives and Records Service, General Services Administration." U.S. GPO, 1958.

———. "Statement by the President Upon Signing the Agricultural Act of 1956: Executive Orders and Presidential Proclamations, Signed by the President; Presidential Signing Statements." May 28, 1956.

Elshtain, Jean Bethke. "Alchemy and the Law: Why This Marriage Can't Be Saved." *UC Davis Law Review* 25, no. 4 (Summer 1992): 1171–87.

Environmental Working Group. Farm Subsidy Database. https://farm.ewg.org/.

———. "Fooling Ourselves: Voluntary Programs Fail to Clean Up Dirty Water." Research, February 7, 2016. https://www.ewg.org/research/fooling-ourselves.

———. "Here Today, Gone Tomorrow: USDA Conservation Program for Sensitive Cropland Wastes Billions of Tax Dollars." June 7, 2017. https://www.ewg.org/news-insights/news-release/here-today-gone-tomorrow-usda-conservation-program-sensitive-cropland.

Erisman, Jan Willem, Mark A. Sutton, James Galloway, Zbigniew Klimont, and Wilfried Winiwarter. "How a Century of Ammonia Synthesis Changed the World." *Nature Geoscience* 1, no. 10 (2008): 636–39. https://www.nature.com/articles/ngeo325.

Ernst, Daniel R. "Mr. Try-it Goes to Washington: Law and Policy at the Agricultural Adjustment Administration." *Fordham Law Review* 87 (2018): 1795–815.

Ervin, David E., William D. Heffernan, and Gary P. Green. "Cross-Compliance for Erosion Control: Anticipating Efficiency and Distributive Impacts." *American Journal of Agricultural Economics* 66, no. 3 (1984): 273–78.

Eskridge, William N., Jr., and John Ferejohn. "The Article I, Section 7 Game." *Georgetown Law Journal* 80 (1991): 530–64.

Eskridge, William N., Jr., Philip P. Frickey, and Elizabeth Garrett. *Legislation and Statutory Interpretation*. 2nd ed. New York: Foundation Press, 2006.

Eubanks, William S. "The Future of Federal Farm Policy: Steps for Achieving a More Sustainable Food System." *Vermont Law Review* 37 (2012): 957–86.

———. "A Rotten System: Subsidizing Environmental Degradation and Poor Public Health Wins Our Nation's Tax Dollars." *Stanford Environmental Law Journal* 28 (2009): 213–310.

———. "The Sustainable Farm Bill: A Proposal for Permanent Environmental Change." *Environmental Law Reporter News & Analysis* 39 (2009): 10493–509.

Even, William J. "Green Payments: The Next Generation of US Farm Programs." *Drake Journal of Agricultural Law* 10 (2005): 173–204.

Ezekiel, Mordecai. "Henry A. Wallace, Agricultural Economist." *American Journal of Agricultural Economics* 48, no. 4, part 1 (1966): 789–802.

Farber, Daniel A. "The Conservative as Environmentalist: From Goldwater and the Early Reagan to the 21st Century." *Arizona Law Review* 59 (2017): 1005–60.

Farhang, Sean, and Ira Katznelson. "The Southern Imposition: Congress and Labor in the New Deal and Fair Deal." *Studies in American Political Development* 19, no. 1 (2005): 1–30.

Feldman, Noah. *The Three Lives of James Madison: Genius, Partisan, President*. New York: Random House, 2017.

Feng, Hongli, Lyubov A. Kurkalova, Catherine L. Kling, and Philip W. Gassman. "Environmental Conservation in Agriculture: Land Retirement vs. Changing Practices on Working Land." *Journal of Environmental Economics and Management* 52, no. 2 (2006): 600–614.

Ferguson, Edwin E. "Nation-Wide Erosion Control: Soil Conservation Districts and the Power of Land-Use Regulation." *Iowa Law Review* 34 (1948): 166–87.

Finney, Bradley R. "Agricultural Law Stifles Innovation and Competition." *Alabama Law Review* 72 (2020): 785–838.

Fite, Gilbert C. *Cotton Fields No More: Southern Agriculture, 1865–1980*. Lexington: University of Kentucky Press, 2014. Online version provided by Project Muse, https://muse.jhu.edu.

Flippen, J. Brooks. *Nixon and the Environment*. Albuquerque: University of New Mexico Press, 2000.

Ford, Aaron L. "The Legislative Reorganization Act of 1946." *American Bar Association Journal* 32, no. 11 (November 1946): 741–44.

Ford, Gerald R. "Safe Drinking Water Act: Statement by the President on Signing the Bill into Law, Executive Orders and Presidential Proclamations." December 17, 1974.

———. "Toxic Substances Control Act: Statement by the President on Signing S.3149 into Law, Executive Orders and Presidential Proclamations." October 12, 1976.

Frarey, Larry C., Ron Jones, and Staci J. Pratt. "Conservation Districts as the Foundation for Watershed-Based Programs to Prevent and Abate Polluted Agricultural Runoff." *Hamline Law Review* 18 (1994): 151–80.

Freeman, Joanne B. *The Field of Blood: Violence in Congress and the Road to Civil War*. New York: Farrar, Straus and Giroux, 2018.

Frey, Fred C., and T. Lynn Smith. "The Influence of the AAA Cotton Program Upon the Tenant, Cropper, and Laborer." *Rural Sociology* 1, no. 4 (1936): 481.

Friedman, Barry. "The History of the Countermajoritarian Difficulty, Part One: The Road to Judicial Supremacy." *New York University Law Review* 73, no. 2 (1998): 333–433.

———. "The History of the Countermajoritarian Difficulty, Part Two: Reconstruction's Political Court." *Georgetown Law Journal* 91, no. 1 (2002): 1–65.

———. "The History of the Countermajoritarian Difficulty, Part Three: The Lesson of *Lochner*." *New York University Law Review* 76, no. 5 (2001): 1383–455.

———. "The History of the Countermajoritarian Difficulty, Part Four: Law's Politics." *University of Pennsylvania Law Review* 148, no. 4 (1999): 971–1064.

———. "The History of the Countermajoritarian Difficulty, Part Five: The Birth of an Academic Obsession." *Yale Law Journal* 112, no. 2 (2002): 153–260.

Froomkin, David, and Ian Shapiro. "The New Authoritarianism in Public Choice." *Political Studies* 70, no. 3 (2021): 00323217211041893, at 2–3. https://journals.sagepub.com/doi/pdf/10.1177/00323217211041893.

Frossard, E., W. E. H. Blum, B. P. Warkentin. *Function of Soils for Human Societies and the Environment*. Special publication 266. London: Geological Society, 2006. https://pubs.geoscienceworld.org/gsl/books/book/1635/Function-of-Soils-for-Human-Societies-and-the.

Frump, Christopher. "Up to Our Ears: Corn Overproduction, Its Environmental Toll, and Using the 2012 US Farm Bill to Limit Corn Subsidies, Increase Environmental Protection Incentives, and Place Accountability on Crop Operations." *Florida A&M University Law Review* 8 (2012): 419–48.

Galloway, George B. "The Operation of the Legislative Reorganization Act of 1946." *American Political Science Review* 45, no. 1 (March 1951): 41–68.

Gao, Lijing, and J. Arbuckle. "Examining Farmers' Adoption of Nutrient Management Best Management Practices: A Social Cognitive Framework." *Agriculture and Human Values* 39, no. 3 (2021): 1–19. https://link.springer.com/article/10.1007/s10460-021-10266-2.

Gardner, Robert. "Trees as Technology: Planting Shelterbelts on the Great Plains." *History and Technology* 25, no. 4 (2009): 325–41.

Garrett, Elizabeth. "The Congressional Budget Process: Strengthening the Party-in-Government." *Columbia Law Review, Symposium: Law and Political Parties* 100, no. 3 (April 2000): 702–30.

———. "Harnessing Politics: The Dynamics of Offset Requirements in the Tax Legislative Process." *University of Chicago Law Review* 65 (1998): 501–70.

Gates, P. W. "Homesteading in the High Plains." *Agricultural History* 51, no. 1 (1977): 109–33.

Gilbert, Jess. "Eastern Urban Liberals and Midwestern Agrarian Intellectuals: Two Group Portraits of Progressives in the New Deal Department of Agriculture." *Agricultural History* 74, no. 2 (Spring 2000): 162–80.

———. *Planning Democracy: Agrarian Intellectuals and the Intended New Deal*. New Haven: Yale University Press, 2015. Online version available from JSTOR.org.

Gilbert, Jess, and Carolyn Howe. "Beyond 'State vs. Society': Theories of the State and New Deal Agricultural Policies." *American Sociological Review* 56, no. 2 (April 1991): 204–20.

Gleason, Robert A., N. H. Euliss Jr., B. A. Tangen, M. K. Laubhan, and B. A. Browne. "USDA Conservation Program and Practice Effects on Wetland Ecosystem Services in the Prairie Pothole Region." *Ecological Applications* 21, no. SP1 (2011): S65–S81.

Glick, Philip M. "The Soil and the Law." In *Soils and Men: Yearbook of Agriculture, 1938*, 296–318. U.S. Department of Agriculture, 1938.

———. "The Soil and the Law: II." *Journal of Farm Economics* 20, no. 3 (1938): 616–40. https://www.jstor.org/stable/pdf/1231280.pdf.

Gluck, Abbe R., and Lisa Schultz Bressman. "Statutory Interpretation from the Inside—An Empirical Study of Congressional Drafting, Delegation, and the Canons: Part I." *Stanford Law Review* 65, no. 5 (June 2013): 901–1026.

Goldstein, Jared A. "The American Liberty League and the Rise of Constitutional Nationalism." *Temple Law Review* 86 (2013): 287–330.

Goodwin, George. "The Seniority System in Congress." In *Legislative Politics U.S.A.*, 2nd ed., edited by Theodore J. Lowi. Boston: Little, Brown, 1962 and 1965.

Gottlieb, Robert. "The Next Environmentalism: How Movements Respond to the Changes that Elections Bring—From Nixon to Obama." *Environmental History* 14, no. 2 (2009): 298–308.

Grofman, Bernard. "Public Choice, Civic Republicanism, and American Politics: Perspectives of a 'Reasonable Choice' Modeler." *Texas Law Review* 71 (1993): 1541–88.

Grossman, Margaret Rosso. "Agriculture and the Polluter Pays Principle: An Introduction." *Oklahoma Law Review* 59, no. 1 (Spring 2006): 1–39.

——. "Good Agricultural Practice in the United States: Conservation and Climate." *Environmental Law Review* 13, no. 4 (2011): 296–317.

Grove, Wayne A. "The Economics of Cotton Harvest Mechanization in the United States, 1920–1970." Thesis, University of Illinois at Urbana-Champaign, 2000.

Grubbs, Donald H. *Cry from the Cotton: The Southern Tenant Farmers' Union and the New Deal*. Fayetteville: University of Arkansas Press, 1971.

Güldner, Dino, Laura Larsen, and Geoff Cunfer. "Soil Fertility Transitions in the Context of Industrialization, 1750–2000." *Social Science History* 45, no. 4 (2021): 785–811.

Hahn, Steven. "Class and State in Postemancipation Societies: Southern Planters in Comparative Perspective." *American Historical Review* 95, no. 1 (1990): 75–98.

Hamilton, Alexander. Library of Congress, "Federalist Papers: Primary Documents in American History." Project Gutenberg. https://guides.loc.gov/federalist-papers/full-text.

——. *Federalist*, no. 1. "General Introduction."

——. *Federalist*, no. 9. "The Utility of the Union as a Safeguard Against Domestic Faction and Insurrection."

——. *Federalist*, no. 30. "Other Defects of the Present Confederation."

——. *Federalist*, no. 31. "The Same Subject Continued: Concerning the General Power of Taxation."

——. *Federalist*, no. 32. "The Same Subject Continued: Concerning the General Power of Taxation."

——. *Federalist*, no. 33. "The Same Subject Continued: Concerning the General Power of Taxation."

——. *Federalist*, no. 34. "The Same Subject Continued: Concerning the General Power of Taxation."

——. *Federalist*, no. 35. "The Same Subject Continued: Concerning the General Power of Taxation."

——. *Federalist*, no. 36. "The Same Subject Continued: Concerning the General Power of Taxation."

Hamilton, Neil D. "Legal Issues in Enforcing Federal Soil Conservation Programs: An Introduction and Preliminary Review." UC Davis Law Review 23 (1989): 637–74.

Hansen, John Mark. *Gaining Access: Congress and the Farm Lobby, 1919–1981*. Chicago: University of Chicago Press, 1991.

Hansen, Zeynep K., and Gary D. Libecap. "Small Farms, Externalities, and the Dust Bowl of the 1930s." *Journal of Political Economy* 112, no. 3 (2004): 665–94.

Hardin, Charles M. *The Politics of Agriculture: Soil Conservation and the Struggle for Power in Rural America*. Glencoe IL: Free Press, 1952.

Hargreaves, Mary W. M. "Dry Farming Alias Scientific Farming." *Agricultural History* 22, no. 1 (1948): 39–56.

Hargreaves, M. W. "Land-Use Planning in Response to Drought: The Experience of the Thirties." *Agricultural History* 50, no. 4 (1976): 561–82.

Harl, Neil E. *The Farm Debt Crisis of the 1980s*. Ames: Iowa State University Press, 1990.

Hathaway, Oona A. "Path Dependence in the Law: The Course and Pattern of Legal Change in a Common Law System." *Iowa Law Review* 86 (2000): 601–66.

Heady, Earl O. "Externalities of American Agricultural Policy." *University of Toledo Law Review* 7 (1976): 795–836.

Held, R. Burnell, and Marion Clawson. *Soil Conservation in Perspective*. Published for Resources for the Future, Inc. Baltimore: Johns Hopkins Press, 1965.

Heller, Walter P., and David A. Starrett. "On the Nature of Externalities." In *Theory and Measurement of Economic Externalities*. New York: Academic Press, 1976.

Hellerstein, Daniel, Nathaniel Alan Higgins, and Michael Roberts. "Options for Improving Conservation Programs: Insights from Auction Theory and Economic Experiments." *Amber Waves*, February 2, 2015. U.S. Department of Agriculture, Economic Research Service. https://www.ers.usda.gov/amber-waves/2015/januaryfebruary/options-for-improving-conservation-programs-insights-from-auction-theory-and-economic-experiments.

Helms, Douglas. "Hugh Hammond Bennett and the Creation of the Soil Erosion Service." *Journal of Soil and Water Conservation* 64, no. 2 (2009): 68A–74A.

———. "Soil and Southern History." *Agricultural History* 74, no. 4 (2000): 723–58.

Hines, N. W. "The Land Ethic and American Agriculture." *Loyola of Los Angeles Law Review* 27 (1993): 841–50.

Hines, N. William. "History of the 1972 Clean Water Act: The Story Behind How the 1972 Act Became the Capstone on a Decade of Extraordinary Environmental Reform." *George Washington Journal of Energy and Environmental Law* 4 (2013): 80–106.

Hoadley, John F. "Easy Riders: Gramm-Rudman-Hollings and the Legislative Fast Track." *PS: Political Science & Politics* 19, no. 1 (1986): 30–36.

Hoffsommer, Harold. "The AAA and the Cropper." *Social Forces* 13, no. 4 (1934): 494–502.

Holland, Austin, David Bennett, and Silvia Secchi. "Complying with Conservation Compliance? An Assessment of Recent Evidence in the US Corn Belt." *Environmental Research Letters* 15, no. 8 (2020): 1–10.

Holleman, Hannah. *Dust Bowls of Empire: Imperialism, Environmental Politics, and the Injustice of "Green" Capitalism*. New Haven: Yale University Press, 2018.

Holmes, William F. "Whitecapping: Agrarian Violence in Mississippi, 1902–1906." *Journal of Southern History* 35, no. 2 (1969): 165–85.

Hooks, Gregory M. "A New Deal for Farmers and Social Scientists: The Politics of Rural Sociology in the Depression Era." *Rural Sociology* 48, no. 3 (1983): 386–408.

Hopkins, Cyril G. "Bread from Stones." Circular No. 168. Urbana: University of Illinois, Agricultural Experiment Station, September 1913.

Hovenkamp, Herbert, "Appraising the Progressive State." *Iowa Law Review* 102, no. 3 (2017): 1063–112.

———. "Legislation, Well-Being, and Public Choice." *University of Chicago Law Review* 57, no. 1 (1990): 63–116.

Howard, Robert L. "The Supreme Court, The Constitution, and the A.A.A." *Kentucky Law Journal* 25, no. 4 (1937): 291–320.

Hudson, Norman W. "A World View of the Development of Soil Conservation." *Agricultural History* 59, no. 2 (1985): 326–39.

Hunter, Robert F. "The AAA Between Neighbors: Virginia, North Carolina, and the New Deal Farm Program." *Journal of Southern History* 44, no. 4 (1978): 537–70.

Hurt, R. D. *The Dust Bowl: An Agricultural and Social History*. Chicago: Nelson-Hall, 1981.

———. "The National Grasslands: Origin and Development in the Dust Bowl." *Agricultural History* 59, no. 2 (1985): 246–59.

Ifft, Jennifer, Deepak Rajagopal, and Ryan Weldzuis. "Ethanol Plant Location and Land Use: A Case Study of CRP and the Ethanol Mandate." *Applied Economic Perspectives and Policy* 41, no. 1 (2019): 37–55.

Intergovernmental Panel on Climate Change. "Climate Change 2022: Impacts, Adaptation and Vulnerability." *IPCC Sixth Assessment Report*. https://www.ipcc.ch/report/ar6/wg2/.

Irons, Peter H. *The New Deal Lawyers*. Princeton NJ: Princeton University Press, 1982.

Janzen, H. Henry. "The Soil Remembers." *Soil Science Society of America Journal* 80, no. 6 (2016): 1429–32.

Jentleson, Adam. *Kill Switch: The Rise of the Modern Senate and the Crippling of American Democracy*. New York: Liveright, 2021.

Johnson, Charles S., Edwin R. Embree, and W. W. Alexander. *The Collapse of Cotton Tenancy: Summary of Field Studies and Statistical Surveys 1933–1935*. Chapel Hill: University of North Carolina Press, 1935.

Jones, Charles O., and Randall Strahan. "The Effect of Energy Politics on Congressional and Executive Organization in the 1970s." *Legislative Studies Quarterly* 10, no. 2 (May 1985): 151–79.

Jones, John W. "Trojan Legend: Who Is Sinon?" *Classical Journal* 61, no. 3 (1965): 122–28.

Jones, Lewis W. "The Negro Farmer." *Journal of Negro Education* 22, no. 3 (1953): 322–32.

Kagan, Robert A. *Adversarial Legalism: The American Way of Law*. Cambridge MA: Harvard University Press, 2001.

Kahn, Paul W. "Gramm-Rudman and the Capacity of Congress to Control the Future." *Hastings Constitutional Law Quarterly* 13 (1985): 185–232.

Karle, Sarah Thomas, and David Karle. "200 Million Trees: Fabricating a Rain-Making Scheme." *Journal of Architectural Education* 69, no. 1 (2015): 54–57.

Katznelson, Ira. *Fear Itself: The New Deal and the Origins of Our Time*. New York: Liveright, 2013.

———. *When Affirmative Action Was White: An Untold History of Racial Inequality in Twentieth-Century America*. New York: W. W. Norton, 2005.

Katznelson, Ira, and John S. Lapinski. "At the Crossroads: Congress and American Political Development." *Perspectives on Politics* 4, no. 2 (2006): 243–60.

Katznelson, Ira, and Quinn Mulroy. "Was the South Pivotal? Situated Partisanship and Policy Coalitions During the New Deal and Fair Deal." *Journal of Politics* 74, no. 2 (2012): 604–20.

Katznelson, Ira, K. Geiger, and D. Kryder. "Limiting Liberalism: The Southern Veto in Congress, 1933–1950." *Political Science Quarterly* 10, no. 2 (1993): 283–306.

Kauffman, George B. "The Role of Gold in Alchemy, Part I." *Gold Bulletin* 18, no. 1 (1985): 31–44.

Kelman, Steven. "Public Choice and Public Spirit." *Public Interest* 87 (1987): 80–94.

Kester, Howard. *Revolt Among the Sharecroppers*. New York: Covici, Friede, 1936.

Key, V. O. *Southern Politics in State and Nation: A New Edition*. Knoxville: University of Tennessee Press, 1949. 8th printing, 2018.

Kile, Orville Merton. *The Farm Bureau Through Three Decades*. Baltimore: Waverly Press, 1948.

King, Ronald F. *Budgeting Entitlements: The Politics of Food Stamps*. Washington DC: Georgetown University Press, 2000.

Kirby, Jack Temple. "The Transformation of Southern Plantations c. 1920–1960." *Agricultural History* 57, no. 3 (1983): 257–76.

Kirkendall, Richard S. "Howard Tolley and Agricultural Planning in the 1930's." *Agricultural History* 39, no. 1 (1965): 25–33.

———. *Social Scientists and Farm Politics in the Age of Roosevelt*. Columbia: University of Missouri Press, 1966.

Knapp, David C. "Congressional Control of Agricultural Conservation Policy: A Case Study of the Appropriations Process." *Political Science Quarterly* 71, no. 2 (1956): 257–81.

Kotz, Nick. *Let Them Eat Promises: The Politics of Hunger in America*. Prentice-Hall, 1969.

Kraft, Michael E. "US Environmental Policy and Politics: From the 1960s to the 1990s." *Journal of Policy History* 12, no. 1 (2000): 17–42.

Kramer, Randall A., and Sandra S. Batie. "Cross Compliance Concepts in Agricultural Programs: The New Deal to the Present." *Agricultural History* 59, no. 2 (1985): 307–19.

Kravitz, Walter. "The Advent of the Modern Congress: The Legislative Reorganization Act of 1970." *Legislative Studies Quarterly* 15, no. 3 (August 1990): 375–99.

Kyvig, David E. "The Road Not Taken: FDR, the Supreme Court, and Constitutional Amendment." *Political Science Quarterly* 104, no. 3 (1989): 463–81.

Lacey, John J. *Farm Bureau in Illinois: History of Illinois Farm Bureau*. Illinois Agricultural Association, 1965.

LaDale Winling, Robert K. Nelson, et al. "Electing the House of Representatives." In *American Panorama*, edited by Robert K. Nelson and Edward L. Ayers. Accessed October 14, 2022. https://dsl.richmond.edu/panorama/congress.

Lane, Dorsey Edw. "The Civil Rights Act of 1957." *Howard Law Journal* 4, no. 1 (1958): 36–49.

Lapinski, John S. *The Substance of Representation: Congress, American Political Development, and Lawmaking*. Princeton NJ: Princeton University Press, 2013.

Lark, Tyler J., J. Meghan Salmon, and Holly K. Gibbs. "Cropland Expansion Outpaces Agricultural and Biofuel Policies in the United States." *Environmental Research Letters* 10, no. 4 (2015): 044003.

Laycock, W. A. "History of Grassland Plowing and Grass Planting on the Great Plains." *Impacts of the Conservation Reserve Program in the Great Plains* (1988): 3–8.

Lee, J. A., and T. E. Gill. "Multiple Causes of Wind Erosion in the Dust Bowl." *Aeolian Research* 19 (2015): 15–36.

Lehrer, Nadine. "Negotiating a Political Path to Agroforestry through the Conservation Security Program." *Agroforestry Systems* 75, no. 1 (2009): 103–16.

Leman, Christopher. "Political Dilemmas in Evaluating and Budgeting Soil Conservation Programs: The RCA Process." In *Soil Conservation Policies, Institutions, and Incentives*, edited by H. G. Halcrow, E. O. Heady, and M. L. Cotner, 58–59. Soil Conservation Society of America, 1982.

Lenihan, Martin H., and Kathryn J. Brasier. "Ecological Modernization and the US Farm Bill: The Case of the Conservation Security Program." *Journal of Rural Studies* 26, no. 3 (2010): 219–27.

Leonard, Charles A. *A Search for a Judicial Philosophy: Mr. Justice Roberts and the Constitutional Revolution of 1937*. Port Washington NY: Kennikat Press, 1971.

Leopold, A. "The Conservation Ethic." *Journal of Forestry* 31, no. 6 (1933): 634–43.

Leopold, Aldo. "The Land Ethic." In *A Sand County Almanac, and Sketches Here and There*. Oxford University Press, 1949 and 2021.

LeRay, Nelson L., George L. Wilber, and Grady B. Crowe. "Plantation Organization, and the Resident Labor Force, Delta Area of Mississippi." Bulletin 606. Mississippi State University, Agricultural Experiment Station, 1960.

Levine, Louis. "Economic Security for Agricultural Labor." *Journal of Farm Economics* 44, no. 2 (1962): 506–17.

Levinson, Daryl J., and Richard H. Pildes. "Separation of Parties, Not Powers." *Harvard Law Review* 119, no. 8 (2006): 2311–86.

Lewis, Jerry M., and T. R. Hensley. "The May 4 Shootings at Kent State University: The Search for Historical Accuracy." Kent State University M4Y. Published in revised form by the *Ohio Council for the Social Studies Review* 34, no. 1 (Summer 1999): 9–21. https://www.kent.edu/may-4-historical-accuracy.

Lewis, M. E. "National Grasslands in the Dust Bowl." *Geographical Review* 79, no. 2 (1989): 161–71.

Libecap, Gary D., and Zeynep Kocabiyik Hansen. "'Rain Follows the Plow' and Dry Farming Doctrine: The Climate Information Problem and Homestead Failure in the Upper Great Plains, 1890–1925." *Journal of Economic History* 62, no. 1 (2002): 86–120.

Libecap, G. D., and Z. K. Hansen. "U.S. Land Policy, Property Rights and the Dust Bowl of the 1930s." Working Paper. Nota di Lavoro, no. 69. Fondazione Eni Enrico Mattei (FEEM), Milano, September 2001. https://www.econstor.eu/bitstream/10419/155232/1/NDL2001-069.pdf.

Linder, Marc. "Farm Workers and the Fair Labor Standards Act: Racial Discrimination in the New Deal." *Texas Law Review* 65 (1986): 1335–93.

Lowitt, Richard. "Henry A. Wallace and the 1935 Purge in the Department of Agriculture." *Agricultural History* 53, no. 3 (July 1979): 607–62.

MacLean, Nancy. *Democracy in Chains: The Deep History of the Radical Right's Stealth Plan for America*. Penguin Random House, 2017.

Madigan, E. R. "Adverse Consequences of Trade Embargoes." *Creighton Law Review* 27 (1994): 941–44.

Madison, James. Library of Congress, "Federalist Papers: Primary Documents in American History." Project Gutenberg. https://guides.loc.gov/federalist-papers/full-text.

———. *Federalist*, no. 10. "The Same Subject Continued: The Union as a Safeguard Against Domestic Faction and Insurrection."

———. *Federalist*, no. 37. "Concerning the Difficulties of the Convention in Devising a Proper Form of Government."

———. *Federalist*, no. 42. "The Powers Conferred by the Constitution Further Considered."

———. *Federalist*, no. 47. "The Particular Structure of the New Government and the Distribution of Power Among Its Different Parts."

———. *Federalist*, no. 48. "These Departments Should Not Be So Far Separated as to Have No Constitutional Control Over Each Other."

———. *Federalist*, no. 51. "The Structure of the Government Must Furnish the Proper Checks and Balances Between the Different Departments."

———. *Federalist*, no. 58. "Objection That the Number of Members Will Not Be Augmented as the Progress of Population Demands Considered."

———. *Federalist*, no. 62. "The Senate."

———. *Federalist*, no. 63. "The Senate Continued."

Malone, Linda A. "Conservation at the Crossroads: Reauthorization of the 1985 Farm Bill Conservation Provisions." *Virginia Environmental Law Journal* 8, no. 2 (1989): 215–34.

———. "A Historical Essay on the Conservation Provisions of the 1985 Farm Bill: Sodbusting, Swampbusting, and the Conservation Reserve." *University of Kansas Law Review* 34 (1985): 577–98.

Mann, Susan A. "Sharecropping in the Cotton South: A Case Study of Uneven Development in Agriculture." *Rural Sociology* 49, no. 4 (1984): 412–29.

Martín, Antonio José, Tatsuya Shinagawa, and Javier Pérez-Ramírez. "Electrocatalytic Reduction of Nitrogen: From Haber-Bosch to Ammonia Artificial Leaf." *Chem* 5, no. 2 (2019): 263–83. https://www.sciencedirect.com/science/article/pii/s245192941830490x.

Massey, Dean T. "Land Use Regulatory Power of Conservation Districts in the Midwestern States for Controlling Nonpoint Source Pollutants." *Drake Law Review* 33, no. 1 (1983): 35–112.

May, Irvin, Jr. "Marvin Jones: Agrarian and Politician." *Agricultural History* 52, no. 2 (1977): 421–40.

Mayer, Michael S. "The Eisenhower Administration and the Civil Rights Act of 1957." *Congress & the Presidency: A Journal of Capital Studies* 16, no. 2 (1989): 137–54.

McConnell, Grant. "The Conservation Movement—Past and Present." *Western Political Quarterly* 7, no. 3 (1954): 463–78.

McCorvie, Mary R., and Christopher L. Lant. "Drainage District Formation and the Loss of Midwestern Wetlands, 1850–1930." *Agricultural History* 67, no. 4 (1993): 13–39.

McCullough, Rose, and Daniel Weiss. "An Environmental Look at the 1985 Farm Bill." *Journal of Soil and Water Conservation* 40, no. 3 (1985): 267–70.

McDean, Harry Carson. "M. L. Wilson and Agricultural Reform in Twentieth Century America." Dissertation, University of California, Los Angeles, 1969. Order no. 7008174, ProQuest.

McFadden, Jonathan, and R. A. Hoppe. "The Evolving Distribution of Payments from Commodity, Conservation, and Federal Crop Insurance Programs." Economic Information Bulletin No. (EIB-184). U.S. Department of Agriculture, Economic Research Service, November 2017. https://www.ers.usda.gov/publications/pub-details/?pubid=85833.

McGuire, Jean M., Lois Wright Morton, J. Gordon Arbuckle Jr., and Alicia D. Cast. "Farmer Identities and Responses to the Social—Biophysical Environment." *Journal of Rural Studies* 39 (2015): 145–55.

McLeman, Robert A., Juliette Dupre, Lea Berrang Ford, James Ford, Konrad Gajewski, and Gregory Marchildon. "What We Learned from the Dust Bowl: Lessons in Science, Policy, and Adaptation." *Population and Environment* 35, no. 4 (2014): 417–40.

McNeill, John R., and Verena Winiwarter. "Breaking the Sod: Humankind, History, and Soil." *Science* 304, no. 5677 (2004): 1627–29.

McSweeny, William T., and Randall A. Kramer. "Farm Level Impacts of Cross Compliance Between Soil Conservation and Commodity Support and Insurance Programs." No. 279124. Annual Meeting, American Agricultural Economics Association, Logan, Utah, August 1–4, 1982.

Mercier, Stephanie A., and Steve A. Halbrook. *Agricultural Policy of the United States: Historic Foundations and 21st Century Issues*. Palgrave MacMillan, 2020.

Metzger, Gillian E. "1930s Redux: The Administrative State Under Siege." *Harvard Law Review* 131, no. 1 (2017): 1–95.

Mickey, Robert. *Paths Out of Dixie: The Democratization of Authoritarian Enclaves in America's Deep South, 1944–1972*. Princeton: Princeton University Press, 2015.

Mikva, Abner J. "Justice Brennan and the Political Process: Assessing the Legacy of Baker v. Carr." *University of Illinois Law Review*, no. 3 (1995): 683–98.

Milde, Karl F. "Roman Contributions to the Law of Soil Conservation." *Fordham Law Review* 19 (1950): 192–96.

Minami, Katsuyuki. "Soil and Humanity: Culture, Civilization, Livelihood and Health." *Soil Science and Plant Nutrition* 55, no. 5 (2009): 603–15. https://www.tandfonline.com/doi/full/10.1111/j.1747-0765.2009.00401.x.

Mitchell, H. L. *Mean Things Happening in This Land: The Life and Times of H. L. Mitchell, Co-Founder of the Southern Tenant Farmers Union*. Montclair NJ: Allanheld, Osmun, 1979.

Mitchell, Timothy. "The Resources of Economics: Making the 1973 Oil Crisis." *Journal of Cultural Economy* 3, no. 2 (2010): 189–204.

Mitchell, T. W. "Destabilizing the Normalization of Rural Black Land Loss: A Critical Role for Legal Empiricism." *Wisconsin Law Review*, no. 2 (2005): 557–616.

———. "From Reconstruction to Deconstruction: Undermining Black Landownership, Political Independence, and Community through Partition Sales of Tenancies in Common." *Northwestern University Law Review* 95 (2001): 505–80.

Montgomery, David R. *Dirt: The Erosion of Civilizations*. Berkeley: University of California Press, 2007.

Morath, Sarah J. "The Farm Bill: A Wicked Problem Seeking a Systematic Solution." *Duke Environmental Law and Policy Forum* 25 (2014): 389–426.

Morgan, Dan. "The Farm Bill and Beyond." *Economic Policy Paper Series*. German Marshall Fund of the United States, 2010.

Morgan, Robert J. *Governing Soil Conservation: Thirty Years of the New Decentralization*. Published for Resources for the Future. Baltimore: Johns Hopkins Press, 1965.

———. "Pressure Politics and Resources Administration." *Journal of Politics* 18, no. 1 (1956): 39–60.

Morgan, R. P. C. *Soil Erosion and Conservation*. New York: John Wiley & Sons, 1986.

National Association of Conservation Districts. "Conservation Compliance Coalition Praises Agreement in Senate Farm Bill." Press Release, May 14, 2013. http://www.nacdnet.org/news/newsroom/2013/conservation-compliance-coalition-praises-agreement-in-senate-farm-bill.

Nelson, Lawrence J. "The Art of the Possible: Another Look at the 'Purge' of the AAA Liberals in 1935." *Agricultural History* 57, no. 4 (1983): 416–35.

———. *King Cotton's Advocate: Oscar G. Johnston and the New Deal*. Knoxville: University of Tennessee Press, 1999.

———. "Oscar Johnston, the New Deal, and the Cotton Subsidy Payments Controversy, 1936–1937." *Journal of Southern History* 40, no. 3 (1974): 399–416.

Nelson, Philip J. "To Hold the Land: Soil Erosion, Agricultural Scientists, and the Development of Conservation Tillage Techniques." *Agricultural History* 71, no. 1 (1997): 71–90.

Newman, William R. "What Have We Learned from the Recent Historiography of Alchemy?" *Isis* 102, no. 2 (2011): 313–21.

Newman, William R., and Lawrence M. Principe. "Alchemy vs. Chemistry: The Etymological Origins of a Historiographic Mistake." *Early Science and Medicine* 3, no. 1 (1998): 32–65.

Nixon, President Richard. "Endangered Species Act of 1973: Statement by the President Upon Signing the Bill into Law, Executive Orders and Presidential Proclamations." December 28, 1973.

———. "Letter Accepting the Resignation of Clifford M. Hardin as Secretary of Agriculture." November 11, 1971. https://www.presidency.ucsb.edu/documents/letter-accepting-the-resignation-clifford-m-hardin-secretary-agriculture.

———. "Remarks Announcing the Resignation of Clifford M. Hardin and Intention to Nominate Earl L. Butz as Secretary of Agriculture." November 11, 1971. https://www.presidency.ucsb.edu/documents/remarks-announcing-the-resignation-clifford-m-hardin-and-intention-nominate-earl-l-butz.

Nourse, Victoria. "A Decision Theory of Statutory Interpretation: Legislative History by the Rules." *Yale Law Journal* 122 (2012): 70–153.

———. "Toward a Due Foundation for the Separation of Powers: The Federalist Papers as Political Narrative." *Texas Law Review* 74 (1996): 447–522.

Ogg, Clayton. "Soil Conservation Under Integrated Farm Program." No. 279125. Annual Meeting, American Agricultural Economics Association, Logan, Utah, August 1–4, 1982.

Ogg, Clayton W., James D. Johnson, and Kenneth C. Clayton. "A Policy Option for Targeting Soil Conservation Expenditures." *Journal of Soil and Water Conservation* 37, no. 2 (1982): 68–72.

Olson, Mancur. *The Logic of Collective Action: Public Goods and the Theory of Groups*. Cambridge MA: Harvard University Press, 1965.

Orden, David, R. Paarlberg, and T. Roe. *Policy Reform in American Agriculture: Analysis and Prognosis*. Chicago: University of Chicago Press, 1999.

Ornstein, Norman J. "The Democrats Reform Power in the House of Representatives." In *America in the Seventies: Problems, Policies, and Politics*, edited by Allan P. Sindler. Boston: Little, Brown, 1977.

———. "The Politics of the Deficit." In *Essays in Contemporary Economic Problems, 1985: The Economy in Deficit*, edited by Phillip Cagan, 311–33. Washington DC: American Enterprise Institute, 1985.

Orren, Karen, and Stephen Skowronek. *The Search for American Political Development*. Cambridge University Press, 2004.

Paarlberg, Don. *Farm and Food Policy: Issues of the 1980s*. Lincoln: University of Nebraska Press, 1980.

Page, Scott E. "Path Dependence." *Quarterly Journal of Political Science* 1, no. 1 (2006): 87–115.

Patch, B. W. "Soil Conservation and Agricultural Adjustment." *Editorial Research Reports*, 1936, vol. 1. http://library.cqpress.com/cqresearcher/cqresrre1936012700.

Patton, Patrick C. "Standing at Thermopylae: A History of the American Liberty League." PhD diss., Temple University, 2015.

Pepper, George Wharton. *Philadelphia Lawyer*. Philadelphia and New York: T. B. Lippincott, 1944.

Perea, Juan F. "The Echoes of Slavery: Recognizing the Racist Origins of the Agricultural and Domestic Worker Exclusion from the National Labor Relations Act." *Ohio State Law Journal* 72 (2011): 95–138.

Phelps, Jess, "A Vision of the New Deal Unfulfilled? Soil and Water Conservation Districts and Land Use Regulation." *Drake Journal of Agricultural Law* 11 (2006): 353–82.

Phillips, Sara T. *This Land, This Nation: Conservation, Rural America, and the New Deal*. Cambridge University Press, 2007.

Pinchot, Gifford. "How Conservation Began in the United States." *Agricultural History* 11, no. 4 (October 1937): 255–65.

Pollans, Margot J. "Drinking Water Protection and Agricultural Exceptionalism." *Ohio State Law Journal* 77 (2016): 1195–260.

Prado, Mariana, and Michael Trebilcock. "Path Dependence, Development, and the Dynamics of Institutional Reform." *University of Toronto Law Journal* 59, no. 3 (2009): 341–80.

Prentiss, Paul F. "Sodbusting in the West: Is Congress Waiting for the Dust to Settle." *William Mitchell Environmental Law Journal* 3 (1985): 67–86.

Prokopy, Linda S., Kristin Floress, J. Gordon Arbuckle, Sarah P. Church, Francis R. Eanes, Yuling Gao, Benjamin M. Gramig, Pranay Ranjan, and Ajay S. Singh. "Adoption of Agricultural Conservation Practices in the United States: Evidence from 35 Years of Quantitative Literature." *Journal of Soil and Water Conservation* 74, no. 5 (2019): 520–34.

Pusey, Merlo J. *Charles Evans Hughes*. Vol. 2. New York: MacMillan, 1951.

Rankin, J. Lee. "The Supreme Court, The Depression, and the New Deal: 1930–1941." *Nebraska Law Review* 40, no. 1 (1960): 35–62.

Rasmussen, Chris. "'Never a Landlord for the Good of the Land': Farm Tenancy, Soil Conservation, and the New Deal in Iowa." *Agricultural History* 73, no. 1 (1999): 70–95.

Reagan, President Ronald. "Statement on Signing H.R. 2100 into Law: Executive Orders and Presidential Proclamations; Presidential Signing Statements." December 23, 1985.

Reynolds, B. "Black Farmers in America, 1865–2000: The Pursuit of Independent Farming and the Role of Cooperatives." RBS Research Report 194. U.S. Department of Agriculture, Rural Business-Cooperative Service, 2002. https://naldc.nal.usda.gov/catalog/20698.

Ribaudo, Marc. "The Limits of Voluntary Conservation Programs." *Choices* 30, no. 2 (2015): 1–5. https://www.jstor.org/stable/choices.30.2.14.

Ribaudo, Marc O., David Zilberman, Rebecca Taylor, and Ben Gordon. "Conservation Programs Can Accomplish More with Less by Improving Cost-Effectiveness." *Choices* 32, no. 4 (2017): 1–6.

Richards, Leonard L. *The Slave Power: The Free North and Southern Domination, 1780–1860*. Baton Rouge: Louisiana State University, 2000.

Richman, Sheldon. "A Matter of Degree, Not Principle: The Founding of the American Liberty League." *Journal of Libertarian Studies* 6, no. 2 (1982): 145–67.

Rieselbach, Leroy N. *Congressional Reform in the Seventies*. Morristown: General Learning Press, 1977.

Rieselbach, Leroy N., and Joseph K. Unekis. "Ousting the Oligarchs: Assessing the Consequences of Reform and Change on Four House Committees." *Congress & the Presidency: A Journal of Capital Studies* 9, no. 1 (1981): 83–117.

Roberts, John C. "Are Congressional Committees Constitutional: Radical Textualism, Separation of Powers, and the Enactment Process." *Case Western Reserve Law Review* 52, no. 2 (2001): 489–572.

Rodell, F. *Nine Men: A Political History of the Supreme Court from 1790 to 1955*. New York: Random House, 1955.

Rodriguez, Divina Gracia P., David S. Bullock, and Maria A. Boerngen. "The Origins, Implications, and Consequences of Yield-Based Nitrogen Fertilizer Management." *Agronomy Journal* 111, no. 2 (2019): 725–35.

Roosevelt, President Franklin Delano. "Statement on Signing the Soil Conservation and Domestic Allotment Act." March 1, 1936. *Executive Orders and Presidential Proclamations: Signed by the President; Presidential Signing Statements*. Available from ProQuest Congressional Universe.

Rose, Carol M. "Joseph Sax and the Idea of the Public Trust." *Ecology Law Quarterly* 25 (1998): 351.

Rosenberg, Nathan A., and Bryce Wilson Stucki. "The Butz Stops Here: Why the Food Movement Needs to Rethink Agricultural History." *Journal of Food Law and Policy* 13 (2017): 12–25.

Ross, John R. "Man over Nature: Origins of the Conservation Movement." *American Studies* 16, no. 1 (1975): 49–62.

Ros-Tonen, M. A., and J. H. van Boxel. "El Nino in Latin America: The Case of Peruvian Fishermen and North-East Brazilian Peasants." *European Review of Latin American and Caribbean Studies* 67 (1999): 5–20.

Rubin, Edward L. "Beyond Public Choice: Comprehensive Rationality in the Writing and Reading of Statutes." *New York University Law Review* 66 (1991): 1–65.

Rudolph, Frederick. "The American Liberty League, 1934–1940." *American Historical Review* 56, no. 1 (1950): 19–33.

Ruhl, James B. "Farms, Their Environmental Harms, and Environmental Law." *Ecology Law Quarterly* 27 (2000): 263–350.

Ruhl, J. B. "Three Questions for Agriculture About the Environment." *Florida State University Journal of Land Use and Environmental Law* 17, no. 2 (2018): 395–408.

Saloutos, Theodore. *The American Farmer and the New Deal.* Vol. 1. Ames: Iowa State Press, 1982.

Sampson, R. Neil. *Farmland or Wasteland: A Time to Choose; Overcoming the Threat to America's Farm and Food Future.* Emmaus PA: Rodale Press, 1981.

———. *For Love of the Land: A History of the National Association of Conservation Districts.* League City TX: National Association of Conservation Districts, 1985.

Sax, Joseph L. "Liberating the Public Trust Doctrine from Its Historical Shackles." *UC Davis Law Review* 14 (1980): 185–94.

———. "The Public Trust Doctrine in Natural Resource Law: Effective Judicial Intervention." *Michigan Law Review* 68, no. 3 (1970): 471–566.

Schapsmeier, Edward L., and Frederick H. Schapsmeier. "Eisenhower and Agricultural Reform: Ike's Farm Policy Legacy Appraised." *American Journal of Economics and Sociology* 51, no. 2 (1992): 147–60.

———. "Eisenhower and Ezra Taft Benson: Farm Policy in the 1950s." *Agricultural History* 44, no. 4 (1970): 369–78.

———. *Ezra Taft Benson and the Politics of Agriculture: The Eisenhower Years, 1953–1961.* Danville IL: Interstate Printers & Publishers, 1975.

———. "Henry A. Wallace: Agrarian Idealist or Agricultural Realist?" *Agricultural History* 41, no. 2 (1967): 127–38.

Schertz, Lyle P., and Otto C. Doering. *The Making of the 1996 Farm Act.* Ames: Iowa State University Press, 1999.

Scholes, Mary C., and Robert J. Scholes. "Dust Unto Dust." *Science* 342, no. 6158 (2013): 565–66. https://www.science.org/doi/10.1126/science.1244579.

Schutz, Anthony B. "Agricultural Discharges Under the CWA: Old Questions and New Insights." *University of the Pacific Law Review* 52 (2020): 567–98.

Seegers, L. Walter. "Cooley, Harold Dunbar." NCPedia.org, from the *Dictionary of North Carolina Biography*, 1979. https://www.ncpedia.org/biography/cooley-harold-dunbar.

Shane, Peter M. *Madison's Nightmare: How Executive Power Threatens American Democracy*. Chicago: University of Chicago Press, 2009.

Shannon, Fred A. "The Homestead Act and the Labor Surplus." *American Historical Review* 41, no. 4 (1936): 637–51.

Shepsle, Kenneth A., and Barry R. Weingast. "The Institutional Foundations of Committee Power." *American Political Science Review* 81, no. 1 (March 1987): 85–104.

———. "When Do Rules of Procedure Matter?" *Journal of Politics* 46, no. 1 (February 1984): 206–21.

Shughart, William F. "Bending Before the Storm: The US Supreme Court in Economic Crisis, 1935–1937." *Independent Review* 9, no. 1 (2004): 55–83.

Sigelman, Lee. "Politics, Economics, and the American Farmer: The Case of 1980." *Rural Sociology* 48, no. 3 (1983): 367–85.

Silverstein, Gordon. *Law's Allure: How Law Shapes, Constrains, Saves, and Kills Politics*. Cambridge University Press, 2009.

Skocpol, Theda. *Protecting Soldiers and Mothers: The Political Origins of Social Policy in the United States*. Boston: Harvard University Press, 1995.

Smith, H. N. "Rain Follows the Plow: The Notion of Increased Rainfall for the Great Plains, 1844–1880." *Huntington Library Quarterly* 10, no. 2 (February 1947): 169–93.

Smith, John P. "The Social and Ecological Correlates of Bankruptcy During the Farm Fiscal Crisis, 1970–1987." *Mid-American Review of Sociology* 12, no. 1 (1987): 35–53.

Smith, Loren M., William R. Effland, Kathrine D. Behrman, and Mari-Vaughn V. Johnson. "Assessing the Effects of USDA Conservation Programs on Ecosystem Services provided by Wetlands." *National Wetlands Newsletter* 37, no. 5 (2015): 10–14.

Smith, Trevor J. "Corn, Cows, and Climate Change: How Federal Agriculture Subsidies Enable Factory Farming and Exacerbate US Greenhouse Gas Emissions." *Washington Journal of Environmental Law and Policy* 9, no. 1 (2019): 26.

Smith, Vincent H., ed. *The Economic Welfare and Trade Relations Implications of the 2014 Farm Bill*. Emerald Group, 2015.

Smith, Vincent H., J. W. Glauber, and B. K. Goodwin, eds. *Agricultural Policy in Disarray*. Vols. 1 and 2. Washington DC: American Enterprise Instituted, 2018.

Soifer, Aviam. "Truisms That Never Will Be True: The Tenth Amendment and the Spending Power." *University of Colorado Law Review* 57, no. 5 (1986): 793–834.

Speigner, Theodore R. "An Analysis of Policy Issues of Small Watershed Acts 1954–1956." *Journal of Geography* 61, no. 9 (1962): 408–14.

Spezio, Teresa Sabol. "The Santa Barbara Oil Spill and Its Effect on United States Environmental Policy." *Sustainability* 10, no. 8 (2018): 2750.

Stam, Jerome M., and Bruce L. Dixon. "Farmer Bankruptcies and Farm Exits in the United States, 1899–2002." Agriculture Information Bulletin No. 788. U.S. Department of Agriculture, Economic Research Service, 2004. https://www.ers.usda.gov/webdocs/publications/42532/17750_aib788_1_.pdf?v=.

Stephens, P. H. "Why the Dust Bowl?" *Journal of Farm Economics* 19, no. 3 (1937): 750–57.

Stith, Kate. "Congress' Power of the Purse." *Yale Law Journal* 97 (1988): 1343–96.

———. "Rewriting the Fiscal Constitution: The Case of Gramm-Rudman-Hollings." *California Law Review* 76 (1988): 593–670.

Stockman, David. *The Triumph of Politics: How the Reagan Revolution Failed*. New York: Harper & Row, 1986.

Stradling, David, and Richard Stradling. "Perceptions of the Burning River: Deindustrialization and Cleveland's Cuyahoga River." *Environmental History* 13, no. 3 (2008): 515–35.

Stuart, Diana, Bruno Basso, Sandy Marquart-Pyatt, Adam Reimer, G. Philip Robertson, and Jinhua Zhao. "The Need for a Coupled Human and Natural Systems Understanding of Agricultural Nitrogen Loss." *BioScience* 65, no. 6 (2015): 571–78.

Subedi, Dipak, A. K. Giri, and T. McDonald. "Commercial Farms Received the Most Government Payments in 2019." *Amber Waves*, July 6, 2021. U.S. Department of Agriculture, Economic Research Service. https://www.ers.usda.gov/amber-waves/2021/july/commercial-farms-received-the-most-government-payments-in-2019/.

Sunbury, Ben. *The Fall of the Farm Credit Empire*. Ames: Iowa State University Press, 1990.

Swidler, Eva-Maria. "The Social Production of Soil." *Soil Science* 174, no. 1 (2009): 2–8.

Switzer, Jacqueline Vaugh. *Green Backlash: The History and Politics of Environmental Opposition in the U.S.* Boulder: Lynne Rienner, 1997.

Talbert, Jeffery C., and Matthew Potoski. "Setting the Legislative Agenda: The Dimensional Structure of Bill Cosponsoring and Floor Voting." *Journal of Politics* 64, no. 3 (August 2002): 864–91.

Talbot, Ross B. "Farm Legislation in the 86th Congress." *Journal of Farm Economics* 43, no. 3 (1961): 582–605.

Talbot, Ross B., and Don F. Hadwiger. *The Policy Process in American Agriculture*. San Francisco: Chandler, 1968.

Taylor, Dorceta E. *The Rise of the American Conservation Movement: Power, Privilege, and Environmental Protection*. Durham: Duke University Press, 2016.

Thaler, Evan A., Jeffrey S. Kwang, Brendon J. Quirk, Caroline L. Quarrier, and Isaac J. Larsen. "Rates of Historical Anthropogenic Soil Erosion in the Midwestern

United States." *Earth's Future* 10, no. 3 (2022): E2021EF002396. https://agupubs.onlinelibrary.wiley.com/doi/pdfdirect/10.1029/2021EF002396.

Tiefer, Charles. "The Reconceptualization of Legislative History in the Supreme Court." *Wisconsin Law Review* 2000, no. 2 (2000): 205–78.

Tolson, Franita. "The Law of Democracy at a Crossroads: Reflecting on Fifty Years of Voting Rights and the Judicial Regulation of the Political Thicket." *Florida State University Law Review* 43, no. 2 (2016): 345–52.

Tomer, M. D., and M. A. Locke. "The Challenge of Documenting Water Quality Benefits of Conservation Practices: A Review of USDA-ARS's Conservation Effects Assessment Project Watershed Studies." *Water Science and Technology* 64, no. 1 (2011): 300–310.

Tomer, M. D., E. J. Sadler, R. E. Lizotte, R. B. Bryant, T. L. Potter, M. T. Moore, T. L. Veith, Claire Baffaut, M. A. Locke, and M. R. Walbridge. "A Decade of Conservation Effects Assessment Research by the USDA Agricultural Research Service: Progress Overview and Future Outlook." *Journal of Soil and Water Conservation* 69, no. 5 (2014): 365–73.

Turner, James Morton. "'The Specter of Environmentalism': Wilderness, Environmental Politics, and the Evolution of the New Right." *Journal of American History* 96, no. 1 (2009): 123–48.

Urban, Michael A. "An Uninhabited Waste: Transforming the Grand Prairie in Nineteenth Century, Illinois, USA." *Journal of Historical Geography* 31 (2005): 647–65.

U.S. Census Bureau. "Historical Population Change Data (1910–2020)." https://www.census.gov/data/tables/time-series/dec/popchange-data-text.html.

U.S. Commission on Civil Rights. "The Decline of Black Farming in America: A Report of the United States Commission on Civil Rights." 1982.

———. "Equal Opportunity in Farm Programs: An Appraisal of Services Rendered by Agencies of the United States Department of Agriculture; a Report." 1965.

U.S. Department of Agriculture, Agricultural Adjustment Administration (AAA). "Agricultural Adjustment, 1937–1938: A Report of the Activities Carried on by the Agricultural Adjustment Administration." Washington DC: GPO, 1939.

———. "Agricultural Adjustment 1938–39: A Report of the Activities Carried on by the Agricultural Adjustment Administration July 1, 1938, through June 30, 1939." Washington DC: GPO, 1939.

———. "Agricultural Conservation 1936: A Report of the Activities of the Agricultural Adjustment Administration." Washington DC: GPO, 1937.

———. "Census of Agriculture: 2017." https://www.nass.usda.gov/Publications/AgCensus/2017/Full_Report/Volume_1,_Chapter_1_US/.

———. "Census of Agriculture Historical Archive: 1930." https://agcensus.library.cornell.edu/census_year/1930-census/.

U.S. Department of Agriculture, Commodity Credit Corporation. "CCC Budget Essentials." https://web.archive.org/web/20220208020323/https://www.fsa.usda.gov/about-fsa/budget-and-performance-management/budget/ccc-budget-essentials/index.

U.S. Department of Agriculture, Economic Research Service (ERS). *Data Files: U.S. and State-Level Farm Income and Wealth Statistics*, "Federal Government Direct Farm Payments by Program, U.S. and States." https://data.ers.usda.gov/reports.aspx?ID=17833.

———. *Data Files: U.S. and State-Level Farm Income and Wealth Statistics*, "Net Cash Income." https://data.ers.usda.gov/reports.aspx?ID=17831.

———. "The Land Utilization Program 1934 to 1964: Origin, Development, and Present Status." Agricultural Economic Report No. 85. Washington DC: GPO, 1965.

———. "Major Land Uses: Cropland Used for Crops." https://www.ers.usda.gov/data-products/major-land-uses/major-land-uses/#Cropland.

U.S. Department of Agriculture, Farm Service Agency (FSA). "Conservation Reserve Program." https://www.fsa.usda.gov/programs-and-services/conservation-programs/conservation-reserve-program/.

———. "Reports and Statistics," Conservation Reserve Program. https://www.fsa.usda.gov/programs-and-services/conservation-programs/reports-and-statistics/index.

U.S. Department of Agriculture, Natural Resources Conservation Service (NRCS). "Conservation Stewardship Program." https://www.nrcs.usda.gov/wps/portal/nrcs/main/national/programs/financial/csp/.

———. "Easements." https://www.nrcs.usda.gov/wps/portal/nrcs/main/national/programs/easements/.

———. "Environmental Quality Incentives Program." https://www.nrcs.usda.gov/wps/portal/nrcs/main/national/programs/financial/eqip/.

———. "Soil Erosion—About the Data." https://www.nrcs.usda.gov/wps/portal/nrcs/detail/soils/survey/geo/?cid=stelprdb1041925.

———. *Summary Report: 2017 National Resources Inventory*. Natural Resources Conservation Service, Washington DC, and Center for Survey Statistics and Methodology, Iowa State University, Ames, Iowa, September 2020. https://www.nrcs.usda.gov/sites/default/files/2022-10/2017NRISummary_Final.pdf.

———. "USDA Celebrates 5 Million Acres Enrolled in Conservation Easements." News Release, April 2, 2021. https://www.nrcs.usda.gov/wps/portal/nrcs/detail/national/newsroom/releases/?cid=NRCSEPRD1760829.

U.S. Department of Agriculture, Risk Management Agency. "Summary of Business Reports." https://www.rma.usda.gov/SummaryOfBusiness.

U.S. Department of State, Office of the Historian. "The 1973 Arab-Israeli War." https://history.state.gov/milestones/1969-1976/arab-israeli-war-1973.

Vance, Rupert B. "Human Factors in the South's Agricultural Readjustment." *Law and Contemporary Problems* 1, no. 3 (1934): 259–74.

Vermeule, Adrian. "The Constitutional Law of Congressional Procedure." *University of Chicago Law Review* 71, no. 2 (Spring 2004): 361–438.

Volanto, Keith J. "The AAA Cotton Plow-Up Campaign in Arkansas." *Arkansas Historical Quarterly* 59, no. 4 (Winter 2000): 388–406.

Volanto, Keith Joseph. "Ordered Liberty: The AAA Cotton Programs in Texas, 1933–1940." PhD diss., Texas A&M University, 1998. Order no. 9903222, ProQuest. https://www-proquest-com.proxy2.library.illinois.edu/dissertations-theses/ordered-liberty-aaa-cotton-programs-texas-1933/docview/304492883/se-2?accountid=14553.

Vos, Nathan. "Agricultural Drainage and the Des Moines Water Works Lawsuit." *Drake Journal of Agricultural Law* 22 (2017): 109–34.

Waldron, Jeremy. *The Dignity of Legislation*. Cambridge University Press, 1999.

———. *Law and Disagreement*. Oxford University Press, 1999.

———. "Legislation, Authority, and Voting." *Georgetown Law Journal* 84 (1996): 2185–214.

Weiss, Harvey, and Raymond S. Bradley. "What Drives Societal Collapse?" *Science* 291, no. 5504 (2001): 609–10. https://www.science.org/doi/full/10.1126/science.1058775.

West, Darrell M. "Gramm-Rudman-Hollings and the Politics of Deficit Reduction." *ANNALS of the American Academy of Political and Social Science* 499, no. 1 (1988): 90–100.

Whatley, Warren C. "Labor for the Picking: The New Deal in the South." *Journal of Economic History* 43, no. 4 (1983): 905–29.

White, Joseph, and Aaron B. Wildavsky. *The Deficit and the Public Interest: The Search for Responsible Budgeting in the 1980s*. Berkeley: University of California Press, 1989.

White, Lynn. "The Historical Roots of Our Ecological Crisis." *Science* 155, no. 3767 (March 10, 1967): 1203–7.

Whittington, Keith E. "The Place of Congress in the Constitutional Order." *Harvard Journal of Law and Public Policy* 40 (2017): 573–602.

Wildavsky, Aaron B. *The New Politics of the Budgetary Process*. Glenview: Scott, Foresman, 1988.

Wilkinson, Charles F. "Soil Conservationists and the Uses of Law." *Journal of Soil and Water Conservation* 42, no. 5 (1987): 304–11.

Williams, Brian Scott. "Perpetual Mobilization and Environmental Injustice: Race and the Contested Development of Industrial Agriculture in the Yazoo-Mississippi Delta." PhD diss., Ohio State University, 2013.

Williams, Craig L. "Soil Conservation and Water Pollution Control: The Muddy Record of the United States Department of Agriculture." *Boston College Environmental Affairs Law Review* 7 (1978): 365–422.

Winquist, Thomas R. "Civil Rights: Legislation: The Civil Rights Act of 1957." *Michigan Law Review* 56, no. 4 (1958): 619–30.

WJW. "Constitutional Law: Agricultural Adjustment Act: The General Welfare Clause and the Tenth Amendment." *Michigan Law Review* 34, no. 3 (1936): 366–84.

Wolfskill, George. *The Revolt of the Conservatives: A History of the American Liberty League, 1934–1940*. Boston: Houghton Mifflin, 1962.

Woodman, Harold D. "Post Civil-War Southern Agriculture and the Law." In "Southern Agriculture Since the Civil War: A Symposium." Special issue, *Agricultural History* 53, no. 1 (January 1979): 319–37.

Woods, Clyde. *Development Arrested: The Blues and Plantation Power in the Mississippi Delta*. New York: Verso, 1998.

Woodward, Donald E. "A Look Back at the Watershed Protection and Flood Prevention Act of 1954 (Pl-566)." In *Watershed Management 2015*, 160–68. American Society of Civil Engineers, 2015.

Worster, D. *Dust Bowl: The Southern Plains in the 1930s*. 25th anniversary ed. Oxford University Press, 2004.

Worster, Donald. "Transformations of the Earth: Toward an Agroecological Perspective in History." *Journal of American History* 76, no. 4 (1990): 1087–106.

Wu, JunJie, Richard M. Adams, Catherine L. Kling, and Katsuya Tanaka. "From Microlevel Decisions to Landscape Changes: An Assessment of Agricultural Conservation Policies." *American Journal of Agricultural Economics* 86, no. 1 (February 2004): 26–41.

Yaalon, Dan H., and Richard W. Arnold. "Attitudes Toward Soils and Their Societal Relevance: Then and Now." *Soil Science* 165, no. 1 (2000): 5–12.

Zelizer, Julian. *On Capitol Hill: The Struggle to Reform Congress and Its Consequences, 1948–2000*. Cambridge University Press, 2006.

Congressional Sources

Due to the large volume of congressional source materials, they are organized in this section of the bibliography by document type and chamber of Congress, except for the Congressional Record citations provided in the notes. For committee hearing records, numbering protocols or reference numbers were added relatively recently. Beginning with the 99th Congress, House and Senate hearings have reference numbers (e.g., House serial numbers or S. Hrg.). For ease of reference and better organization of the notes, I have applied the modern numbering system to the hearing records prior to the 99th Congress, using the chamber (H. Hrg. or S. Hrg.), Congress and year (H. Hrg. 74-1935 or S. Hrg. 74-1935), and additional summary information after a comma (e.g., H. Hrg. 83-1953, "Long Range, part 8"). All congressional source materials are arranged numerically within categories.

Public Laws

P.L. 73-10. Agricultural Adjustment Act of 1933. 73rd Congress, 1st Session, May 12, 1933.

P.L. 74-46. "An Act: To Provide for the Protection of Land Resources against Soil Erosion, and for Other Purposes." 74th Congress, 1st Session, April 27, 1935.

P.L. 74-461. Soil Conservation and Domestic Allotment Act of 1936. 74th Congress, 2nd Session, February 29, 1936.

P.L. 75-210. The Bankhead-Jones Farm Tenant Act. 75th Congress, 1st Session, July 22, 1937.

P.L. 75-430. Agricultural Adjustment Act of 1938. 75th Congress, 3rd Session, February 16, 1938.

P.L. 79-601. The Legislative Reorganization Act of 1946. 79th Congress, 2nd Session, August 2, 1946.

P.L. 80-104. Federal Insecticide, Fungicide, and Rodenticide Act. 80th Congress, 1st Session, June 25, 1947.

P.L. 80-845. Water Pollution Control Act. 80th Congress, 2nd Session, June 30, 1948.

P.L. 83-566. Watershed and Flood Prevention Act of 1954. 83rd Congress, 2nd Session, August 4, 1954.

P.L. 83-690. Agricultural Act of 1954. 83rd Congress, 2nd Session, August 28, 1954.

P.L. 84-40. Department of Agriculture and Farm Credit Administration Appropriation Act, 1956. 84th Congress, 1st Session, H.R. 5239, May 23, 1955.

P.L. 84-540. Agricultural Act of 1956. 84th Congress, 2nd Session, May 28, 1956.

P.L. 84-554. Department of Agriculture and Farm Credit Administration Appropriation Act, 1957. 84th Congress, 2nd Session, H.R. 11177, June 4, 1956.

P.L. 85-118. Department of Agriculture and Farm Credit Administration Appropriation Act, 1958. 85th Congress, 1st Session, H.R. 7441, August 2, 1957.

P.L. 85-459. Department of Agriculture and Farm Credit Administration Appropriation Act, 1959. 85th Congress, 2nd Session, H.R. 11767, June 13, 1958.

P.L. 87-5. "An Act: To Provide a Special Program for Feed Grains for 1961." 87th Congress, 2nd Session, H.R. 4510, March 22, 1961.

P.L. 87-128. Agricultural Act of 1961. 87th Congress, 1st Session, S. 1634, August 8, 1961.

P.L. 87-703. Food and Agriculture Act of 1962. 87th Congress, 2nd Session, H.R. 12391, September 27, 1962.

P.L. 88-26. Feed Grain Act of 1963. 88th Congress, 1st Session, H.R. 4997, May 20, 1963.

P.L. 89-321. Food and Agriculture Act of 1965. 89th Congress, 1st Session, H.R. 9811, November 3, 1965.

P.L. 91-190. National Environmental Policy Act of 1969. 91st Congress, 2nd Session, S. 1075, January 1, 1970.

P.L. 91-510. The Legislative Reorganization Act of 1970. 91st Congress, October 26, 1970.

P.L. 91-524. Agricultural Act of 1970. 91st Congress, 2nd Session, November 30, 1970.

P.L. 91-604. Clean Air Act Amendments of 1970. 91st Congress, 2nd Session, H.R. 17255, December 31, 1970.

P.L. 93-86. Agriculture and Consumer Protection Act of 1973. 93rd Congress, 1st Session, August 10, 1973.

P.L. 93-159. Emergency Petroleum Allocation Act of 1973. 93rd Congress, 1st Session, November 27, 1973.

P.L. 93-205. Endangered Species Act of 1973. 93rd Congress, 1st Session, December 28, 1973.

P.L. 93-319. Energy Supply and Environmental Coordination Act of 1974. 93rd Congress, 2nd Session, June 22, 1974.

P.L. 93-344. Congressional Budget and Impoundment Control Act of 1974. 93rd Congress, 2nd Session, July 12, 1974.

P.L. 93-511. To Extend the Emergency Petroleum Allocation Act of 1973 until August 31, 1973. 93rd Congress, 2nd Session, December 5, 1974.

P.L. 93-523. Safe Drinking Water Act of 1974. 93rd Congress, 2nd Session, December 16, 1974.

P.L. 94-469. Toxic Substances Control Act. 94th Congress, 2nd Session, October 11, 1976.

P.L. 95-113. Food and Agriculture Act of 1977. 95th Congress, 1st Session, September 29, 1977.

P.L. 95-192. Soil and Water Resources Conservation Act of 1977. 95th Congress, 1st Session, November 18, 1977.

P.L. 97-98. Agriculture and Food Act of 1981. 97th Congress, 1st Session, S. 884, December 22, 1981.

P.L. 99-177. Balanced Budget and Emergency Deficit Control Act of 1985. 99th Congress, 1st Session, December 12, 1985.

P.L., 99-198. The Food Security Act of 1985. 99th Congress, 1st Session, December 23, 1985.

P.L. 101-508. Omnibus Budget Reconciliation Act of 1990. 101st Congress, 2nd Session, November 5, 1990.

P.L. 101-624. Food, Agriculture, Conservation and Trade Act of 1990. 101st Congress, 2nd Session, S. 2830, November 28, 1990.

P.L. 104-127. Federal Agriculture Improvement and Reform Act of 1996. 104th Congress, 2nd Session, H.R. 2854, April 4, 1996.

P.L. 105-33. Balanced Budget Act of 1997. 105th Congress, 1st Session, August 5, 1997.

P.L. 107-171. Farm Security and Rural Investment Act of 2002. 107th Congress, 2nd Session, H.R. 2646, May 13, 2002.

P.L. 110-234. Food, Conservation, and Energy of 2008. 110th Congress, 2nd Session, H.R. 2419, May 22, 2008.

P.L. 112-25. Budget Control Act of 2011. 112th Congress, 1st Session, August 2, 2011.

P.L. 113-79. Agricultural Act of 2014. 113th Congress, 2nd Session, H.R. 2642, February 7, 2014.

P.L. 115-334. Agriculture Improvement Act of 2018. 115th Congress, 2nd Session, December 20, 2018.

P.L. 117-169. The Inflation Reduction Act of 2022. 117th Congress, 2nd Session, August 16, 2022.

Legislation

U.S. House of Representatives

73 H.R. 9967. "A Bill: To protect and aid tenants, sharecroppers, and operating owners of farms . . ." 73rd Congress, 2nd Session, June 18, 1934.

74 H.J. Res. 270. "Joint Resolution: Authorizing the appointment of a committee to investigate living and working conditions of share-tenants, share-croppers, and agricultural laborers." 74th Congress, 1st Session, April 30, 1935.

74 H.R. 10500. "A Bill: To make further provision for the conservation and proper utilization of the soil resources of the Nation." *Introduced in the House*. 74th Congress, 2nd Session, January 22, 1936.

83 H. Con. Res. 189. "Concurrent Resolution. . . . That the Secretary of Agriculture is requested to prepare and develop a new and comprehensive soil conservation program . . ." 83rd Congress, 2nd Session, January 6, 1954.

83 H.R. 1809. "A Bill: To provide for the deduction from gross income for income-tax purposes of expenses incurred by farmers for the purpose of soil and water conservation." *Introduced in the House*. 83rd Congress, 1st Session, January 16, 1953.

83 H. Res. 454. "Resolution . . . for the consideration of the bill (H.R. 6788) . . ." 83rd Congress, 2nd Session, March 2, 1954.

84 H.R. 2646. "A Bill: To establish a Conservation Acreage Reserve, to promote conservation and improvement of agricultural soil and water resources, to stabilize

farmers' income, to adjust total agricultural production to consumer and export needs, to maintain an abundant and even flow of farm commodities in interstate commerce, and for other purposes." *Introduced in the House.* 84th Congress, 1st Session, January 20, 1955.

84 H.R. 3912. "A Bill: To establish a conservation acreage reserve, to promote conservation and improvement of agricultural soil and water resources, to stabilize farmers' income, to adjust total agricultural production to consumer and export needs, to maintain an abundant and even flow of farm commodities in interstate commerce, and for other purposes." *Introduced in the House.* 84th Congress, 1st Session, February 10, 1955.

84 H.R. 5942. "A Bill: To establish a conservation acreage reserve, to promote conservation and improvement of agricultural soil and water resources, to stabilize farmers' income, to adjust total agricultural production to consumer and export needs, to maintain an abundant and even flow of farm commodities in interstate commerce, and for other purposes." *Introduced in the House.* 84th Congress, 1st Session, May 2, 1955.

84 H.R. 6052. "A Bill: To establish a conservation acreage reserve, to promote conservation improvement of agricultural soil and water resources in relation to watershed development, to stabilize farmers income, to adjust total agricultural production to consumer and export needs, to maintain an abundant and even flow of farm commodities in interstate commerce, and for other purposes." *Introduced in the House.* 84th Congress, 1st Session, May 5, 1955.

84 H.R. 8137. "A Bill: To establish a conservation acreage, reserve, to promote conservation improvement of agricultural soil, water, and related resources, to stabilize farmers' income, to adjust total agricultural production to consumer and export needs, to maintain an abundant and even flow of farm commodities in interstate commerce, and for other purposes." *Introduced in the House.* 84th Congress, 2nd Session, January 5, 1956.

84 H.R. 8151. "A Bill: To establish a conservation acreage reserve, to promote conservation improvement of agricultural soil, water, and related resources, to stabilize farmers' income, to adjust total agricultural production to consumer and export needs, to maintain an abundant and even flow of farm commodities in interstate commerce, and for other purposes." *Introduced in the House.* 84th Congress, 2nd Session, January 5, 1956.

84 H.R. 8156. "A Bill: To establish a Conservation Acreage Reserve, to promote conservation and improvement of agricultural soil and water resources, to stabilize farmers' income, to adjust total agricultural production to consumer and export needs, to maintain an abundant and even flow of farm commodities in interstate

commerce, and for other purposes." *Introduced in the House.* 84th Congress, 2nd Session, January 5, 1956.
84 H.R. 8369. "A Bill: To establish a conservation acreage reserve, to promote conservation improvement of agricultural soil and water resources in relation to watershed development, to stabilize farmers' income, to adjust total agricultural production to consumer and export needs, to maintain an abundant and even flow of farm commodities in interstate commerce, and for other purposes." *Introduced in the House.* 84th Congress, 2nd Session, January 12, 1956.
91 H. Res. 1209. "Resolution: That the House of Representatives does not favor the Reorganization Plan Number 3 of 1970 . . ." 91st Congress, 2nd Session, September 23, 1970.
92 H.R. 11896. "A Bill: To amend the Federal Water Pollution Control Act." *Introduced in the House.* 92nd Congress, 1st Session, November 19, 1972.
94 H.R. 10456. "Land and Water Resource Conservation Act of 1975." 94th Congress, 1st Session, October 30, 1975.
94 H.R. 14912. "Land and Water Resource Conservation Act of 1976." 94th Congress, 2nd Session, July 28, 1976.
95 H.R. 75. "Land and Water Resources Conservation Act of 1977." 95th Congress, 1st Session, January 4, 1977.
96 H.R. 3681. "A Bill: To amend the Agricultural Act of 1949 to provide that farmers may not participate in the price support program for the 1980 and 1981 crops of feed grains and wheat unless they implement plans for conservation and sustained use of soil and water on their farms." *Introduced in the House.* 96th Congress, 1st Session, April 24, 1979.
99 H.R. 1656. "Agricultural Recovery Act of 1985." 99th Congress, 1st Session, March 21, 1985.
99 H.R. 2100-IH. "Food Security Act of 1985." *Introduced in the House.* 99th Congress, 1st Session, April 17, 1985.
99 H.R. 2100-PCS. "Food Security Act of 1985." *Placed on calendar in the Senate.* 99th Congress, 1st Session, October 16, 1985.
99 H.R. 2108. "Soil Conservation Act of 1985." 99th Congress, 1st Session, April 18, 1985.
99 H.R. 2112. "Agricultural Trade, Food, and Fiber Act of 1985." 99th Congress, 1st Session, April 18, 1985.
106 H.R. 5511. "Conservation Security Act of 2000." 106th Congress, 2nd Session, October 19, 2000.
107 H.R. 1949. "Conservation Security Act of 2001." 107th Congress, 1st Session, May 22, 2001.

107 H.R. 2646. "Farm Security and Rural Investment Act of 2002." *Reported in the House*. 107th Congress, 1st Session, September 10, 2001.

U.S. Senate

74 S.3780. "A Bill: To make further provision for the conservation and proper utilization of the soil resources of the Nation." *Introduced in the Senate*. 74th Congress, 2nd Session, January 16 (calendar day, January 22, 1936).

74 S.3780. "A Bill: To make further provision for the conservation and proper utilization of the soil resources of the Nation." *Reported in the Senate with amendments*. S. Rept. 1481, Calendar No. 1545. 74th Congress, 2nd Session, January 30, 1936.

74 S.3780. "An Act: To promote the conservation and profitable use of agricultural land resources by temporary Federal aid to farmers and by providing for a permanent policy of Federal aid to States for such purposes." *Ordered to lie on the table in the House*. 74th Congress, 2nd Session, February 18, 1936.

74 S.3780. "An Act: To promote the conservation and profitable use of agricultural land resources by temporary Federal aid to farmers and by providing for a permanent policy of Federal aid to States for such purposes." *Ordered printed in the Senate* (showing differences between the bill and the House amendment). 74th Congress, 2nd Session, February 24, 1936.

83 S.379. "A Bill: To permit the deduction deduction from gross income for income-tax purposes of expenses incurred by farmers for the purpose of soil and water conservation." *Introduced in the Senate*. 83rd Congress, 1st Session, January 13, 1953.

83 S.3036. "A Bill: To require the Secretary of Agriculture to establish programs for the use of acreage diverted from production by the establishment of acreage allotments." *Introduced in the Senate*. 83rd Congress, 2nd Session, February 26, 1954.

84 S.1396. "A Bill: To establish a conservation acreage reserve, to promote conservation and improvement of agricultural soil and water resources, to stabilize farmers' income, to adjust total agricultural production to consumer and export needs, to maintain an abundant and even flow of farm commodities in interstate commerce, and for other purposes." *Introduced in the Senate*. 84th Congress, 1st Session, March 10, 1956.

84 S.3183. "A Bill: To provide an improved farm program." *Introduced in the Senate*. 84th Congress, 2nd Session, February 10, 1956.

92 S.2770. "A Bill: To amend the Federal Water Pollution Control Act." *Reported in the Senate*. 92nd Congress, 1st Session, October 28, 1971.

94 S.2081. "Land and Water Resource Conservation Act of 1975." 94th Congress, 1st Session, July 10, 1975.

95 S.106. "Land and Water Resources Conservation Act of 1977." 95th Congress, 1st Session, January 10, 1977.

97 S.1825. "A Bill: To prohibit price support for crops produced on certain lands in the western part of the United States which have not been used in the past 10 years for agricultural purposes, and for other purposes." *Introduced in the Senate.* 97th Congress, 1st Session, November 9, 1981.

98 S.663. "A Bill: A bill to prohibit the payment of certain agriculture incentives to persons who produce certain agricultural commodities on highly erodible land." *Introduced in the Senate.* 98th Congress, 1st Session, March 3, 1983.

99 S.1050. "Agricultural Export Enhancement and Soil Conservation Act of 1985." 99th Congress, 1st Session, April 15, 1985.

106 S.1426. "Conservation Security Act of 1999." 106th Congress, 1st Session, July 22, 1999.

106 S.3260. "Conservation Security Act." 106th Congress, 2nd Session, October 27, 2000.

Committee Hearings and Reports

U.S. House of Representatives

"Constitutional Amendments and Major Civil Rights Acts of Congress Referenced in *Black Americans in Congress*." https://history.house.gov/Exhibitions-and-Publications/BAIC/Historical-Data/Constitutional-Amendments-and-Legislation/.

H. Ex. Doc. 1, 53-1895. "Report of the Secretary of Agriculture." Ex. Doc. 1, Part 6. 53rd Congress, 3rd Session, 1895.

H. Ex. Doc. 37-78, 1863. "Report of the Commissioner of Agriculture for the Year 1862." 37th Congress, 3rd Session, March 3, 1863.

H. Ex. Doc. 75-149, 1937. "Farm Tenancy: Message from the President of the United States Transmitting The Report of the Special Committee on Farm Tenancy." 75th Congress, 1st Session, February 16, 1937.

H. Ex. Doc. 85-149, 1957. "Amendment to the Budget Involving a Reduction in the Amount for the Department of Agriculture." Communication from the President of the United States. 85th Congress, 1st Session, April 11, 1957.

H. Ex. Doc. 85-266, 1959 Budget. "The Budget of the United States Government, for the Fiscal Year Ending June 30, 1959." H. Doc. No. 266, Part 1. 85th Congress, 2nd Session, 1958.

H. Ex. Doc. 87-15, 1962 Budget. "The Budget of the United States Government for the Fiscal Year Ending June 30, 1962." 87th Congress, 1st Session, January 16, 1961.

H. Ex. Doc. 91-364, 1970, and H. Ex. Doc. 9-366, 1970. "Reorganization Plan No. 3 of 1970." Message from the President of the United States. 91st Congress, 2nd Session, July 9, 1970.

H. Ex. Doc. 355, 53-1895. "Yearbook of the United States Department of Agriculture." 53rd Congress, 3rd Session. Washington DC: GPO, 1895.

H. Ex. Doc. 399, 56-1900. "Field Operations of the Division of Soils, 1899." U.S. Dept. of Agric., Report No. 64. 56th Congress, 1st Session, February 5, 1900.

H. Ex. Doc. 655, 57-1902. "Field Operations of the Bureau of Soils, 1901." U.S. Dept. of Agric., Bureau of Soils. 57th Congress, 1st Session, February 15, 1902.

H. Hrg. 71-1929, "Farm Relief." Committee on Agriculture. "Farm Relief Legislation." Hearings, Part 6. 71st Congress, 1st Session, April 5–6, 1929.

H. Hrg. 74-1935, H.R. 557. Committee on Agriculture. "Amend Cotton Control Act." Hearings on H.R. 557 (H.R. 6424). 74th Congress, 1st Session, February 20, 21, 22, 25, 1935.

H. Hrg. 74-1935, H.R. 7054. Subcommittee of the Committee on Public Lands. "Soil Erosion Program." Hearing on H.R. 7054—Reported. 74th Congress, 1st Session, March 20, 21, 22, 25, 1935.

H. Hrg. 75-1937, H.R. 8. Committee on Agriculture. "Farm Tenancy." Hearing on H.R. 8. 75th Congress, 1st Session, January and February 1937.

H. Hrg. 83-1953, "Long Range, part 8." Committee on Agriculture. "Long Range Farm Program, Part 8." Hearings. 83rd Congress, 1st Session, October 17, 1953, Bloomington and Quincy IL.

H. Hrg. 83-1954, H.R. 7339. Committee on Banking and Currency. "Increase Borrowing Power of the Commodity Credit Corporation." Hearings on H.R. 7339. 83rd Congress, 2nd Session, February 25 and 26, 1954.

H. Hrg. 83-1954, "Long Range," part 21. Committee on Agriculture. "Long Range Farm Program." Hearings, Part 21. 83rd Congress, 2nd Session, April 15, 1954.

H. Hrg. 84-1955, 1956 Appropriations. Committee on Appropriations, Subcommittee. "Department of Agriculture Appropriations for 1956." Hearings. 84th Congress, 1st Session, July 29, 1955.

H. Hrg. 84-1956, H.R. 11958. Committee on Agriculture. "Soil Bank Act Amendment—Acreage Reserve." Hearings on H.R. 11958. 84th Congress, 2nd Session, July 11, 1956.

H. Hrg. 85-1957, "Acreage Reserve." Committee on Agriculture. "Acreage Reserve Allotment for Corn." Hearings. 85th Congress, 1st Session, January 30, 31 and February 1, 4, 5, 6, 7, and 8, 1957.

H. Hrg. 85-1957, Appropriations, part 4. Subcommittee of the Committee on Appropriations. "Department of Agriculture Appropriations for 1958, Part 4." 85th Congress, 1st Session, April 16, 1957.

H. Hrg. 85-1957, Appropriations, part 6. Subcommittee of the Committee on Appropriations. "Department of Agriculture Appropriations for 1958, Part 6." Hearings. 85th Congress, 1st Session, March 29, 1957.

H. Hrg. 85-1957, "Present Conditions." Subcommittee of the Committee on Appropriations. "Present Conditions in Agriculture." 85th Congress, 1st Session, February 26, 1957.

H. Hrg. 85-1957, "Soil Bank." Committee on Agriculture. "Soil Bank Program." Hearing. 85th Congress, 1st Session, January 7, 8, 9, and 10, 1957.

H. Hrg. 85-1958, "Acreage Reserve." Subcommittee of the Committee on Government Operations. "The Soil Bank: 1956 Acreage Reserve Program." Hearing. 85th Congress, 2nd Session, September 30, 1958.

H. Hrg. 85-1958, Appropriations, part 1. Subcommittee of the Committee on Appropriations. "Department of Agriculture Appropriations for 1959, Part 1." 85th Congress, 2nd Session, February 1958.

H. Hrg. 85-1958, Appropriations, part 4. Subcommittee of the Committee on Appropriations. "Department of Agriculture Appropriations for 1959," part 4. 85th Congress, 2nd Session, February 4, 1958.

H. Hrg. 86-1959, Appropriations, part 3. Subcommittee of the Committee on Appropriations. "Department of Agriculture Appropriations for 1960," part 3. 86th Congress, 1st Session, 1959.

H. Hrg. 94-1976, "Resource Conservation." Committee on Agriculture, Subcommittee on Conservation and Credit. "Resource Conservation." Hearings on H.R. 14912 and S.2081. 94th Congress, 2nd Session, August 3 and 4, 1976.

H. Hrg. 95-1977, "General Farm Bill." Committee on Agriculture. "General Farm Bill." Hearings, part 2, Serial No. 95-H. 95th Congress, 1st Session, March 1–4, 1977.

H. Hrg. 96-1979, "Conservation." Committee on Agriculture, Subcommittee on Conservation and Credit. "Federal Conservation and Farm Credit Act Amendments." Hearings. 96th Congress, 1st Session, September through November 1979.

H. Hrg. 98-1983, "Conservation." Committee on Agriculture, Subcommittee on Conservation, Credit, and Rural Development. "Miscellaneous Conservation." Hearings. 98th Congress, 1st Session, May 4, 1983, and September 20, 1983.

H. Hrg. 99-2, 1985, "Economic Conditions." Committee on Agriculture. "Economic Conditions in Agriculture and Fiscal Year 1986 Department of Agriculture Budget Proposals (Secretary John R. Block)." Hearings, Serial No. 99-2. 99th Congress, 1st Session, February 21, 1985.

H. Hrg. 99-5, 1985, "General Farm Bill." Committee on Agriculture, Subcommittee on Conservation, Credit, and Rural Development. "General Farm Bill of 1985 (Conservation, Credit, and Rural Development Programs)." Hearings, Serial No. 99-5, part 2. 99th Congress, 1st Session, March 26 and 29, April 4, and May 13, 1985.

H. Hrg. 103-91, 1994, "Review." Committee on Agriculture, Subcommittee on Environment, Credit, and Rural Development. "Review of the Budget and Policy Consequences of Extending the Conservation Reserve Program." Hearing, Serial No. 103-91. 103rd Congress, 2nd Session, August 2, 1994.

H. Hrg. 103-92, 1994, "Future." Committee on Agriculture, Subcommittee on Environment, Credit, and Rural Development and Committee on Agriculture, Nutrition, and Forestry, Subcommittee on Agricultural Research, Conservation, Forestry and General Legislation. "Future of the Conservation Reserve Program." Hearings, Serial No. 103-92. 103rd Congress, 2nd Session, September 1, 1994, Aberdeen SD.

H. Hrg. 109-31, 2006, "Review." Committee on Agriculture, Subcommittee on Conservation, Credit, Rural Development, and Research. "Review of Federal Farm Policy." Hearings, Serial No. 109-31. 109th Congress, 2nd Session, May 1, 2006.

H. Hrg. 110-12, 2007, "Review." Committee on Agriculture, Subcommittee on Conservation, Credit, Energy, and Research. "Review of USDA Farm Bill Conservation Programs." Hearing, Serial No. 110-12. 110th Congress, 1st Session, April 19, 2007.

H. Rept. 49-848. Committee on Agriculture. "Agricultural Experiment Stations." 49th Congress, 1st Session, March 3, 1886.

H. Rept. 74-528. Committee on Agriculture. "To Provide for Protection of Land Resources against Soil Erosion, and for Other Purposes." Report to accompany H.R. 7054. 74th Congress, 1st Session, March 29, 1935.

H. Rept. 74-1973. Committee on Agriculture. "Soil Conservation Act." Report to accompany H.R. 10835. 74th Congress, 2nd Session, February 6, 1936.

H. Rept. 74-2079. Committee of Conference. "Soil Conservation and Domestic Allotment Act." Conference Report to accompany S.3780. 74th Congress, 2nd Session, February 26, 1936.

H. Rept. 75-1767. Committee of Conference. "Agricultural Adjustment Act of 1938." Conference Report. 75th Congress, 3rd Session, February 7, 1938.

H. Rept. 84-203. Committee on Agriculture. "Price Support Programs for Basic Commodities, Wheat, and Dairy Products." Report to accompany H.R. 12. 84th Congress, 1st Session, March 10, 1955.

H. Rept. 84-1986. Committee of Conference. "Agricultural Act of 1956." Conference Report. 84th Congress, 2nd Session, April 6, 1956.

H. Rept. 84-2197. Committee of Conference. "Agricultural Act of 1956." Conference Report. 84th Congress, 2nd Session, May 22, 1956.

H. Rept. 85-438. Committee on Appropriations. "Department of Agriculture and Farm Credit Administration Appropriation Bill, 1958." Report to accompany H.R. 7441. 85th Congress, 1st Session, May 10, 1957.

H. Rept. 85-682. Committee of Conference. "Department of Agriculture and Farm Credit Administration Appropriation Bill, 1958." Conference Report. 85th Congress, 1st Session, July 5, 1957.

H. Rept. 85-1584. Committee on Appropriations. "Department of Agriculture and Farm Credit Administration Appropriation Bill, 1959." Report to accompany H.R. 11767. 85th Congress, 2nd Session, March 28, 1958.

H. Rept. 85-1776. Committee of Conference. "Department of Agriculture and Farm Credit Administration Appropriation Bill, 1959." Conference Report. 85th Congress, 2nd Session, May 26, 1958.

H. Rept. 86-2219. Committee on Government Operations. "Price Support and Production Adjustment Activities." 86th Congress, 2nd Session, August 31, 1960.

H. Rept. 87-170. Committee of Conference. "1961 Feed Grain Program." Conference Report to accompany H.R. 4510. 87th Congress, 1st Session, March 17, 1961.

H. Rept. 87-448. Committee on Appropriations. "Department of Agriculture and Related Agencies Appropriation Bill, 1962." 87th Congress, 1st Session, June 2, 1961.

H. Rept. 91-1329. Committee on Agriculture. "Agricultural Act of 1970." Report to accompany H.R. 18546. 91st Congress, 2nd Session, July 23, 1970.

H. Rept. 91-1464. Committee on Government Operations. "Approving Reorganization Plan No. 3 of 1970." 91st Congress, 2nd Session, September 23, 1970.

H. Rept. 91-1594. Committee of Conference. "Agricultural Act of 1970." Conference Report to accompany H.R. 18546. 91st Congress, 2nd Session, October 9, 1970.

H. Rept. 94-1744. Committee on Agriculture. "Agricultural Resources Conservation Act of 1976." Report on S.2081. 94th Congress, 2nd Session, September 30, 1976.

H. Rept. 95-348. Committee on Agriculture. "Agricultural Act of 1977." Report on H.R. 7171. 95th Congress, 1st Session, May 16, 1977.

H. Rept. 95-599. Committee of Conference. "Food and Agriculture Act of 1977." Conference Report. 95th Congress, 1st Session, September 12, 1977.

H. Rept. 97-106. Committee on Agriculture. "Food and Agriculture Act of 1981." Report to accompany H.R. 3603. 97th Congress, 1st Session, May 19, 1981.

H. Rept. 97-377. Committee of Conference. "Agriculture and Food Act of 1981." Conference Report. 97th Congress, 1st Session, December 9, 1981.

H. Rept. 98-696. Committee on Agriculture. "Soil Conservation Act of 1984." Report on H.R. 3457. 98th Congress, 2nd Session, April 24, 1984.

H. Rept. 99-271. Committee on Agriculture. "Food Security Act of 1985." Report on H.R. 2100, H. Rept. 99-271, part 1. 99th Congress, 1st Session, September 13, 1985.

H. Rept. 99-447. Committee of Conference. "Food Security Act of 1985." Conference Report. 99th Congress, 1st Session, December 17, 1985.

H. Rept. 101-916. Committee of Conference. "Food, Agriculture, Conservation, and Trade Act of 1990." Conference Report to accompany S. 2830. 101st Congress, 2nd Session, October 22, 1990.

H. Rept. 102-617. Committee on Appropriations. "Agriculture, Rural Development, Food and Drug Administration, and Related Agencies Appropriations Bill, 1993." Report to accompany H.R. 5487. 102nd Congress, 2nd Session, June 25, 1992.

H. Rept. 103-153. Committee on Appropriations. "Agriculture, Rural Development, Food and Drug Administration, and Related Agencies Appropriations Bill, 1994." Report to accompany H.R. 2493. 103rd Congress, 1st Session, June 23, 1993.

H. Rept. 103-542. Committee on Appropriations. "Agriculture, Rural Development, Food and Drug Administration, and Related Agencies Appropriations Bill, 1995." Report to accompany H.R. 4554. 103rd Congress, 2nd Session, June 9, 1994.

H. Rept. 104-172. Committee on Appropriations. "Agriculture, Rural Development, Food and Drug Administration, and Related Agencies Appropriations Bill, 1996." Report to accompany H.R. 1976. 104th Congress, 1st Session, June 30, 1995.

H. Rept. 104-462. Committee on Agriculture. "Agricultural Market Transition Act." Report accompanying H.R. 2854, H. Rept. 104-462, part 1. 104th Congress, 2nd Session, February 9, 1996.

H. Rept. 104-494. Committee of Conference. "Federal Agriculture Improvement and Reform Act of 1996." Conference Report accompanying H.R. 2854. 104th Congress, 2nd Session, March 25, 1996.

H. Rept. 104-726. Committee of Conference. "Making Appropriations for Agriculture, Rural Development, Food and Drug Administration, and Related Agencies Programs for the Fiscal Year Ending September 30, 1997." Conference Report to accompany H.R. 3603. 104th Congress, 2nd Session, July 30, 1996.

H. Rept. 107-191, parts 1–3. Committee on Agriculture. "Farm Security Act of 2001." Report to accompany H.R. 2646. 107th Congress, 1st Session, August 2, 2001.

H. Rept. 107-424. Committee of Conference. "Farm Security and Rural Investment Act of 2002." Conference Report to accompany H.R. 2646. 107th Congress, 2nd Session, May 1, 2002.

H. Rept. 110-256. Committee on Agriculture. "Farm, Nutrition, and Bioenergy Act of 2007." Report to accompany H.R. 2419, H. Rept. 110-256, part 1. 110th Congress, 1st Session, July 23, 2007.

H. Rept. 110-627. Committee of Conference. "Food, Conservation, and Energy Act of 2008." Conference Report to accompany H.R. 2410. 110th Congress, 2nd Session, May 13, 2008.

H. Rept. 113-333. Committee of Conference. "Agricultural Act of 2014." Conference Report to accompany H.R. 2642. 113th Congress, 2nd Session, January 27, 2014.

H. Rept. 115-661. Committee on Agriculture. "Agriculture and Nutrition Act of 2018." Report to accompany H.R. 2. 115th Congress, 2nd Session, May 3, 2018.

H. Rept. 115-1072. Committee of Conference. "Agriculture Improvement Act of 2018." Conference Report to accompany H.R. 2. 115th Congress, 2nd Session, December 10, 2018.

U.S. Senate

S. Com. Print., 94-1975, "Conservation." Committee on Agriculture and Forestry. "Conservation of the Land, and the Use of Waste Materials for Man's Benefits." *Committee Print.* 94th Congress, 1st Session, March 25, 1975. Prepared by the Council for Agricultural Science and Technology (CAST).

S. Ex. Doc. 85-67, 1957. "Appropriations, Budget Estimates, etc.; Statements." S. Doc. 85-67. 85th Congress, 1st Session, January 1957.

S. Ex. Doc. 87-12, 1961. "Special Feed Grain Program for 1961." Communication from the President of the United States. 87th Congress, 1st Session, February 20, 1961.

S. Ex. Doc. 92-93, 1972. "Federal Water Pollution Act—Veto Message." Message from the President of the United States. 92nd Congress, 2nd Session, October 17, 1972.

S. Hrg. 70-1928, S.3484. Subcommittee of the Committee on Agriculture and Forestry. "Conservation of Rainfall." Hearing on S.3484. 70th Congress, 1st Session, May 3, 1928.

S. Hrg. 74-1935, "Protection." Subcommittee of the Committee on Agriculture and Forestry. "Protection of Land Resources Against Soil Erosion." Hearings on S.2149, S.2418, and H.R. 7054. 74th Congress, 1st Session, April 2 and 3, 1935.

S. Hrg. 74-1935, S.1800. Subcommittee of the Committee on Agriculture and Forestry. "To Create the Farm Tenant Homes Corporation." Hearing on S.1800. 74th Congress, 1st Session, March 5, 1935.

S. Hrg. 74-1936, "Substitute." Committee on Agriculture and Forestry. "Substitute Legislation for the Invalidated Agricultural Adjustment Act." Hearing. 74th Congress, 2nd Session, January 14, 1936.

S. Hrg. 83-1953, "Commodity Inventories." Committee on Agriculture and Forestry. "Commodity Inventories of the Commodity Credit Corporation." Hearing. 83rd Congress, 1st Session, February 23, 1953.

S. Hrg. 83-1954, S.3052. Committee on Agriculture and Forestry. "General Farm Program." Hearings on S.3052, part 2. 83rd Congress, 2nd Session, April 2, 1954.

S. Hrg. 84-1955, S.2604. Committee on Agriculture and Forestry. "To Increase the Borrowing of the Commodity Power of the Commodity Credit Corporation." Hearings on S.2604. 84th Congress, 1st Session, July 28, 1955.

S. Hrg. 84-1956, "Price-Support," part 8. Committee on Agriculture and Forestry. "Price-Support Program." Hearings, part 8. 84th Congress, 2nd Session, January 12, 1956.

S. Hrg. 84-1956, S.3280. Committee on Agriculture and Forestry. Hearing on S.3280. June 20, 1956, Executive Session, transcript.

S. Hrg. 85-1957, Appropriations. Subcommittee of the Committee on Appropriations. "Agricultural Appropriations for 1958." Hearings on H.R. 7441. 85th Congress, 1st Session, April 30, 1957.

S. Hrg. 85-1958, "Farm Program," part 2. Subcommittee of the Committee on Agriculture and Forestry. "Farm Program: Cotton Price Supports and Acreage Allotments." Hearings, part 2. 85th Congress, 2nd Session, February 17 and 18, 1958.

S. Hrg. 85-1958, "Farm Program," part 3. Subcommittee of the Committee on Agriculture and Forestry. "Farm Program, Price Supports and Acreage Allotments for Corn and Feed Grains." Hearings, part 3. 85th Congress, 2nd Session, February 20 and 21, 1958.

S. Hrg. 87-1961, S.J. Res. 98. Committee on the Judiciary, Subcommittee on Federal Charters, Holidays, and Celebrations. "Homestead Act Centennial." Hearing, S.J. Res. 98. 87th Congress, 1st Session, July 26, 1961.

S. Hrg. 93-1974, "Implementation." Committee on Agriculture and Forestry, Subcommittee on Agricultural Production, Marketing, and Stabilization of Prices. "Implementation of Conservation Programs—REAP and RECP." Hearing. 93rd Congress, 2nd Session, June 27, 1974.

S. Hrg. 94-1976, "Cost-Sharing." Subcommittee on Environment, Soil Conservation, and Forestry of the Committee on Agriculture and Forestry. "Federal Soil Conservation Cost-Sharing Program." Hearing. 94th Congress, 2nd Session, July 6, 1976.

S. Hrg. 95-1977, "Protection." Committee on Agriculture, Nutrition, and Forestry, Subcommittee on Environment, Soil Conservation, and Forestry. "Protection and Enhancement of Soil and Water Resources." Hearing. 95th Congress, 1st Session, August 2 and 4, 1977.

S. Hrg. 97-1981, "Recommendations." Committee on Agriculture, Nutrition, and Forestry. "Administration's Recommendations for a Comprehensive Soil and Water Conservation Program." Hearing. 97th Congress, 1st Session, October 28, 1981.

S. Hrg. 97-1982, "Oversight." Committee on Agriculture, Nutrition, and Forestry, Subcommittee on Soil and Water Conservation. "Oversight on Proposed Modifications in Soil and Water Conservation Policy." Hearing. 97th Congress, 2nd Session, March 30, 1982.

S. Hrg. 97-1982, S.1825. Committee on Agriculture, Nutrition, and Forestry, Subcommittee on Soil and Water Conservation and Subcommittee on Agricultural Production, Marketing, and Stabilization of Prices. "Prohibition of Price Supports for Crops Produced on Certain Lands." Hearing on S.1825. 97th Congress, 2nd Session, May 6, 1982.

S. Hrg. 97-1982, S.3074. Committee on Agriculture, Nutrition, and Forestry, Subcommittee on Agricultural Production, Marketing, and Stabilization of Prices. "Agricultural Act of 1982." Hearing on S.3074. 97th Congress, 2nd Session, December 9, 1982.

S. Hrg. 98-254, "Oversight." Committee on Agriculture, Nutrition, and Forestry, Subcommittee on Soil and Water Conservation, Forestry, and Environment. "Oversight on USDA Soil and Water Conservation Programs." 98th Congress, 1st Session, March 9, 1983.

S. Hrg. 98-1983, S.663. Committee on Agriculture, Nutrition, and Forestry. "Prohibition of Incentive Payments for Crops Produced on Highly Erodible Land." Hearing on S.663. 98th Congress, 1st Session, April 19, 1983.

S. Hrg. 99-55, 1985, "Framework." Committee on Agriculture, Nutrition, and Forestry. "Framework and Analysis for Agricultural Policy in 1985." 99th Congress, 1st Session, March 12, 14, 19, 20, and May 1, 1985.

S. Hrg. 104-359, 1995, "Conservation." Committee on Agriculture, Nutrition, and Forestry. "Conservation, Wetlands, and Farm Policy." Hearing. 104th Congress, 1st Session, March 14, 1995.

S. Hrg. 107-225, 2001, "Conservation." Committee on Agriculture, Nutrition, and Forestry. "Conservation." Hearing. 107th Congress, 1st Session, February 28 and March 1, 2001.

S. Hrg. 107-828, 2001, "Working Lands." Committee on Agriculture, Nutrition, and Forestry. "Conservation on Working Lands for the New Federal Farm Bill." Hearing. 107th Congress, 1st Session, July 31, 2001.

S. Hrg. 108-564, 2004, "Oversight." Committee on Agriculture, Nutrition, and Forestry, Subcommittee on Forestry, Conservation, and Rural Revitalization. "Oversight of Conservation Programs of the 2002 Farm Bill." Hearings. 108th Congress, 2nd Session, May 11, 2004.

S. Hrg. 110-18, 2007, "Working Lands." Committee on Agriculture, Nutrition, and Forestry. "Working Lands Conservation: Conservation Security Program and Environmental Quality Incentives Program." Hearings. 110th Congress, 1st Session, January 17, 2007.

S. Hrg. 110-155, 2007, "Recommendations." Committee on Agriculture, Nutrition, and Forestry. "Conservation Policy Recommendations for the Farm Bill." Hearings. 110th Congress, 1st Session, May 1, 2007.

S. Hrg. 115-592, 2017, "Perspectives." Committee on Agriculture, Nutrition, and Forestry. "Conservation and Forestry: Perspectives on the Past and Future Direction for the 2018 Farm Bill." Hearings. 115th Congress, 1st Session, June 29, 2017.

S. Rept. 70-1211. Committee on Agriculture and Forestry. "Conservation of Rainfall." Report to accompany S.3484. 71st Congress, 1st Session, May 3, 1928.

S. Rept. 71-1784. Committee on Agriculture and Forestry. "Conservation of Rainfall." Report to accompany S.1444. 71st Congress, 3rd Session, February 17, 1931.

S. Rept. 74-466. Committee on Agriculture and Forestry. "Protection of Land Resources Against Soil Erosion." Report to accompany H.R. 7054. 74th Congress, 1st Session, April 11, 1935.

S. Rept. 74-1481. Committee on Agriculture and Forestry. "Conservation and Utilization of the Soil Resources." Report to accompany S.3780. 74th Congress, 2nd Session, January 30, 1936.

S. Rept. 84-1484. Committee on Agriculture and Forestry. "Agricultural Act of 1956." Report to accompany S.3183. 84th Congress, 2nd Session, February 16, 1956.

S. Rept. 85-1438. Committee on Appropriations. "Agricultural and Farm Credit Administration Appropriation Bill, 1959." Report to accompany H.R. 11767. 85th Congress, 2nd Session, April 18, 1958.

S. Rept. 91-1154. Committee on Agriculture and Forestry. "Agricultural Act of 1970." Report to accompany H.R. 18546. 91st Congress, 2nd Session, September 4, 1970.

S. Rept. 91-1250. Committee on Government Operations. "Reorganization Plan No. 3 of 1970, Providing for the Establishment of the Environmental Protection Agency." 91st Congress, 2nd Session, September 29, 1970.

S. Rept. 93-173. Committee on Agriculture and Forestry. "Agriculture and Consumer Protection Act of 1973." Report on S.1888. 93rd Congress, 1st Session, May 23, 1973.

S. Rept. 93-1033. Committee on Government Operations, Permanent Subcommittee on Investigations. "Russian Grain Transactions." 93rd Congress, 2nd Session, July 29, 1974.

S. Rept. 94-895. Committee on Agriculture and Forestry. "Land and Water Resource Conservation Act of 1976." Report on S.2081. 94th Congress, 2nd Session, May 14, 1976.

S. Rept. 95-180. Committee on Agriculture, Nutrition, and Forestry. "Food and Agriculture Act of 1977." Report on S.275. 95th Congress, 1st Session, May 16, 1977.

S. Rept. 95-418. Committee of Conference. "Food and Agriculture Act of 1977." Conference Report. 95th Congress, 1st Session, September 9, 1977.

S. Rept. 97-126. Committee on Agriculture, Nutrition, and Forestry. "Agriculture and Food Act of 1981." Report to accompany S.884. 97th Congress, 1st Session, May 27, 1981.

S. Rept. 97-290. Committee of Conference. "Agriculture and Food Act of 1981." Conference Report. 97th Congress, 1st Session, December 10, 1981.

S. Rept. 98-296. Committee on Agriculture, Nutrition, and Forestry. "Highly Erodible Land Conservation Act of 1983." Report on S.663. 98th Congress, 1st Session, November 2, 1983.

S. Rept. 98-1984, S.663, unpublished conference transcript. Committee of Conference. *Unpublished Transcript* on S.663. 98th Congress, 2nd Session, June 27, 1984.

S. Rept. 99-145. Committee on Agriculture, Nutrition, and Forestry. "Agriculture, Food, Trade, and Conservation Act of 1985." Report on S.1714. 99th Congress, 1st Session, September 30, 1985.

S. Rept. 102-334. Committee on Appropriations. "Agriculture, Rural Development, Food and Drug Administration, and Related Agencies Appropriations Bill, 1993." Report to accompany H.R. 5487. 102nd Congress, 2nd Session, July 23, 1992.

S. Rept. 103-102. Committee on Appropriations. "Agriculture, Rural Development, Food and Drug Administration, and Related Agencies Appropriations Bill, 1994." Report to accompany H.R. 2493. 103rd Congress, 1st Session, July 20, 1993.

S. Rept. 103-290. Committee on Appropriations. "Agriculture, Rural Development, Food and Drug Administration, and Related Agencies Appropriations Bill, 1995." Report to accompany H.R. 4554. 103rd Congress, 2nd Session, June 23, 1994.

S. Rept. 104-142. Committee on Appropriations. "Agriculture, Rural Development, Food and Drug Administration, and Related Agencies Appropriations Bill, 1996." Report to accompany H.R. 1976. 104th Congress, 1st Session, September 14, 1995.

S. Rept. 104-317. Committee on Appropriations. "Agriculture, Rural Development, Food and Drug Administration, and Related Agencies Appropriations Bill, 1997." Report to accompany H.R. 3603. 104th Congress, 2nd Session, July 11, 1996.

S. Rept. 107-117. Committee on Agriculture, Nutrition, and Forestry. "Agriculture, Conservation, and Rural Enhancement Act of 2001." Report to accompany S.1731. 107th Congress, 1st Session, December 7, 2001.

S. Rept. 110-220. Committee on Agriculture, Nutrition, and Forestry. "Food and Energy Security Act of 2007." Report on S.2302. 110th Congress, 1st Session, November 2, 2007.

S. Rept. 112-203. Committee on Agriculture, Nutrition, and Forestry. "Agriculture Reform, Food and Jobs Act of 2012." Report to accompany S.3240. 112th Congress, 2nd Session, August 28, 2012.

S. Rept. 113-88. Committee on Agriculture, Nutrition, and Forestry. "Agriculture Reform, Food and Jobs Act of 2013." Report to accompany S.954. 113th Congress, 1st Session, September 4, 2013.

Congressional Budget Office (CBO)

"The Budget and Economic Outlook: 2022 to 2023," May 2022. https://www.cbo.gov/publication/58147.

"Cost Estimate for the Conference Agreement on H.R. 2646 Relative to the March 2002 Baseline," May 6, 2002. https://www.cbo.gov/publication/13637.

"Details about Baseline Projects for Selected Programs: USDA Mandatory Farm Programs." https://www.cbo.gov/data/baseline-projections-selected-programs #25. [CBO, "Baseline."]

"Details about Baseline Projections for Selected Programs: USDA Mandatory Farm Programs; March 2007 CBO Baseline," February 23, 2007. https://www.cbo.gov/sites/default/files/recurringdata/51317-2007-03-ccc.pdf.

"Details about Baseline Projections for Selected Programs: USDA's Mandatory Farm Programs—CBO's May 2019 Baseline," May 2, 2019. https://www.cbo.gov/system/files/2019-05/51317-2019-05-usda_0.pdf.

"Details about Baseline Projections for Selected Programs: USDA's Mandatory Farm Programs—CBO's Baseline as of March 6, 2020," March 6, 2020. https://www.cbo.gov/system/files/2020-03/51317-2020-03-usda.pdf.

"Details about Baseline Projections for Selected Programs: USDA Mandatory Farm Programs," May 2022. https://www.cbo.gov/system/files/2022-05/51317-2022-05-usda.pdf.

"Estimated Budgetary Effects of Public Law 117-169, to Provide for Reconciliation Pursuant to Title II of S. Con. Res. 14," September 7, 2022. https://www.cbo.gov/publication/58455.

"H.R. 2419, Food, Conservation, and Energy Act of 2008: Direct Spending and Revenue Effects of the Conference Agreement," May 13, 2008. https://www.cbo.gov/publication/41696.

"H.R. 2642, Agricultural Act of 2014: Effects on Direct Spending and Revenues of Conference Agreement on H.R. 2642, as Reported on January 27, 2014," Cost Estimate, January 28, 2014. https://www.cbo.gov/publication/45049.

"H.R. 2646, Farm Security Act of 2001: Cost Estimate for the Bill as Reported by the House Committee on Agriculture on August 2, 2001," August 23, 2001. https://www.cbo.gov/publication/13246.

Congressional Quarterly Almanac (CQ Almanac)

"Farm Bill Delayed a Year." *CQ Almanac 2001*, 57th ed., 3-3 through 3-8. Washington DC: Congressional Quarterly, 2002. http://library.cqpress.com/cqalmanac/cqal01-106-6384-328622.

"Farm Bill Granted a Limited 'Win' to All Sides." *CQ Almanac 1985*, 41st ed., 517–39. Washington DC: Congressional Quarterly, 1986. http://library.cqpress.com/cqalmanac/cqal85-1147043.

"Farm Price Supports." *CQ Almanac 1955*, 11th ed., 169–72. Washington DC: Congressional Quarterly, 1956. http://library.cqpress.com/cqalmanac/cqal55-1352801.

"Farm Program Funds Included in Stopgap Bill." CQ *Almanac 1984*, 40th ed., 383–87. Washington DC: Congressional Quarterly, 1985. http://library.cqpress.com/cqalmanac/cqal84-1151538.

"'02 Farm Bill Revives Subsidies." CQ *Almanac 2002*, 58th ed., 4-3 through 4-7. Washington DC: Congressional Quarterly, 2003. http://library.cqpress.com/cqalmanac/cqal02-236-10375-664394.

"President Outlines 9-Point Farm Program." CQ *Almanac 1956*, 12th ed., 10-52 through 10-58. Washington DC: Congressional Quarterly, 1957. http://library.cqpress.com/cqalmanac/cqal56-1348268.

"'Sodbuster' Conservation Bill." CQ *Almanac 1984*, 40th ed., 364–65. Washington DC: Congressional Quarterly, 1985. http://library.cqpress.com/cqalmanac/cqal84-1153217.

"Soil Bank Enacted After Farm Bill VETO." CQ *Almanac 1956*, 12th ed., 02-375 through 02-392. Washington DC: Congressional Quarterly, 1957. http://library.cqpress.com/cqalmanac/cqal56-1348982.

"$34 Billion Voted for Farm, Food Programs." CQ *Almanac 1982*, 38th ed., 255–62. Washington DC: Congressional Quarterly, 1983. http://library.cqpress.com/cqalmanac/cqal82-1164192.

"Work Begins on New Farm Bill; Conference Put Off Until 2008." CQ *Almanac 2007*, 63rd ed., edited by Jan Austin, 3-3 through 3-7. Washington DC: Congressional Quarterly, 2008. http://library.cqpress.com/cqalmanac/cqal07-1006-44912-2047970.

Congressional Research Service (CRS)

CRS Report 80-144S. Dallavalle, Rita S., and Leo V. Mayer. "Soil Conservation in the United States: The Federal Role; Origins, Evolution and Current Status," September 1980.

CRS Report 98-175. "House Committee Jurisdiction and Referral: Rules and Practice," updated October 21, 2014. https://crsreports.congress.gov/product/pdf/RS/98-175.

CRS Report 98-242. "Committee Jurisdiction and Referral in the Senate," updated April 15, 2008. https://crsreports.congress.gov/product/pdf/RS/98-242.

CRS Report 98-244. "Markup in Senate Committee: Choosing A Text," updated September 10, 2007. https://crsreports.congress.gov/product/pdf/RS/98-244.

CRS Report 98-310. "Senate Unanimous Consent Agreements: Potential Effects on the Amendment Process," updated May 17, 2017. https://crsreports.congress.gov/product/pdf/RS/98-310.

CRS Report 98-612. "Special Rules and Options for Regulating the Amending Process," updated November 17, 2010. https://crsreports.congress.gov/product/pdf/RS/98-612.

CRS Report 98-853. "The Amending Process in the Senate," updated September 16, 2015. https://crsreports.congress.gov/product/pdf/RL/98-853.

CRS Report 98-995. "The Amending Process in the House of Representatives," updated September 16, 2015. https://crsreports.congress.gov/product/pdf/RL/98-995.

CRS Report IF10657. "Budgetary Effects of the BCA as Amended: The 'Parity Principle,'" In Focus, updated February 23, 2018. https://crsreports.congress.gov/product/pdf/IF/IF10657/4.

CRS Report IF12024. Stubbs, Megan. "Farm Bill Primer: Conservation Title," In Focus, January 25, 2022. https://crsreports.congress.gov/product/pdf/IF/IF12024/IF12024.pdf/.

CRS Report R41083. "House Committee Markups: Manual of Procedures and Procedural Strategies," updated March 27, 2018. https://crsreports.congress.gov/product/pdf/R/R41083.

CRS Report R42459. Stubbs, Megan. "Conservation Compliance and U.S. Farm Policy," updated October 16, 2016. https://crsreports.congress.gov/product/pdf/R/R42459.

CRS Report R43424. Rybicki, Elizabeth. "Considering Legislation on the House Floor: Common Practices in Brief," updated January 13, 2017. https://crsreports.congress.gov/product/pdf/R/R43424.

CRS Report R45698. Stubbs, Megan. "Agricultural Conservation in the 2018 Farm Bill," April 18, 2019. https://crsreports.congress.gov/product/pdf/R/R45698.

CRS Report R46727. Johnson, Renee. "Defining a Socially Disadvantaged Farmer or Rancher (SDFR): In Brief," updated March 21, 2021. https://crsreports.congress.gov/product/pdf/R/R46727.

CRS Report RL30360. "Filibuster and Cloture in the Senate," updated April 7, 2017. https://crsreports.congress.gov/product/pdf/RL/RL30360.

CRS Report RL31195. "The 2002 Farm Bill: Overview and Status," 3, table 2, last updated September 3, 2002. https://www.everycrsreport.com/reports/RL31195.html or https://www.everycrsreport.com/files/20020903_RL31195_40b8544604db896699d6cfe569a7dec811e3rd46c.pdf.

CRS Report RL33939. "The Rise of Senate Unanimous Consent Agreements," updated March 14, 2008. https://crsreports.congress.gov/product/pdf/RL/RL33939.

CRS Report RS20594. "How Unanimous Consent Agreements Regulate Senate Floor Action," updated April 12, 2007. https://crsreports.congress.gov/product/pdf/rs/rs20594.

CRS Report RS21740. Cowan, Tadlock. "Conservation Security Program: Implementation and Current Issues," July 24, 2006.

U.S. Government Accountability Office/General Accounting Office (GAO)

"Agricultural Conservation: USDA Needs to Better Ensure Protection of Highly Erodible Cropland and Wetlands." GAO-03-418, April 21, 2003. https://www.gao.gov/products/gao-03-418.

"Agricultural Conservation: USDA's Environmental Quality Incentives Program Could Be Improved to Optimize Benefits." GAO-17-225, April 13, 2017. https://www.gao.gov/products/gao-17-225.

"Conservation Security Program: Despite Cost Controls, Improved USDA Management is Needed to Ensure Proper Payments and Reduce Duplication with Other Programs." Report to the Chairman, Committee on Appropriations, U.S. Senate (Senator Thad Cochran [R-MS]). GAO-06-312, April 2006. https://www.gao.gov/products/gao-06-312.

"Farm Programs: Conservation Compliance Provisions Could Be Made More Effective." RCED-90-206, September 24, 1990. https://www.gao.gov/products/rced-90-206.

"Farm Programs: USDA Should Take Additional Steps to Ensure Compliance with Wetland Conservation Provisions." GAO-21-241, April 2, 2021. https://www.gao.gov/products/gao-21-241.

"Greater Conservation Benefits Could Be Attained Under the Rural Environmental Assistance Program." Report to Congress by the Comptroller General of the United States. GAO/B-114833, February 16, 1972. https://www.gao.gov/assets/210/204487.pdf.

"Preserving America's Farmland—A Goal the Federal Government Should Support." Report to Congress by the Comptroller General of the United States. GAO-CED-79-109, September 20, 1979. https://www.gao.gov/products/ced-79-109.

"To Protect Tomorrow's Food Supply, Soil Conservation Needs Priority Attention." Report to Congress by the Comptroller General of the United States. GAO/CED-77-30, February 14, 1977. https://www.gao.gov/products/CED-77-30.

INDEX

AAA. *See* Agricultural Adjustment Administration (AAA)
ACEP. *See* Agricultural Conservation Easement Program (ACEP)
acreage-based production controls: Farm Bill of 1935 and, 44–45, 46, 47–48; Farm Bill of 1938 and, 85–86; of parity system, 87–89, 103; resurrection in 1960s of, 137–38; Soil Conservation and Domestic Allotment Act and, 57, 58, 60, 67, 81; surpluses and, 159. *See also* acreage reserve program of Soil Bank; allotments, acreage
acreage reserve program of Soil Bank: background and overview of, 91, 94, 95–96, 103; congressional process and, 95–100; final 1956 Farm Bill and, 100, 101, 103, 106; investigative reports and, 118–19, 120–23, 124; operation and enrollment of, 107, 115–16, 117; sabotage and demise of, 105, 108–15, 117–18, 123–24, 131, 137
Agricultural Act of 1949, 86
Agricultural Act of 1954, 91–92
Agricultural Act of 1956, 85, 99–100, 105, 129, 130, 247n43. *See also* Soil Bank
Agricultural Act of 1970, 139, 254n6
Agricultural Act of 2014, 176, 177, 181, 183–84, 190, 192, 213

Agricultural Adjustment Act of 1933, 16, 44–45, 52–54, 57–58, 108
Agricultural Adjustment Act of 1938, 73–74, 85–86, 98
Agricultural Adjustment Administration (AAA), 49, 53, 57–58, 59, 62, 63, 71, 73–74
Agricultural Conservation Easement Program (ACEP), 9, 182, 193, 218, 266n26
Agricultural Conservation Program, 137–38, 183
Agricultural Improvement Act of 2018, 7, 177, 181, 182, 190–92, 213
Agricultural Resources Conservation Act of 1976, 149–50
Agricultural Water Quality Incentives Program, 182–83
Agriculture and Consumer Protection Act of 1973, 144–45, 146–47, 148, 152, 205
Agriculture and Food Act of 1981, 156–59, 213
Aiken, George, 115–16
allotments, acreage: Farm Bills and, 85–86, 98; parity system and, 86, 87, 88–89; Soil Bank and, 98, 110–11, 121, 122–23, 125; Soil Conservation and Domestic Allotment Act and, 61, 63–64, 74

327

amendment limits, legislative, 23
American Farm Bureau, 59, 74, 80, 91, 94, 98, 103, 131
American Liberty League, 53
ammonia, 86
anchovies, 144
Anderson, Clinton, 97
Andresen, August H., 113
anticommunism, 128
appropriations: in 1980s, 158, 162; for CSP, 184, 188, 191; for Soil Bank, 105–6, 107, 111–15, 117–18; as a weapon, 204. *See also* budgets, federal
Arkansas, 119, 120–21, 127
Armstrong, William, 156–57, 158, 160, 161

Balanced Budget Act of 1977, 208
Bankhead, John H., 49, 57, 59–60, 61, 64, 67, 202
Bankhead Cotton Control Act of 1934, 45
Bankhead-Jones Farm Tenant Act, 49, 232n48
bankruptcies, farm, 155–56, 158, 168
Barlow, Tom, 149, 151, 152, 153, 170, 257n17
Benedict, Murray R., 73
Bennett, Hugh Hammond, 37–38, 41, 45, 48, 72–73, 79
Benson, Ezra Taft: Soil Bank demise and, 108, 112, 114, 119; Soil Bank emergence and, 92, 94, 95–96, 98, 99
Bergland, Robert, 153–54
Berry, Marion, 30
Bessette, Joseph, 21
bicameral approval, 20, 21–22, 24
Biden, Joe, 218
bill passage process, 22–26, 33
bill reconciling, 23–24, 193
Block, John, 159, 160, 182

Boehlert, Sherwood, 27–28
Boileau, Gerald, 66–68
Boll Weevil Democrats, 170
Bosso, Christopher, 33–34
Brown v. Board of Education, 92, 127–28
Budget Control Act of 2011, 208
Budget Control and Impoundment Act of 1974, 147, 207
budget discipline, 156, 163, 178, 179–80, 208–9, 211, 212
Budget Enforcement Act of 1990, 208
budgets, federal: Budget Control Act of 2011 and, 208; Budget Control and Impoundment Act of 1974 and, 147, 207; CRP and, 179–81; CSP and, 187–88, 189, 190; Farm Bill of 1985 and, 162–63, 169; Farm Bills of 1990s and, 178, 180, 186; Farm Bills of 2000s and, 27, 183–84, 187, 189; Inflation Reduction Act and, 193, 218–19; RCPP and, 192; under Reagan administration, 156, 159, 162–63, 166, 169, 178–79, 207–13; reform and, 147, 217–18, 271n40; Soil Bank and, 112–13, 114, 117, 120. *See also* appropriations; Congressional Budget Office (CBO)
Bush, George H. W., 207
Bush, George W., 190
Butler, William A., 53, 233n66, 234n68. See also *U.S. v. Butler*
Butz, Earl, 139, 144, 145, 149
byproducts of war, 86, 87

Carson, Rachel, 142
Carter, Jimmy, 149, 155
CBO. *See* Congressional Budget Office (CBO)
CCC. *See* Commodity Credit Corporation (CCC)

INDEX

AAA. *See* Agricultural Adjustment Administration (AAA)
ACEP. *See* Agricultural Conservation Easement Program (ACEP)
acreage-based production controls: Farm Bill of 1935 and, 44–45, 46, 47–48; Farm Bill of 1938 and, 85–86; of parity system, 87–89, 103; resurrection in 1960s of, 137–38; Soil Conservation and Domestic Allotment Act and, 57, 58, 60, 67, 81; surpluses and, 159. *See also* acreage reserve program of Soil Bank; allotments, acreage
acreage reserve program of Soil Bank: background and overview of, 91, 94, 95–96, 103; congressional process and, 95–100; final 1956 Farm Bill and, 100, 101, 103, 106; investigative reports and, 118–19, 120–23, 124; operation and enrollment of, 107, 115–16, 117; sabotage and demise of, 105, 108–15, 117–18, 123–24, 131, 137
Agricultural Act of 1949, 86
Agricultural Act of 1954, 91–92
Agricultural Act of 1956, 85, 99–100, 105, 129, 130, 247n43. *See also* Soil Bank
Agricultural Act of 1970, 139, 254n6
Agricultural Act of 2014, 176, 177, 181, 183–84, 190, 192, 213
Agricultural Adjustment Act of 1933, 16, 44–45, 52–54, 57–58, 108
Agricultural Adjustment Act of 1938, 73–74, 85–86, 98
Agricultural Adjustment Administration (AAA), 49, 53, 57–58, 59, 62, 63, 71, 73–74
Agricultural Conservation Easement Program (ACEP), 9, 182, 193, 218, 266n26
Agricultural Conservation Program, 137–38, 183
Agricultural Improvement Act of 2018, 7, 177, 181, 182, 190–92, 213
Agricultural Resources Conservation Act of 1976, 149–50
Agricultural Water Quality Incentives Program, 182–83
Agriculture and Consumer Protection Act of 1973, 144–45, 146–47, 148, 152, 205
Agriculture and Food Act of 1981, 156–59, 213
Aiken, George, 115–16
allotments, acreage: Farm Bills and, 85–86, 98; parity system and, 86, 87, 88–89; Soil Bank and, 98, 110–11, 121, 122–23, 125; Soil Conservation and Domestic Allotment Act and, 61, 63–64, 74

amendment limits, legislative, 23
American Farm Bureau, 59, 74, 80, 91, 94, 98, 103, 131
American Liberty League, 53
ammonia, 86
anchovies, 144
Anderson, Clinton, 97
Andresen, August H., 113
anticommunism, 128
appropriations: in 1980s, 158, 162; for CSP, 184, 188, 191; for Soil Bank, 105–6, 107, 111–15, 117–18; as a weapon, 204. *See also* budgets, federal
Arkansas, 119, 120–21, 127
Armstrong, William, 156–57, 158, 160, 161

Balanced Budget Act of 1977, 208
Bankhead, John H., 49, 57, 59–60, 61, 64, 67, 202
Bankhead Cotton Control Act of 1934, 45
Bankhead-Jones Farm Tenant Act, 49, 232n48
bankruptcies, farm, 155–56, 158, 168
Barlow, Tom, 149, 151, 152, 153, 170, 257n17
Benedict, Murray R., 73
Bennett, Hugh Hammond, 37–38, 41, 45, 48, 72–73, 79
Benson, Ezra Taft: Soil Bank demise and, 108, 112, 114, 119; Soil Bank emergence and, 92, 94, 95–96, 98, 99
Bergland, Robert, 153–54
Berry, Marion, 30
Bessette, Joseph, 21
bicameral approval, 20, 21–22, 24
Biden, Joe, 218
bill passage process, 22–26, 33
bill reconciling, 23–24, 193
Block, John, 159, 160, 182

Boehlert, Sherwood, 27–28
Boileau, Gerald, 66–68
Boll Weevil Democrats, 170
Bosso, Christopher, 33–34
Brown v. Board of Education, 92, 127–28
Budget Control Act of 2011, 208
Budget Control and Impoundment Act of 1974, 147, 207
budget discipline, 156, 163, 178, 179–80, 208–9, 211, 212
Budget Enforcement Act of 1990, 208
budgets, federal: Budget Control Act of 2011 and, 208; Budget Control and Impoundment Act of 1974 and, 147, 207; CRP and, 179–81; CSP and, 187–88, 189, 190; Farm Bill of 1985 and, 162–63, 169; Farm Bills of 1990s and, 178, 180, 186; Farm Bills of 2000s and, 27, 183–84, 187, 189; Inflation Reduction Act and, 193, 218–19; RCPP and, 192; under Reagan administration, 156, 159, 162–63, 166, 169, 178–79, 207–13; reform and, 147, 217–18, 271n40; Soil Bank and, 112–13, 114, 117, 120. *See also* appropriations; Congressional Budget Office (CBO)
Bush, George H. W., 207
Bush, George W., 190
Butler, William A., 53, 233n66, 234n68. See also *U.S. v. Butler*
Butz, Earl, 139, 144, 145, 149
byproducts of war, 86, 87

Carson, Rachel, 142
Carter, Jimmy, 149, 155
CBO. *See* Congressional Budget Office (CBO)
CCC. *See* Commodity Credit Corporation (CCC)

Citizens' Councils, 92–93
Civilian Conservation Corps, 44
Civil Rights Act of 1957, 204
Civil Rights Act of 1964, 102, 204–5
civil rights movement, 93, 128, 129, 245n26
Clark, Dick, 147, 149
Clean Air Act (CAA) Amendments of 1970, 140–41
Clean Air Act (CAA) Amendments of 1990, 207
Clean Water Act, 141–42, 256n14
climate change, 5, 14, 193, 207, 215–20
Clinton, Bill, 180, 213
cloture, 23, 217
coalition-building, 17, 25–26, 205–6, 211, 213, 215, 217, 219
Cobb, James, 93
Cochran, Thad, 187–88
Coleman, Thomas, 164
Colorado River Basin Salinity Control Program, 183
Combest, Larry, 27, 29, 186
committees, congressional, 18, 22, 23–24, 148, 226n35. *See also specific committees*
Commodity Credit Corporation (CCC): CRP and, 167, 180; CSP and, 184; Soil Bank and, 100, 106–7, 111; Soviet Union loans and, 143; surpluses and, 88, 96, 137
competition of factions: about, 20–21, 25, 26; budgets and, 207, 209–10, 211–13, 215; climate change and, 216–17; Farm Bill of 1985 and, 170–72; public interest and, 219–20; Reform Amendment of 2001 and, 28; Soil Bank and, 102, 104, 132, 133; Soil Conservation and Domestic Allotment Act and, 80–81; southern political power and, 205–6; working theory on Congress and, 197–99, 200–202

compliance, conservation. *See* conservation compliance
Conaway, K. Michael, 191
Congress, U.S.: background and overview of, 16–17; control of, 43, 56, 94, 169, 178, 193; CRP and, 167–68; energy-oil crisis and, 146; environmental movement of 1970s and, 139, 140–43, 255n9, 256n14; erosion in late 1800s and, 40; legislative process and, 20, 22–26, 226n35; overview of farm legislation and, 10–14; resurrection of acreage reserve program and, 137–38; Richard Nixon and, 141; sodbuster and swampbuster and, 160–62, 163–66; southern Republicans in, 189, 217; wetlands and, 177. *See also* House of Representatives, U.S.; legislative process; Senate, U.S.; southern Democrats in Congress; working theory on Congress; *and specific committees and legislation*
Congressional Budget and Impoundment Control Act of 1974, 147, 207
Congressional Budget Office (CBO), 176, 187, 189, 190, 207, 208–10, 271n24
congressional districts, 217
consequences of agriculture, 4–6
conservation background and overview, 6–8, 9, 11–13
conservation compliance: about, 151, 153–54, 168, 176; budgets and, 158–59; crop insurance and, 176, 177;

conservation compliance (*cont.*)
crossing of ideological lines of, 156–57; CRP and, 167; highly erodible lands and, 165–66, 167, 176; Reagan administration and, 159–60; sodbuster and swampbuster and, 160–62, 163–66; wetlands and, 176–77. *See also* Food Security Act of 1985
conservation easements, 9, 165, 182
conservation interests, 102, 132, 160, 163, 170, 171, 214
conservation question: background and overview of, 13, 14; Congress and legislative power and, 16–17, 20, 27, 31–34; working theory on Congress and, 195–96, 197, 207, 215
conservation reserve authority, 105, 137
Conservation Reserve Program (CRP): in 1990s, 178–80; in 2000s, 180–81, 185, 186; acreage caps of, 12, 165, 178, 180–81, 214; background and overview of, 9, 30, 223n15; Farm Bill of 1985 and, 162–63, 164, 165, 166–67; funding of, 175, 179–81, 214; Soil Bank and, 95–96, 101–5, 117, 118, 120, 126, 131; Working Lands Programs and, 182
Conservation Security Act effort, 185
Conservation Security Program (CSP): about, 12, 184–85, 186–87; budget and, 187–88, 189–90, 193, 215; Farm Bill of 2018 and, 190–92; resurrection in 2008 of, 188–90
Conservation Stewardship Program (CSP), 9, 189, 218
Constitution, U.S.: countermajoritarian features of, 26; farming regulation and, 53–54; framing of, 17–22, 33, 196; Tenth Amendment and, 53–54, 65, 79

contracts: cotton, 46–47, 48; CSP and, 184, 188, 191; expiration of, 178–79, 180, 181; production control and, 57; prohibition of, 59, 64–65, 67; RCCP and, 192; Soil Bank and, 114, 117, 120, 129; Special Areas Conservation Program and, 157
Cooley, Harold, 78–79, 94, 105, 113, 138, 148
corn: in 1960s, 137–38; ethanol and, 181; feed grains and, 89; prices in 1970s of, 143–44; Soil Bank and, 94, 96–99, 107–11, 118, 124, 125, 137, 249n11; southern cotton and, 68, 89, 96–97, 138, 203, 206; surpluses of, 44, 88–89
Corn Belt, 5, 98, 110, 118, 120
corn faction and interests, 94, 96–97, 103, 138, 203
cost-share payments, 9, 29, 177, 183, 186
cotton: acres in 1940s and 1950s of, 86; Farm Bill of 1933 and, 44–45, 46; feed grains programs of 1960s and, 138; prices of, 143; Reform Amendment of 2001 and, 28–29; sodbuster bill and, 163; Soil Bank and, 96, 99, 103, 107–11, 116–17, 119, 121–25, 128; Soil Conservation and Domestic Allotment Act and, 68, 69–71, 73, 242n78; surpluses of, 44, 89. *See also* cotton ginners; cotton industries; cotton planters; southern cotton faction
cotton ginners, 52, 119–20, 121, 123
cotton harvester, mechanical, 86, 129
cotton industries, 111, 116–17, 119–20, 121–23
cotton planters: production control of 1930s and, 46; race and, 93; sharecropping and tenant farmers and,

46–48, 69–70, 76, 101, 129; Soil Bank demise and, 109–10, 116, 119–20, 121, 123–26, 128, 129–30; Soil Bank emergence and, 92, 93–94, 98–99, 100, 102, 103; southern political power and, 44, 71, 94, 202, 206
Cotton Section of the Agricultural Adjustment Administration, 71
Council on Environmental Quality, 140
countercyclical payment programs, 187, 189
crop insurance, 7–9, 16, 176, 177, 215
CRP. *See* Conservation Reserve Program (CRP)
CSP. *See* Conservation Security Program (CSP); Conservation Stewardship Program (CSP)
Culkin, Francis, 68
cultivation methods of homesteaders, 40
Cuyahoga River fire, 139–40

dairy faction and interests, 62, 66–69
dairy surpluses, 88
Daniel, Pete, 93, 129
Daschle, Tom, 163
Davis, Chester, 58
dead zones, 5, 6, 195
debt: of farmers in 1970s and 1980s, 150–51, 155, 158; national, 147, 179, 208–9; sharecropping and, 46–47
deficits, national, 179, 208–9, 210, 271n24
de la Garza, Eligio "Kika," 161–62, 170
Democrats: 1930s political power of, 43, 44, 56, 60, 66; 1950s political power of, 91, 92, 105; 1970s political power of, 148; 1980s political power of, 169; Inflation Reduction Act and, 193–94, 218; southerners and, 205. *See also* southern Democrats in Congress
direct payments, 184, 187, 188, 191
disaster assistance, 109, 188
discretionary funding, 106–7
discrimination, 31, 160
diverting acres. *See* acreage-based production controls
Division of Agricultural Soils, 40
domestic allotment, 61, 63–64. *See also* allotments, acreage
double majority requirement, 21–22, 23, 25, 197–98, 205
Doxey, Wall, 65, 70
drainage systems, 5, 177
drought: of 1949–56, 86–87; about, 39, 40; emergency legislation and, 187; production control and, 77; Soil Bank and, 104, 108–10, 118, 119, 120. *See also* Dust Bowl
Dust Bowl: background and overview of, 10, 38–39, 42–43, 55, 56; lessons of, 87, 200, 215; policy impact of, 72; Soil Conservation and Domestic Allotment Act of 1936 and, 57, 73–75
dust storms, 37, 39, 43, 48, 50, 74–75, 87

Earth Day, 140
easements, conservation, 9, 165, 182
Eisenhower, Dwight D.: about, 92, 127; Civil Rights Act and, 127; on conservation, 90; Farm Bill of 1956 veto and passage and, 99–100, 102, 105, 130; Soil Bank and, 11, 85, 92, 94–95, 97, 104–5, 117, 204
Ellender, Allen J., 95, 97, 102, 105, 202, 246n27
El Nino, 144
embargoes, 144, 146, 155

Index 331

enclosure movement, 129
enclosures, 205–6
Endangered Species Act, 146
energy-oil crisis, 146–47, 150–51
environmental movement: of 1950s, 90, 102; of 1970s, 139–43, 146–47, 149, 200, 257n17; of 1980s, 155, 163, 170, 171, 200
Environmental Protection Agency (EPA), 140–41, 147, 255n10
Environmental Quality Incentives Program (EQIP), 9, 12, 183–84, 186, 191, 193, 214, 218
erosion, soil. *See* soil erosion
ethanol, 181
"ever-normal granary," 74
exporting crops, 95, 143–44, 145, 155

FACT. *See* Food, Agriculture, Conservation, and Trade Act of 1990 (FACT)
factions: climate change and, 216, 217, 219; Farm Bill of 1985 and, 168, 169–72; legislative process and, 16, 17, 18–19, 20–21, 25–26, 32–34; political power and, 26, 33–34, 55, 197, 201–2, 204, 206; Reform Amendment of 2001 and, 28, 31; Soil Bank and, 91, 94, 102, 104, 125, 127, 130, 133; Soil Conservation and Domestic Allotment Act and, 68–69, 76, 78, 80–81, 82; working theory on Congress and, 197–204, 206, 209–10, 211–15. *See also* competition of factions; *and specific factions*
farm bill coalition, 17, 26, 163, 175, 200, 213
Farm Bill of 1933, 16, 44–45, 52–54, 57–58, 108

Farm Bill of 1935, 10, 37, 38–39, 50–55, 57, 59, 72
Farm Bill of 1938, 73–74, 85–86, 98
Farm Bill of 1949, 86
Farm Bill of 1954, 91–92
Farm Bill of 1956. *See* Agricultural Act of 1956; Soil Bank
Farm Bill of 1962 effort, 138–39
Farm Bill of 1970, 139, 254n6
Farm Bill of 1973, 144–45, 146–47, 148, 152, 205
Farm Bill of 1977, 152, 156, 205
Farm Bill of 1981, 156–59, 213
Farm Bill of 1985. *See* Food Security Act of 1985
Farm Bill of 1990, 176–77, 178, 182–83, 201, 207
Farm Bill of 1995 reauthorization effort, 180
Farm Bill of 1996, 176, 177, 178, 180–81, 184, 186
Farm Bill of 2002. *See* Farm Security and Rural Investment Act of 2002
Farm Bill of 2008, 181, 182, 183, 188, 190
Farm Bill of 2014, 176, 177, 181, 183–84, 190, 192, 213
Farm Bill of 2018, 7, 177, 181, 182, 190–92, 213
farm bills background and overview, 7–9, 16, 17
Farm Bureau, 59, 74, 80, 91, 94, 98, 103, 131
farm coalition: acreage allotments and, 88–89; factions of 1950s of, 91, 94, 102, 125, 133, 204; Soil Conservation and Domestic Allotment Act and, 62, 68, 69
Farm Credit System, 155–56, 168
farm depression, post–World War I, 42

farm economy: of 1930s, 72–73, 75; of 1950s, 87; of 1970s, 143, 150; of 1980s, 155–56, 158, 167–69; of 2000s, 28, 185; post–World War I, 42

Farmers Home Administration, 160

farm interest faction: of 1950s, 91, 94, 102, 125, 133, 204; of 1970s, 142, 153, 156, 205; CSP and, 186, 188, 189, 191; Farm Bill of 1985 and, 154, 170–72; Jeffords bill and, 154; pesticides and, 142–43; public interest and, 195; Soil Bank and, 102, 103, 132, 200; Soil Conservation and Domestic Allotment Act and, 80–81. *See also* midwestern farm factions and interests; southern cotton faction

"Farmland Protection Policy Act," 158

Farmland Protection Program, 182

farm loans. *See* loans, farm

farm program payments: background and overview of, 7–9, 12, 13, 16; CSP and, 187, 188; Farm Bill of 1985 and, 168; limiting, 78; Reform Amendment of 2001 and, 27, 28–29, 31, 227n46; working theory on Congress and, 196, 207, 213, 214

Farm Security and Rural Investment Act of 2002, 27, 180–84, 187, 189. *See also* Reform Amendment of 2001

Farm Service Agency, 223n15

farm size: Soil Bank and, 100, 120–24, 126, 128–29; Soil Conservation and Domestic Allotment Act and, 70, 78

farms today, 9

farm tenancy. *See* tenant farmers

Farm Tenant Homes Corporation, 49

The Fault Lines of Farm Policy (Coppess), 10

Federal Agriculture Improvement and Reform Act of 1996, 176, 177, 178, 180–81, 184, 186

The Federalist Papers, 17–18, 32, 33, 199

feed grains: acreage increase in 1950s of, 89; New Deal Era and, 45; programs of 1960s for, 137–38, 167, 254n3; Soil Bank and, 103, 124; southern farmers and, 205; surpluses of, 88–89, 98, 137

fertilizers: 1970s and, 146, 148, 150, 158; introduction of synthetic, 86; production and, 149; Soil Bank and, 114, 119, 124, 131; water quality and, 5, 41, 149, 150

filibusters, 23, 217

Fite, Gilbert, 128

floods, 41, 50, 90

Foley, Tom, 148, 170

Food, Agriculture, Conservation, and Trade Act of 1990 (FACT), 176–77, 178, 182–83, 201, 207

Food, Conservation, and Energy Act of 2008, 181, 182, 183, 188, 190

Food and Agricultural Act of 1977, 152, 156, 205

food assistance programs, 7, 139, 148, 180, 191, 205, 217

Food Security Act of 1985: background and overview of, 9, 11, 137, 151, 153–56; conclusions on, 166–72, 175; congressional and presidential consensus and, 162–66; faction competition and, 200–201; Farm Bill of 1981 and, 156–59; Ronald Reagan's retreat from compliance opposition and, 159–60; sodbuster bill and, 160–62; Title XII of, 172; wetlands and, 177

Food Stamp Act of 1964, 139, 148, 205
food stamps, 139, 148, 169, 171, 200, 205
Ford, Gerald, 146, 149–50
foreign markets and competitors, 95, 102, 122, 179, 186
forfeited commodities, 95
fossil fuels, 5, 217
"Four Horsemen of Reaction," 52
free land movement, 39
Fulmer, Hampton, 78
funding caps, 187–88, 215
furnish system, 47–48

GAO. *See* Government Accountability Office (GAO)
Garrett, Elizabeth, 210
Gehlbach, Melvin P., 87–88, 89
general public. *See* society
Gilchrist, Fred C., 68
Gingrich, Newt, 180
Glickman, Dan, 164
Goodlatte, Bob, 29
Government Accountability Office (GAO), 150, 151, 153, 187
grains, 124, 137, 143–44, 155, 179. *See also* corn; feed grains; wheat
Gramm-Rudman-Hollings balanced budget legislation, 208
Grasslands Reserve Program (GRP), 182
Gray, L. C., 49
Great Depression: about, 10, 38, 42, 43, 48, 56; agriculture policy impact of, 71–72, 75, 82
Great Plains: drought of 1940s and 1950s and, 86–87, 104, 109; Dust Bowl and, 37, 38, 42–43, 72, 73, 74–75; farming of, 39–40, 42; tree planting in, 44; wetlands compliance in, 177

Great Plains Conservation Program, 157, 183
green ticket concept, 154
GRP. *See* Grasslands Reserve Program (GRP)

Hamilton, Alexander, 17–20, 33, 197, 199
Hansen, John Mark, 91, 93
Hardin, Clifford, 139
Harkin, Tom, 149, 161, 170, 185, 186–87, 188, 189–90
Hatch, Carl A., 61
Hatch Act of 1887, 40
herbicides, 86
Hickenlooper, Bourke, 99
highly erodible lands compliance, 165–66, 167, 176
Hill, Lister, 115–16
Homestead Act of 1862, 39, 43–44
Hoosac Mills, 52–53
Hoover, Herbert, 43
hostage-taking, legislative, 26, 100, 199, 212
House Agriculture Committee: in 1970s, 148, 149; CSP and, 189, 191; Farm Bill of 1985 and, 170; Farm Bill of 2002 and, 27; pesticide reform and, 143; power of, 22–23; RCA and, 154; RCCP and, 192; Sheppard bill and, 41; sodbuster and swampbuster and, 161–62, 163–64; Soil Bank and, 91, 94, 108, 109; Soil Conservation and Domestic Allotment Act and, 60
House of Representatives, U.S., 22, 23, 66–69, 163–65. *See also* Congress, U.S.; House Agriculture Committee; *and specific legislation*
Hovenkamp, Herbert, 33, 198–99
Humphrey, Hubert H., 91

Hurt, Douglas, 87

inaction, favor of, 25
Inflation Reduction Act of 2022, 193–94, 218–19
insecticides, 86
insurance, crop, 7–9, 16, 176, 177, 215
interest groups. *See* factions
interest rates, 155
international markets and competitors, 53–54, 95, 102, 122, 179, 186
Iowa, 119, 145, 147
irrigation, 6, 50, 87

Jackson, Henry M. "Scoop," 140
Jeffords, Jim, 153, 154, 162, 170, 186
Jeffords bill, 154–55
Jim Crow system, 44, 76, 79, 82, 92, 94
Johnson, Lyndon B., 204–5
Johnson, Paul, 177
Johnston, Oscar, 46
Jones, Ed, 164
Jones, Marvin, 59–60, 62, 66, 67, 70, 71

Katznelson, Ira, 38, 81
Kennedy, John F., 137
Kind, Ron, 27, 28

LaHood, Ray, 30
land, federal purchase of, 44, 51
land rental by government: Farm Bureau and, 91; sharecropping and, 48; Soil Bank and, 104, 106, 107, 119, 120, 124. *See also* Conservation Reserve Program (CRP)
land retirement conservation programs, 9, 91, 95, 181–82, 185. *See also* Conservation Reserve Program (CRP)

land use regulation: in 1930s, 45–46, 50, 51, 59–60, 63, 76–77, 79–80, 82; Jeffords bill and, 154–55; Reform Amendment of 2001 and, 29–30, 31; sodbuster and swampbuster and, 161–62, 164. *See also* private property rights; *and specific legislation*
legislative hostage-taking, 26, 100, 199, 212
legislative power, 16–17, 20, 23–24, 66, 171, 199
legislative process: congressional operations, rules, and organization and, 22–26; framing of the Constitution and, 17–22; minority interests and, 21, 25–26; working theory on Congress and, 197–98
Leopold, Aldo, 6, 220
loan rates, 85–86, 88, 91, 96, 102, 130, 154
loans, farm: Agriculture Adjustment Act of 1938 and, 85–86; Farm Bill of 1949 and, 86; Farm Bill of 1985 and, 153, 160, 167; legislation of 1970s and, 139, 144; Soil Bank and, 87, 88, 96, 111, 139. *See also* loan rates
local conservation districts, 79–80
local economies: CRP and, 167, 179; Soil Bank and, 120, 123–24, 125
Lucas, Frank, 31
Lugar, Richard, 182

Madison, James: about, 17–18; factions and, 18, 19, 32, 130, 132, 197–99, 201, 202; majority control and, 21–23, 25–26; priorities and, 72; separation of powers and, 19–20
majorities: congressional, 43, 56, 66, 92, 169; double requirement of,

Index 335

majorities (*cont.*)
21–22, 23, 25, 197–98, 205; factions and, 19, 21, 25, 198, 199; legislative process and, 19, 21–26; supermajorities and, 20, 23, 24–25
mandatory funding: conservation policy growth and, 175; CRP and, 167, 180; CSP and, 184; EQIP and, 183; RCPP and, 192; Soil Bank and, 100, 106, 111, 120
Marlanee, Ron, 155, 161, 164
Maverick, Maury, 69
McCarthy, Joseph, 128
McCarthyism, 128, 253n51
McNary, Charles, 57, 61–62
Middle East, 146
midwestern farmers: conflict with South and, 68, 87, 89; Soil Bank and, 87, 89–92, 97–99, 109, 119, 125, 205
midwestern farm factions and interests, 96, 99, 115, 200, 203, 206; of corn, 102–3; of dairy, 68
Miller, George, 27
Mississippi, 119, 121
Mississippi Delta, 120–21
modernization of agriculture, 5, 48, 93–94, 131. *See also* technological revolution in agriculture
Moran, Jerry, 29
Muskie, Edmund, 141, 142

National Association of Conservation Districts (NACD), 79, 154
National Climate Assessment, 207
National Environmental Policy Act (NEPA), 140, 142
Natural Resources Conservation Service (NRCS), 4, 177, 179, 192
negative power, 25, 26

New Deal: about, 16, 38, 43–44, 45, 56, 71–72, 76–77; conclusions on, 80–82; cotton and, 45–46, 48; Farm Bill of 1933 and, 16, 44–45, 52–54, 57–58, 108; Farm Bill of 1935 and, 10, 37, 38–39, 50–55, 59, 72; post–New Deal era opposition to, 128, 156; Supreme Court and, 52, 53, 54. *See also* Soil Conservation and Domestic Allotment Act of 1936
nitrogen, 5–6, 86, 193
Nixon, Richard M.: about, 139, 141, 145, 147; Agriculture and Consumer Protection Act and, 144, 145; budgets and, 147, 207; environmental policy and, 140–42, 146, 147, 220, 255n10; Soviet grain deal and, 143; Vietnam War and, 140
Norris, George, 59, 62
Nourse, Victoria, 19
nutrient loss, 5–6

Obama, Barack, 213
Office of Management and Budget (OMB), 159, 162–63, 169, 187, 207–8
offsets, budget, 180, 187–88, 189, 208, 209, 214
oil crisis, 146–47, 150–51
oil spill (January 28, 1969), 139
Olson, Mancur, 33, 198
OMB. *See* Office of Management and Budget (OMB)
O'Neill, Thomas "Tip," 170
overriding vetoes, 20, 24–25, 100, 142

parity system: farm bloc and, 91; Soil Bank and, 87–88, 94–96, 98, 103, 122–23, 125–26, 137; Soil Conservation and Domestic Allotment Act and, 74; technological revolution and, 86

partisanship, 91–92, 103, 125, 131–33, 188, 210, 216
path dependency, 32–33, 71, 103, 202, 206
Payment-in-Kind emergency program, 159
payment limits, 78, 100, 113, 114, 213–14
payments-in-kind, 95, 96, 138, 159
Pence, Mike, 30
Pepper, George Wharton, 53, 234n68
Perdue, Sonny, 191
pesticides, 90, 141, 142–43, 150, 158
Peterson, Collin, 189
Pinchot, Gifford, 38
planters, cotton. *See* cotton planters
Plessy v. Ferguson, 92
plowing, 4, 5, 40, 46, 109–10, 145, 150
Poage, W. R., 105, 108, 110, 113, 126, 148, 246n27
polarization, 194, 210, 216
political power: budgets and, 209, 212–13; competition and, 20–21, 205–6; factions and, 26, 33–34, 55, 197, 201–2, 204, 205–6; of House Agriculture Committee, 22–23; of Republicans, 43, 92, 169–70, 178; southern, 44, 46, 71, 94, 202–6; Watergate scandal and, 148. *See also* Democrats
pollution: 1970s and, 140, 141–43; 1980s and, 157; 1990s and, 182–83, 207
Pope, James, 65
population, farm, 195
precedents, 33, 188
prices, crop: of 1970s, 143–45; of 1980s, 155–56, 166; conservation assistance and, 29; during Great Depression, 44, 73, 74–75, 77; post–World War I, 42; RFS and, 181; Soil Bank and, 87, 112, 116, 122, 124. *See also* price support programs

price support programs: 1940s and, 86; 1980s and, 153, 156; background of, 8; CRP and, 167; of New Deal, 74; Soil Bank and, 87–88, 96–97, 99–100, 104, 113, 115, 118; surpluses and, 101–2, 112
private property rights: in 2000s, 192; background and overview of, 30–31, 32, 203; Jeffords bill and, 154–55; neighboring farms and, 156; New Deal reform and, 45–46, 51, 75, 77, 79; RCA and, 154; during Reagan era, 157, 158, 160, 164, 166, 168; sodbuster bill and, 160–62, 164; Soil Bank and, 94, 128, 132–33
processing taxes, 52–53
production: in 1950s, 86, 103, 123–24, 131; in 1970s, 143–45, 146–48, 152; in 1980s, 153, 157, 159, 165, 166–67; background and overview of, 6–7; fertilizer and, 149; Reform Amendment of 2001 and, 29–30; Soil Conservation and Domestic Allotment Act and, 6–7, 76, 77. *See also* acreage-based production controls; *and specific legislation*
"promised land" proclamation, 144, 145, 150–51, 167
public interest: background and overview of, 12, 14; budgets and, 210–11, 212, 217–18; competing factions and, 197–99, 200–202, 215, 219; conclusions on, 195, 196; Farm Bill of 1985 and, 171; legislative process and, 19, 21, 26, 32, 33; Reform Amendment of 2001 and, 28, 29; Soil Bank and, 132, 204; Tom Barlow and, 152. *See also* society
public-private partnership, 9, 192

Index 337

race: farm loan programs and, 31, 60, 160; farm policy and, 92–93, 94, 130, 245n26
Rasmussen, Chris, 73
Rayburn, Sam, 246n27
RCA. *See* Soil and Water Resources Conservation Act of 1977 (RCA)
RCPP. *See* Regional Conservation Partnership Program (RCPP)
Reagan, Ronald, 155, 157, 166, 168, 169
Reagan administration, 156–57, 159–60, 168–69, 171, 200–201, 207–8
reconciling bills, 23–24, 193
referendum of Soil Bank, corn farmer, 98, 110–11, 249n11
Reform Amendment of 2001, 27–32, 227n46, 227n50, 227n69
Regional Conservation Partnership Program (RCPP), 9, 192, 193, 218
renewable fuels, 181
Renewable Fuels Standard (RFS), 181, 189
renting acres, federal government. *See* land rental by government
Republicans: in 1990s, 179–80; budgets and, 147, 156, 208, 211; CSP and, 188; environmental movement and, 140–41, 142; Jeffords bill and, 155; political power of, 43, 92, 169–70, 178
Resettlement Administration, 49
resettlement of farm families, 48–49
retirement of land programs. *See* land retirement conservation programs
RFS. *See* Renewable Fuels Standard (RFS)
right-to-work, 93
river fires, 140, 142
Roberts, Owen, 53–54, 82
Roman Empire, 14–15
Roosevelt, Franklin Delano: election and reelections of, 43, 54; on erosion, 72; Farm Bill of 1935 and, 37–38, 50; Farm Bill of 1938 and, 73, 74; Soil Conservation and Domestic Allotment Act and, 58, 63, 65; southern political power and, 203
row crop farming, 39, 87, 95, 104, 126, 146
Ruckelshaus, William, 140
Rules Committee, 23
Russell, Richard B., 99, 106, 117

sabotage, congressional, 188, 191. *See also* Soil Bank sabotage and demise
Safe Drinking Water Act, 146
school integration, 92–93
self-correction, 219
self-government system: about, 18, 20, 25–26, 196; budgeting and, 210; factions and, 18, 197, 199, 205, 211, 219; New Deal and, 38; self-correction and, 219
self-renewal, 6, 220
Senate, U.S., 22, 23, 25, 165–66. *See also* Congress, U.S.; Senate Agriculture Committee; *and specific legislation*
Senate Agriculture Committee: in 1970s, 142, 148, 150; conservation bills of 1950s and, 91; CSP and, 184–85, 186; Farm Bill of 2002 and, 27; RCCP and, 192; Sheppard bill and, 41; sodbuster and, 160–61; Soil Bank and, 95, 97–98, 109–10; Soil Conservation and Domestic Allotment Act and, 59–61, 64, 82
seniority, congressional, 22, 148, 206
separation of powers, 20
sequestration, 208
set-aside policy, 139, 254n6
sharecroppers: about, 47–48; displacement of, 48, 129, 131; New Deal

policy and, 46, 48, 49, 69–71, 76, 77, 240n65; Soil Bank and, 101, 119, 121, 129–30, 131; southern power and, 94
Sheppard bill, 41, 43
Silent Spring (Carson), 142
Smith, Earl, 59
Smith, Ellison D. "Cotton Ed," 59–60, 61, 65, 70–71, 76, 79, 240n65
Smith, Gordon, 185
SNAP. *See* Supplemental Nutrition Assistance Program (SNAP)
"socially disadvantaged farmers," 31, 229n69
social reform, 39, 76–77
society: factions and, 197, 199, 200, 206; New Deal era and, 45, 51; soil and, 3–4, 6–7, 14, 49, 72–73, 195, 220; Soil Bank and, 125. *See also* public interest
sod-breaking, 40
sodbuster bill, 160–62, 163–65, 262n21
soil and society, 3–4, 6–7, 14, 49, 72–73, 195, 220
Soil and Water Conservation Districts (SWCD), 79–80
Soil and Water Resources Conservation Act of 1977 (RCA), 152, 153–54, 156, 157
Soil Bank: about, 95–96, 100–102; background and overview of, 11, 12, 85–86, 89, 94–95; competing factions and, 200; Conservation Reserve Program of, 95–96, 101–5, 117, 118, 120, 126, 131; Farm Bill veto and passage and, 99–100; feed grains programs of 1960s comparison to, 138; fight for, 96–99, 100–104; funding for, 106–8, 111–15, 250n14; political power and, 204; political terrain of, 91–94; and what could have been, 104. *See also* acreage reserve program of Soil Bank; Soil Bank sabotage and demise
Soil Bank sabotage and demise: acreage reserve program and, 105, 108–15, 117–18, 123–24, 125, 131, 137; consequences of, 130–33, 200, 204; cotton industry and, 116–17; funding and, 106–8, 111–15, 117–18, 120, 250n14, 251n32; Jamie Whitten's investigations and, 118–23, 124, 126, 128; overview of, 105–6; program timing and operations and, 108, 109–11, 115–16, 118–19, 127–29; reasons for, 123–30; weather and, 108–10, 116, 118–19, 120, 121, 122; welfare programs and, 126–27
Soil Conservation and Domestic Allotment Act of 1936: background and overview of, 10, 56–57, 63–64; congressional blame for, 76–79, 82; contracting powers and, 64–65, 67; dairy interests and, 66–69; farm interests and, 80–81; judgment on, 72–73, 74–75; results of, 73–74, 242n78; sharecroppers and, 69–71; state-led program of, 60, 61, 62–64, 73, 74, 78–79, 203; writing, debating, and signing of, 57–63, 66–69, 76–79
Soil Conservation Service (SCS), 51, 73, 147
soil erosion: of 1970s, 145–46, 147, 148–52; of 1980s, 157, 159; background and overview of, 4–6, 222n4; Dust Bowl and, 37–38, 39; farming challenges of, 176; farm tenancy and, 49; society and, 49. *See also specific legislation*
Soil Erosion Act of 1935, 10, 37, 38–39, 50–55, 57, 59, 72

soil loss tolerance rate (T), 4, 222n4
soil surveys of USDA, 40–41
southern cotton faction: climate change and, 217; Farm Bill of 1985 and, 169–70; feed grains programs of 1960s and, 138; in New Deal era, 68–70; power of, 148, 202–6; Soil Bank and, 95, 97, 99, 102, 115–16. *See also* southern Democrats in Congress
southern cotton planters. *See* cotton planters
southern Democrats in Congress: in 1960s, 138, 142; in 1970s, 148; Farm Bill of 1985 and, 169–70; Jeffords bill and, 155; in New Deal era, 44–45, 46, 48, 76, 79, 81–82; pesticide law and, 142; political power of, 44, 46, 81–82, 102, 105, 130, 148, 202–6; Soil Bank and, 91, 92, 93, 94–101, 102. *See also* Soil Bank sabotage and demise; southern cotton faction; *and specific southern Democrats*
"Southern Manifesto," 127
southern Republicans in Congress, 189, 217
Southern Tenant Farmers Union (STFU), 48, 71, 76
Soviet Union grain deal, 143–44, 146, 155
soybeans, 143, 144
"Special Areas Conservation Program," 157–58
spending, federal. *See* budgets, federal
Stabenow, Debbie, 192
states: legislative process and, 20, 22, 23, 24; Soil Conservation and Domestic Allotment Act and, 60–64, 67, 71, 73, 76, 78–80, 203; soil conservation districts of, 80; Supreme Court ruling and, 53–54, 65

status quo protections, 17, 21, 32–34, 175, 199, 202–3, 209, 216
Stenholm, Charlie, 27, 28, 163
STFU. *See* Southern Tenant Farmers Union (STFU)
Stockman, David, 155, 168–69, 207–8, 213
Stone, Harlan, 54
subsidies. *See* farm program payments
supermajorities, 20, 23, 24–25
Supplemental Nutrition Assistance Program (SNAP), 7, 26, 213
supply controls, 85–86, 98. *See also* acreage-based production controls
support for farmers, background and overview of, 7–9
Supreme Court, U.S.: *Brown v. Board of Education* and, 92, 127–28; Farm Bill of 1985 and, 168; legislative power and, 20; *Plessy v. Ferguson* and, 92; Soil Conservation and Domestic Allotment Act and, 52, 57, 63, 203; southern political power and, 203, 204; *U.S. v. Butler* and, 52–54, 65, 76; *U.S. v. Butler* ruling impact and, 57, 64–66, 67, 68, 70, 73, 75–76, 82
surplus crops: in 1960s, 137–38; crop prices and, 144; in New Deal era, 44, 68, 74; Soil Bank demise and, 108, 109, 112, 115, 116, 124; Soil Bank emergence and, 85, 86, 87, 89, 91, 95–99, 102, 204
swampbuster, 163–65

Talmadge, Herman, 148
Tarver, Malcolm, 69–70
taxes, processing, 52–53
technological revolution in agriculture: 1970s and, 146, 148; about, 86,

87, 90; Soil Bank and, 94, 97, 98, 103, 114, 131

tenant farmers: in 1970s, 149; displacement of, 129, 131; Jim Crow system and, 94; New Deal era and, 46–47, 48–50, 69–71, 76, 77, 232n48; Soil Bank and, 101, 119, 121, 123–30, 131

Tenth Amendment of U.S. Constitution, 53–54, 65, 79

Texas, 108, 120, 122

Thye, Edward, 90

Till, Emmit, 92, 127

tillage, 4, 5, 40, 46, 109–10, 145, 150

Title I of Farm Bill. *See* farm program payments

Tolley, Howard R., 58, 59, 77

"To Protect Tomorrow's Food Supply . . ." GAO report, 150

Truman, Harry S., 141

U.S. Department of Agriculture (USDA): ACEP and, 182; Agriculture Resources Conservation Act and, 150; cotton crop and, 44, 46, 48; CRP and, 162–63, 167, 178, 179, 180, 181; CSP and, 186–87, 188, 189, 191; discrimination and, 160; funding and, 12, 105–6, 214, 249n5; in New Deal era, 51, 57, 58–62, 65, 73, 78, 79–80; NRCS and, 4, 177, 179, 192; pesticide eradication campaign of, 142; RCA and, 152, 153–54, 156–57; set-aside policy and, 139; Soil Bank demise and, 105–11, 114, 117–19, 129, 249n5, 250n17; Soil Bank emergence and, 88, 96, 98, 204; Soil Conservation Service and, 51, 73, 147; soil surveying of, 40–41; statistics from, 4, 8, 73, 143, 175, 178; SWCD and, 79–80

U.S. v. Butler, 52–54, 65, 76; ruling impact of, 57, 64–66, 67, 68, 70, 73, 75–76, 82

vetoes: Agricultural Resources Conservation Act and, 149–50; Clean Water Act and, 141–42, 256n14; CSP and, 190; Farm Bill of 1956 and, 97, 99–100, 102, 204; legislative process and, 20, 24–25, 26; southern Democrats and, 44; spending reductions and, 213

vetogates, 24, 26

Vietnam War, 140, 141

Volcker, Paul, 155

Voting Rights Act of 1965, 102, 205

Waldron, Jeremy, 21

Wallace, Henry A.: Farm Bill of 1933 and, 44; Soil Conservation and Domestic Allotment Act and, 57, 58, 71, 74, 77, 79, 80; tenant farming and, 48, 49, 71

war byproducts, 86, 87

water: erosion and, 4–6, 37, 73; irrigation and, 6, 50, 87; Soil Bank and, 101; Watershed and Flood Prevention Act and, 90. *See also* water quality; waterways and water bodies

Watergate scandal, 148

water quality: Agricultural Water Quality Incentives Program and, 182–83; Clean Water Act and, 141–42, 256n14; fertilizers and, 5–6, 41, 149, 150, 222n4; funding and, 28; RCA and, 152, 153–54, 156, 157; Safe Drinking Water Act and, 146; state laws and, 80

Watershed and Flood Prevention Act of 1954, 90

waterways and water bodies, 4, 5, 41, 150–51, 195

Index   341

weather: climate change and, 5, 14, 193, 207, 215–20; conservation program concerns about, 30; erosion and, 5; Soil Bank and, 108–10, 116–17, 118–19, 120, 121–22; technology and, 148; wetlands compliance and, 177. *See also* drought

wetlands, 4, 163–66, 167, 176–77, 181, 182, 207

Wetlands Reserve Program (WRP), 181

wheat: Dust Bowl and, 43; Farm Bill effort of 1962 and, 138–39; homesteading and, 39; loans and, 86, 88; New Deal era and, 44, 73, 242n78; prices in 1970s of, 143–44, 145; Soil Bank and, 94, 96, 99, 107–8, 124; surpluses of, 88–89, 137; technological revolution and, 86–87

Whelchel, Frank, 69

Whitten, Jamie: 1980s appropriations and, 158, 169–70; Soil Bank and, 105–7, 111–14, 117–19, 123, 127, 130, 249n5, 250n14

Wildavsky, Aaron, 210

wind erosion, 42–43, 73, 87

Wolpe, Howard, 163, 164, 170

working lands conservation: Agricultural Water Quality Incentives Program and, 182–83; background and overview of, 182; EQIP and, 183–84. *See also* Conservation Security Program (CSP)

working theory on Congress: climate change and, 215–18; congressional design and, 197–202; new budget regime and, 207–15; southern power and, 202–6

WRP. *See* Wetlands Reserve Program (WRP)

Young, Milton, 97, 114